Selected Titles in This Series

20 Kolmogorov in perspective, 2000

19 **Hermann Grassmann,** Extension theory, 2000

18 **Joe Albree, David C. Arney, and V. Frederick Rickey,** A station favorable to the pursuits of science: Primary materials in the history of mathematics at the United States Military Academy, 2000

17 **Jacques Hadamard (Jeremy J. Gray and Abe Shenitzer, Editors),** Non-Euclidean geometry in the theory of automorphic functions, 1999

16 **P. G. L. Dirichlet (with Supplements by R. Dedekind),** Lectures on number theory, 1999

15 **Charles W. Curtis,** Pioneers of representation theory: Frobenius, Burnside, Schur, and Brauer, 1999

14 **Vladimir Maz′ya and Tatyana Shaposhnikova,** Jacques Hadamard, a universal mathematician, 1998

13 **Lars Gårding,** Mathematics and mathematicians: Mathematics in Sweden before 1950, 1998

12 **Walter Rudin,** The way I remember it, 1997

11 **June Barrow-Green,** Poincaré and the three body problem, 1997

10 **John Stillwell,** Sources of hyperbolic geometry, 1996

9 **Bruce C. Berndt and Robert A. Rankin,** Ramanujan: Letters and commentary, 1995

8 **Karen Hunger Parshall and David E. Rowe,** The emergence of the American mathematical research community, 1876–1900: J. J. Sylvester, Felix Klein, and E. H. Moore, 1994

7 **Henk J. M. Bos,** Lectures in the history of mathematics, 1993

6 **Smilka Zdravkovska and Peter L. Duren, Editors,** Golden years of Moscow mathematics, 1993

5 **George W. Mackey,** The scope and history of commutative and noncommutative harmonic analysis, 1992

4 **Charles W. McArthur,** Operations analysis in the U.S. Army Eighth Air Force in World War II, 1990

3 **Peter L. Duren et al., Editors,** A century of mathematics in America, part III, 1989

2 **Peter L. Duren et al., Editors,** A century of mathematics in America, part II, 1989

1 **Peter L. Duren et al., Editors,** A century of mathematics in America, part I, 1988

KOLMOGOROV
IN PERSPECTIVE

History of Mathematics • Volume 20

KOLMOGOROV
IN PERSPECTIVE

American Mathematical Society
London Mathematical Society

Translated from the Russian by Harold H. McFaden.

2000 *Mathematics Subject Classification.* Primary 01A70.

Library of Congress Cataloging-in-Publication Data

Kolmogorov in perspective / [translated from the Russian by Harold H. McFaden].
p. cm. — (History of mathematics ; v. 20)
Includes bibliographical references.
ISBN 0-8218-0872-9 (hard cover; acid-free paper)
ISBN 0-8218-2918-1 (soft cover; acid-free paper)
1. Kolmogorov, A. N. (Andrei Nikolaevich), 1903– 2. Mathematicians—Russia (Federation)—Biography. I. Series.
QA29.K627 K65 2000
510′.92—dc21
[B] 00-044162

♾ The paper used in this book is acid-free and falls within the guidelines
established to ensure permanence and durability.
· The London Mathematical Society is incorporated under Royal Charter
and is registered with the Charity Commissioners.
Visit the AMS home page at `http://www.ams.org/`

10 9 8 7 6 5 4 3 2 1   11 10 09 08 07 06

Contents

Preface

The Editorial Board of the History of Mathematics series has selected for translation a series of writings from the two volumes in Russian, "Kolmogorov in Remembrance" (A. N. Shiryaev, ed., "Nauka", Moscow, 1993) and "Mathematics and its Historical Development" (V. A. Uspenskiĭ, ed., "Nauka", Moscow, 1991). The former comprises articles by students and colleagues of Andreĭ Nikolaevich Kolmogorov on his life and works. The pieces in the second volume were written during Kolmogorov's lifetime, in the early to mid 1980's. Here Kolmogorov's memories of Aleksandrov were irresistible, sketching the early years of a lifelong mathematical and personal friendship. Shared experiences are supplemented by quotations from letters, as they visited mathematical centers in Germany, France and the United States in the early 1930's. Their comments on the Western societies they encountered are fresh, frank and charming. This article was previously published in "Uspekhi Matematicheskikh Nauk", and the translation included in the present volume was made for "Russian Mathematical Surveys" under the auspices of the London Mathematical Society. The second of the articles by Kolmogorov is a 1982 reissuing of a piece in a collection honoring the tricentennial of Newton's birth. He argues that Newton's understanding of the fundamental concepts of analysis really followed lines of thought much closer to those of our time than is generally conceded.

In the obituary volume, A. N. Shiryaev presents a quite comprehensive account of Kolmogorov's life and works. This is a mathematician's history of mathematics, in that he prefaces many of his subject's achievements with a sketch of his precursors' work, then follows the explanation of Kolmogorov's work with an account of advances made by his successors. In particular, one may recommend his treatment of the tentatives that preceded the famous "Grundbegriffe der Wahrscheinlichkeitsrechnung" and of the growth of the subject following its publication. Reminiscences of V. I. Arnol′d quote liberally from Kolmogorov's own letters and conversations, displaying his mathematical tastes and prejudices. The selections are rounded out by accounts of his activities with students at Moscow State University and of his continuing influence on generations of Soviet mathematicians.

Kolmogorov's bibliography is overwhelming, both in its volume and its range. For the reader who wants an overview of that range, it might be a good place to start, as would the reprinted table of contents from his three-volume "Selected Works" (Kluwer, 1991–1993). An effort here to give even a partial catalogue of the areas of his contributions would be inadequate.

The Editorial Board wish to thank Vladimir Drobot who have advised on the translation, Hal McFaden who translated most of the articles for this volume, and A. N. Shiryaev for sending additional photographs reproduced in the book.

Editorial Board

Andreĭ Nikolaevich Kolmogorov
(April 25, 1903 to October 20, 1987)
A Biographical Sketch of His Life and Creative Paths

A. N. SHIRYAEV

Andreĭ Nikolaevich Kolmogorov occupies a unique place in contemporary mathematics, indeed, in the world of science as a whole. In the breadth and diversity of his scientific pursuits he recalls the classicists in the natural sciences of past centuries.

(N. N. Bogolyubov, B. V. Gnedenko, and S. L. Sobolev, *Russian Math. Surveys* **38** (1983), no. 4, p. 24)

In 1985, 1986, and 1987 the publisher "Nauka" issued three volumes (in Russian) of selected works of A. N. Kolmogorov: *Mathematics and mechanics*, *Probability theory and mathematical statistics*, and *Information theory and the theory of algorithms.**

The papers in these volumes (referred to below as [MM], [PS], and [IA]) were selected by Kolmogorov himself, and he wrote some of the commentaries. At his suggestion a number of commentaries were prepared by his students and successors.

A comparison of the contents of these volumes (see Bibliography, Section IX) with the list of Kolmogorov's fundamental papers (Section I) shows that the papers included (60, 53, and 13 papers, respectively) far from exhaust what he did. And a simple acquaintance with the titles of the papers, not to mention their contents, an acquaintance with the list of his addresses at meetings of the Moscow Mathematical Society (Section VIII), and an acquaintance with his articles in encyclopedias, in the periodicals *Matematika v shkole* and *Kvant*, and in newspapers and popular science publications (Sections II to VII) strike one by their extraordinary breadth and fundamental character.

Here is an incomplete list of areas in which Kolmogorov obtained basic results and developed fundamentally important concepts that have determined the face and paths of development of many branches of twentieth century mathematics and other branches of science and knowledge: the theory of trigonometric series, measure theory and set theory, investigations in the theory of integration, approximation theory, constructive logic, topology, the theory of superposition of functions and Hilbert's thirteenth problem; classical mechanics, ergodic theory, the theory of turbulence, diffusion and models of population dynamics; the foundations of probability theory, limit theorems, the general theory of random processes, the theory

*Translated into English as vols. I, II, and III of *Selected works of A. N. Kolmogorov*, Kluwer, Dordrecht, 1991, 1992, 1993. The page references below are to the translations.

of Markov, stationary, and branching processes, mathematical statistics, the theory of algorithms, information theory, the theory of automata and applications of mathematical methods in the humanitarian sciences (including work in the theory of poetry, the statistics of text, and history); the history and methodology of mathematics, the teaching of school mathematics; school textbooks, popular science publications for school children and teachers of school mathematics.

The biographical sketch of Kolmogorov's life and activities being presented to the reader follows a chronological scheme. The main attention is given to the activities in his fundamental area of interests: mathematics. Connected with this, of course, is the fact that along with "life" and "activities" there is a lot of "mathematics" in the sketch.[1] It is intended that the reader can easily choose from the text the material that interests him.

A reading of this sketch presumes the use of the Bibliography, which includes the following sections:

I. General list of the main publications by A. N. Kolmogorov (pp. 177–203).
II. Encyclopedia articles by Kolmogorov (pp. 204–206).
III. Articles by Kolmogorov in the journal "Matematika v Shkole" (pp. 207–209).
IV. Articles by Kolmogorov in the journal "Kvant" (p. 210).
V. Articles by Kolmogorov on the theory of poetry and the statistics of text (p. 211).
VI. Articles by Kolmogorov on popular science (pp. 212–213).
VII. Newspaper articles by Kolmogorov (pp. 214–215).
VIII. Addresses of Kolmogorov at meetings of the Moscow Mathematical Society (pp. 216–218).
IX. Contents of Selected Works of Kolmogorov, Vols. I–III (pp. 219–222).
X. Publications about Kolmogorov (pp. 223–225).
XI. Publications (by various authors) cited in the biographical sketch (pp. 226–230).

The references to publications in Section I are given in the form [1921;1], ... , [1925;3], ... , where the first number indicates the year and the second the number of a publication for that year. The references to publications in different sections are preceded by the section number: [II-1], ... , [XI-121]. When possible, references to publications of Kolmogorov are given in two variants: to the originals, many of which were written in foreign languages (Section I), and to the respective volume [MM], [PS], or [IA].

Childhood and school years (1903–1920)

A. N. Kolmogorov was born on April 25, 1903 in the town of Tambov, where his mother Mariya Yakovlevna Kolmogorova was staying with a friend of hers on the way from the Crimea. Mariya Yakovlevna died at the birth of her son, who was taken to Yaroslavl at the age of ten days, and then to the home of his maternal

[1]This sketch is based on the author's paper "Andreĭ Nikolaevich Kolmogorov (25/4/1903–20/10/1987): In Memorial," in the memorial volume of the journal *Theory of Probability and its Applications* (**34** (1989), 5–118).

Kolmogorov's parents: Nikolaĭ Matveevich Kataev and Mariya Yakovlevna Kolmogorova.

grandfather in the village of Tunoshna (17 *versts* down the Volga from Yaroslavl). All responsibilities for the care and upbringing of the boy were assumed by Mariya's sister Vera Yakovlevna Kolmogorova, who adopted him and was indeed a mother to him.

Andreĭ Nikolaevich's father, Nikolaĭ Matveevich Kataev, was at the time exiled to Yaroslavl as a Zemstvo (elective district council) statistician, an agronomist by education. After the October Revolution he was head of the training department of Narkomzem (People's Commissariat of Agriculture); he was killed in 1919 on the southern front during Denikin's offensive.

The Kolmogorov sisters—Mariya Yakovlevna, Vera Yakovlevna, and Nadezhda Yakovlevna—were independent women possessing high social ideals. In [X-6] there is information about a secret printshop (a hectograph) in the Kolmogorovs' home in Tunoshna. Vera Yakovlevna helped the Yaroslavl revolutionary underground. The Kolmogorovs' address in Tunoshna was used for communications outside the country. Andreĭ Nikolaevich himself wrote in one of his letters (to V. I. Andrianov) that " ... according to a family story, during one of the searches some illegal literature was saved by putting it under my cradle. Both Vera Yakovlevna and Mariya Yakovlevna were arrested and stayed several months at a Petersburg center for detention pending trial."

Andreĭ Nikolaevich spent his first years of life, up to 1910, at the house near Yaroslavl. In the article "How I became a mathematician" ([1963;8]; see also [1988;1]) he writes:

Mariya Yakovlevna Kolmogorova.

> I experienced the joy of mathematical "discovery" early in life, when I noticed at the age of five or six the pattern
>
> $$1 = 1^2$$
> $$1 + 3 = 2^2$$
> $$1 + 3 + 5 = 3^2$$
> $$1 + 3 + 5 + 7 = 4^2 \qquad \text{and so on.}$$
>
> In our home near Yaroslavl my aunts established a small school in which they attended to ten children of various ages according to the latest prescriptions of the pedagogy of the time. A periodical, *The swallows of spring*, was produced at the school. My discovery was published in it. There I also published arithmetic problems I made up.

(Among them was, for example, in Kolmogorov's words: "There is a button with four holes. To attach it one need only pull the thread through at least two holes. How many ways can the button be attached?")

In 1910 Vera Yakovlevna moved with her adopted son to Moscow, where he studied at the E. A. Repman private gymnasium, which became the 23rd secondary school after the October Revolution.

Kolmogorov mentioned repeatedly the beneficial atmosphere of this gymnasium, which was founded by a circle of democratic intellectuals and was one of the least expensive private gymnasiums of that educational quality. He recalled:

> The classes were small (15–20 students). A considerable portion of the teachers were themselves attracted by science. Sometimes there were university teachers, and our geography teacher took part herself in interesting expeditions. Many pupils competed with each other in independent study of extra material, sometimes even with insidious intentions of using their knowledge to shame the less experienced teachers. An attempt was made to introduce into tradition a public defense of a final student graduation composition

Vera Yakovlevna with Andreĭ (6 years old).

> (of the same type as a university diploma thesis). In mathematics I was one of the first in my class, but the principal more serious scientific passions for me in school were initially biology and then Russian history.

In his answer to a questionnaire of the journal *Rovesnik* Andreĭ Nikolaevich commented in the same connection:

> The general purpose of finding a serious and useful pursuit for myself I owe to family tradition (especially to my aunt V. Ya. Kolmogorova, who raised me) and to the prevailing atmosphere at the wonderful E. A. Repman private gymnasium where I studied. My enthusiasm for science stemmed from the teachers of this school, but it was cultivated with special ardor in a circle of friends, of whom I mention the Seliverstov brothers (Gleb—a mathematician, and Nikolaĭ—an historian).

Recalling the last years of study at the school, Kolmogorov wrote ([1963;8], [1988;1]):

> In 1918–1920 life in Moscow was not easy. Only the most persistent studied seriously in school. At the time I had to leave together with other senior students to work on the construction of the Kazan–Ekaterinburg railroad. Simultaneously with the work I continued to study independently, preparing to take the high school examination. On returning to Moscow I experienced a disappointment: they gave me a graduation certificate without even bothering to give me an examination.

The years as a student and graduate student (1920–1925, 1925–1929)

In 1920, after receiving his high school diploma and not without some hesitation in his choice of a path for the future, Andreĭ Nikolaevich enrolled in the Physics and Mathematics Department of Moscow University, which at the time admitted anyone who wished, without examinations. In his "Memories of P. S. Aleksandrov" ([1986;2]) Andreĭ Nikolaevich writes about this time:

> I arrived at Moscow University with a fair knowledge of mathematics. Thanks to the book "*Novye idei v matematike*" (New ideas in mathematics) I knew in particular the beginnings of set theory. I studied many questions in articles in the Encyclopedia of Brockhaus and Efron, filling out for myself what was presented too concisely in these articles.

In parallel with the university Andreĭ Nikolaevich enrolled in the mathematics section of the D. I. Mendeleev Institute of Chemical Engineering (where an entrance examination in mathematics was required) and studied there for some time. ("I did not discount the idea of a career in engineering; for some reason, metallurgy fascinated me. Engineering was then perceived as something more serious and necessary than pure science," recalled Kolmogorov.) At the same time he maintained his serious enthusiasm for history, participating as an outside student in Professor S. V. Bakhrushin's History Department seminar on ancient Russian history. In 1920 he made his first scientific report in this seminar, treating land relations in Novgorod on the basis of an analysis of property registers of the fifteenth and sixteenth centuries. In his analysis he "made use of ... some devices in mathematical theory" ([1963;8], [1988;1]).

Soon though, his interest in mathematics outweighed all his doubts about the relevance of the profession of a mathematician.

> After taking the first-year examinations during the first months, I received the right, as a second-year student, to 16 kilograms of bread and 1 kilogram of butter per month. According to the notions of the time, that meant complete material well-being. I had clothes, and I made myself wooden-soled shoes.

Andreĭ Nikolaevich describes his student years ([1963;8], [1988;1]):

> However, in 1922–1925 my need of extra income to supplement the very modest stipend of that time led me to a secondary school. I now recall with great pleasure my work at the Potylikha experimental-model school of the Narkompros (People's Commissariat of Education) of the RSFSR. I taught mathematics and physics (they were not then afraid to entrust two subjects at once to nineteen-year-old teachers) and took a very active part in the life of the school (I was secretary of the school council and a tutor in the school boarding house).

As a university student Kolmogorov usually attended only special courses and seminars. During his first year (1920–1921) he attended N. N. Luzin's lectures on the theory of analytic functions and A. K. Vlasov's lectures on projective geometry.

At one of the lectures devoted to a proof of Cauchy's theorem Luzin used the following assertion. "Let a square be partitioned into finitely many squares. Then for any constant C there is a number C' such that for every curve of length at most C the sum of the perimeters of the squares touching the curve does not exceed C'." Luzin posed this as a problem for his listeners to prove. "I was able to show that this assertion is actually erroneous," recalled Andreĭ Nikolaevich ([1921;1], [1988;1]). "Nikolaĭ Nikolaevich at once saw the idea of the example disproving the supposition. It was decided that I should report the counterexample at the student mathematical circle."

Pavel Samuilovich Uryson undertook to check all the constructions and proofs. Thus originated the manuscript "Report to the mathematical circle on square pavings," dated January 4, 1921. The paper had been regarded as lost, but was found recently and appears as Appendix 1 in the third volume [IA] of Kolmogorov's selected works.

In the fall of 1921 Kolmogorov continued to attend Luzin's and Vlasov's lectures, and he started going to the lectures of P. S. Aleksandrov and Uryson. "Incidentally," recalled Aleksandrov [X-2], "at one of Uryson's lectures Andreĭ Nikolaevich noted a mistake in complicated constructions of Uryson as part of his proof of a dimension theorem in three-dimensional space. On another day Uryson corrected the error, but the sharpness of mathematical perception shown by the eighteen-year-old student Kolmogorov made a great impression on him."

Uryson invited Andreĭ Nikolaevich to start coming to him for mathematics studies. Kolmogorov later recalled:

> The Moscow mathematics of that time was very rich in brilliant and talented individuals. But Pavel Samuilovich Uryson stood out even on this background by the universality of his interests in combination with a clarity of purpose in his choice of research topics, the intelligibility of his statements of problems (in particular, to me when he regarded himself as responsible for the direction of my work), and a clear evaluation of his own and others' achievements together with a benevolence toward the accomplishments of the very young.

(From [1972;5], an article of Kolmogorov in L. Neĭman's book *The joy of discovery*, Moscow, 1972, devoted to P. S. Uryson (1898–1924); see also [1988;1].)

At the same time, under the influence of Aleksandrov's lectures, Andreĭ Nikolaevich began to occupy himself with general questions in descriptive set theory, and he arrived at the conception of a very general "theory of operations on sets" continuing and generalizing the investigations of E. Borel, Baire, Lebesgue, Aleksandrov, and Suslin. This work was finished at the very beginning of 1922 (January 3), but its first part was published (with a delay through no fault of the author) only in 1928 ([1928;1], [MM-13]). (Kolmogorov himself remarked: "My descriptive papers lay there on Luzin's desk; he had found them to be methodologically incorrect, and they lay there untouched until 1926.") The second part of the manuscript was accessible to a number of investigators in descriptive set theory, but it was first published only in 1987 as Appendix 2 in the third volume [IA] of Kolmogorov's selected works (English pp. 266–274).

In [1928;1] and [MM-13] Kolmogorov first introduced the concept of a δS-operation X on sets, which is defined as follows. For the basic class of elementary sets let us take the closed subsets of the interval $(0,1)$, including the empty set. The operation X is given by two objects: a collection $\{U^X\}$ of number chains $U^X = \{n_1, n_2, \dots\}$ that are subsets of the natural number series $\{1, 2, \dots\}$, and a sequence of sets $E_1, E_2, \dots$. Corresponding to each U^X is the chain of sets $E_{n_1}, E_{n_2}, \dots$ and the kernel $\bigcap_k E_{n_k}$. The union of the kernels of all the chains corresponding to a given collection $\{U^X\}$ and a given sequence of sets $E_1, E_2, \dots$ is the result of applying the operation X to this sequence of sets.

Next, Kolmogorov defined for this operation X the concept of the complementation operation $\overline{X}$ and proved the following remarkable result: there exists an X-set (on $(0,1)$) whose complement is not an X-set. Specialists in descriptive set theory saw clearly the significance of this result generalizing, in particular, the Suslin theorem on the existence of A-sets (analytic sets, introduced by Aleksandrov) that are not B-sets (Borel sets).

As a second-year student in the fall of 1921, Andreĭ Nikolaevich began to work also in V. V. Stepanov's seminar on trigonometric series, where he solved a problem that had interested Luzin, that of constructing a Fourier series with coefficients tending arbitrarily slowly to zero. In a paper on this topic dated June 2, 1922, Kolmogorov gave the following formulation of his main result about the order of magnitude of the Fourier coefficients ([1923;2], [MM-2]):

It is known that the Fourier coefficients of an integrable function tend to zero. In this note I prove the following proposition about cosine series.

1. *For any sequence* $\{a_n\}_{n=1}^{\infty}$ *converging to zero there is a sequence* $\{a'_n\}_{n=1}^{\infty}$ *such that*

1) $|a_n| < a'_n$,

2) $\sum_{n=1}^{\infty} a'_n \cos nx$ *is the Fourier series of an integrable function.*

Andreĭ Nikolaevich recalled ([1988;1]) that "when Luzin heard about this, he approached me (I remember this was on the stairs of the university) and suggested that I come regularly to him for studies."

In this way Kolmogorov became a student of Luzin, whose form of lessons with students consisted in scientific discussions with them one evening a week on the day of the week specified for them. Such "intensive work with students was one of the novelties cultivated by Nikolaĭ Nikolaevich," recalled Kolmogorov [1988;1].

In 1922 Andreĭ Nikolaevich obtained his most famous result in the area of trigonometric series: the construction of an example of a Fourier–Lebesgue series diverging almost everywhere. In a paper dated June 2, 1922 [1923;1] he writes the following about his result:

> The purpose of this note is to give an example of an integrable (that is, integrable in the Lebesgue sense) function whose Fourier series diverges almost everywhere (that is, everywhere except at the points of a set of measure zero). The function constructed is not square-integrable, and I do not know anything about the order of magnitude of the coefficients of its Fourier series.

Here he points out that his methods do not allow him to construct a Fourier series diverging everywhere. However, later (in 1926) Kolmogorov altered his construction somewhat and constructed an example of an integrable function whose Fourier series diverges everywhere ([1926;2], [MM-11]).

Both these examples were completely unexpected[2] for specialists and made an enormous impression. Kolmogorov maintained his interest in the theory of trigonometric functions and orthogonal series his whole life, returning from time to time to problems in this area of investigation and bequeathing to the young many problems to work on. He published about ten papers on this topic, each of which actually served as a starting point for subsequent investigations that have continued to the present time. (P. L. Ul'yanov's paper, "A. N. Kolmogorov and divergent Fourier series," *Uspekhi Mat. Nauk* **38** (1983), no. 4, 51–91, gives a survey both of Kolmogorov's series work and of further investigations in this direction.)

Along with his interest in the theory of trigonometric series and descriptive set theory Andreĭ Nikolaevich simultaneously studied a number of problems in classical analysis: differentiation, integration, measure theory, as well as in mathematical logic.

In several of his papers during the 1920's he tried various generalizations of the concept of "differentiation". His hope was to give so general a definition for derivative that any measurable (or at least any continuous) function would in general be differentiable in some natural sense. As a rule, however, for any given definition he was able to construct a counterexample with a continuous function that was not differentiable in the proposed sense. Kolmogorov looked at the problem that had arisen in its most general form ([1925;3], [MM-7]). Namely, after formulating a number of requirements that ought naturally to be satisfied by the "generalized derivative" $f'(x)$ of a function $f(x)$ (for example: the generalized derivative coincides with the ordinary derivative if the latter exists; if $\varphi(x) = af(x)$, then the function $\varphi(x)$ has a generalized derivative $\varphi'(x) = af'(x)$; $\dots$), Kolmogorov showed that if the function

$$f(x) = \sum_{n=1}^{\infty} \frac{\cos 3^n x}{3^n}$$

has a finite or infinite "generalized derivative" on a set of positive measure, then it is a *nonmeasurable* function. The example of this function thereby shows that trying to find an effective definition of the derivative for too general a class of continuous functions inevitably leads to the same difficulties as trying to construct a nonmeasurable set. Analogous results in this paper were formulated in connection with summation of divergent series and a general definition of integral. Somewhat later, in a paper published in 1930 ([1930;3], [MM-16]), Kolmogorov carried out a fundamental analysis of the known definitions and some new constructive definitions of integral, introducing orderliness and clarity into the general theory of integration, where the results had been scattered, and the relations between them had been left unclear.

In the introduction to this paper Andreĭ Nikolaevich sets his goal as follows:

> ... to clarify the logical nature of integration. And while, besides a unification of different approaches, a generalization emerged of the concept of integral, the point here apparently is that a generalization of a concept is often useful for comprehending its essence ...

---

[2]In connection with these examples see his diary entry for September 14, 1943 on pages 50–51 below.

A. N. Kolmogorov (1925).

> It is not impossible that all these generalizations may be of interest also for applications, though I see the merits of the more general approach first and foremost in the simplicity and clearness introduced by the new concepts.

Included in the same cycle are the 1925 paper "An axiomatic definition of the integral" ([1925;1], [MM-5]) and also "On the limits of generalizing the integral" ([1925;6], [MM-6]), which was first published in Russian in the first volume *Mathematics and mechanics* of the selected works and which contains proofs of results in the paper [1925;1], [MM-5], along with the 1928 paper "On the process of Denjoy integration" ([1928;3], [MM-14]).

In 1925 Kolmogorov's first paper on intuitionistic logic appeared: "On the *tertium non datur* principle" ([1925;5], [MM-9]). (The second paper was [1932;4], [MM-19].) In commentaries on these papers (English p. 451 of [MM]) written for *Mathematics and mechanics* Kolmogorov defines their plan:

> I conceived of [1925;5] as the introductory part of a broader plan. The construction, in the framework of intuitionistic mathematics, of models of diverse areas of classical mathematics was to serve to prove their consistency (the consistency of intuitionistic mathematics was then regarded as a consequence of its intuitive persuasiveness). In proving the consistency of the classical propositional logic such an approach is unnecessary, of course, but it was thought that the method would be applicable also to a proof of the consistency of classical arithmetic (cf. Gödel's 1933 paper).
>
> The paper [1932;4] was written in the hope that problem-solving logic could eventually be made a permanent part of a course in logic. It was proposed that there be created a single logical apparatus involving objects of two types: propositions and problems.

The development of the ideas put forth by Kolmogorov is traced in a long commentary in [MM] by V. A. Uspenskiĭ and V. E. Plisko about these papers.

In 1924, in his fourth year at the university, Andreĭ Nikolaevich began to be interested in the area of science in which his authority is especially great: probability theory.

His first paper, "On convergence of series whose terms are determined by chance" ([1925;6], [PS-1]) in this new area for him was written jointly with A. Ya. Khinchin (also a student of Luzin). "The whole of my work in probability theory together with Khinchin," recalled Andreĭ Nikolaevich ([1988;1], pp. 19–22), "in general the whole first period of my work in this theory was marked by the fact that we employed methods worked out in the metric theory of functions. Such topics as conditions for the applicability of the law of large numbers or a condition for convergence of a series of independent random variables essentially involved methods forged in the general theory of trigonometric series, that is, methods worked out by Luzin and his students."

The paper [1925;6] consists of four sections. The first was written by Khinchin and the remaining three by Kolmogorov. The results there have the following form in contemporary notation.

Let $\xi_1, \xi_2, \dots$ be a sequence of independent random variables. Then:

I. *The series $\sum \xi_k$ converges almost surely if the* **two series** *$\sum \mathbf{E}\xi_k$ and $\sum \mathbf{D}\xi_k$ converge.*

II. *If the variables $\xi_1, \xi_2, \dots$ are uniformly bounded ($\mathbf{P}(|\xi_k| \le c) = 1$ for $k \ge 1$, where $c < \infty$), then convergence of the* **two series** *$\sum \mathbf{E}\xi_k$ and $\sum \mathbf{D}\xi_k$ is not only sufficient but also necessary for the convergence of the series $\sum \xi_k$ almost surely.*

III. *If $\xi^c = \xi \mathrm{I}(|\xi| \le c)$, then for convergence of the series $\sum \xi_k$ almost surely it suffices that for some $c > 0$ the* **three series**

$$\sum \mathbf{E}\xi_k^c, \quad \sum \mathbf{D}\xi_k^c, \quad \sum \mathbf{P}(|\xi_k| \ge c)$$

converge, and if the series $\sum \xi_k$ converges almost surely, then these three series necessarily converge for any $c > 0$.

In the paper [1925;6], dated December 3, 1925, result I was proved by Khinchin (§ 1) and Kolmogorov (§ 2) by different methods. Khinchin used a generalization of a method of Rademacher (1922), who had considered the case when the quantities ξ_k take only two values c_k and $-c_k$, each with probability 1/2. Kolmogorov obtained his proof in a quite different way, based essentially on the same ideas as in the proof of the by now classical "Kolmogorov inequality":

If $\eta_1, \dots, \eta_n$ are independent random variables with $\mathbf{E}\eta_i = 0$, and if $S_k = \eta_1 + \dots + \eta_k$, then

$$\mathbf{P}\Big(\max_{1 \le k \le n} |S_k| \ge \varepsilon\Big) \le \frac{\mathbf{E}S_n^2}{\varepsilon^2}. \tag{1}$$

(In precisely this form the inequality was given in a subsequent paper of Kolmogorov: "On sums of independent random variables" ([1928;4], [PS-4]) written at the end of 1927; here he also reproved result I).

The proof of assertion II given in § 2 of [1925;6], [PS-4] is based on a lower estimate of the probability $\mathbf{P}\big(\max_{1 \le k \le n} |S_k| \ge \varepsilon\big)$ for bounded random variables η_k, $1 \le k \le n$. The corresponding inequality was given in the now common form in [1928;4], [PS-4] and is formulated in the following way:

If $\eta_1, \dots, \eta_n$ are independent random variables, $\mathbf{E}\eta_k = 0$, and $\mathbf{P}(|\eta_k| \le c) = 1$ for $1 \le k \le n$, where $c < \infty$, then

$$\mathbf{P}\Big(\max_{1\le k\le n} |S_k| \ge \varepsilon\Big) \ge 1 - \frac{(c+\varepsilon)^2}{\mathbf{E}S_n^2}.$$

Thus, this short paper [1925;6]—Kolmogorov's first in probability theory, written jointly with Khinchin—already contains the "Kolmogorov–Khinchin two series theorem", the "Kolmogorov three series theorem", and the "Kolmogorov–Khinchin criterion" for convergence almost surely of a series $\sum \xi_n$ of independent random variables with $\mathbf{E}\xi_n = 0$ for $n \ge 1$, which is simple to state and is now found in all probability texts:

If $\sum \mathbf{E}\xi_n^2 < \infty$, then the series $\sum \xi_n$ converges almost surely.

As the subsequent development of probability theory has shown, the significance of this paper [1925;6] is by no means exhausted by only the fact that it completely solved an important problem. It established new methods that were later used repeatedly and carried over to random elements of more general nature.

In 1925 Andreĭ Nikolaevich graduated from Moscow University and began postgraduate work, with Luzin as his advisor as before. In connection with the graduate school of that time Kolmogorov recalled ([1988;1]) that "graduate study did not then end with a dissertation as now: academic degrees were introduced first in 1934." (The degree of Doctor of the Physical and Mathematical Sciences was awarded to Kolmogorov in 1935 for his accumulated work, without a thesis defense.)

Kolmogorov's fundamental work on conditions for applicability of the law of large numbers and the strong law of large numbers was carried out in 1927–1929. By the end of 1927 he had succeeded in practically finishing the investigations begun by Bernoulli and Poisson and continued by Chebyshev and Markov on finding sufficient conditions, as well as necessary and sufficient conditions, for the law of large numbers to be valid.

In the preface of the book [1986;4] Kolmogorov wrote:

> The cognitive value of probability theory is due to the fact that large-scale random phenomena in their cumulative effect create strict regularities. The very concept of mathematical probability would be barren if it did not find its realization as the frequency of occurrence of some result when uniform conditions are repeated many times. Therefore, the work of Pascal and Fermat can be regarded as only the prehistory of probability theory, while its genuine history begins with Bernoulli's law of large numbers.

This law is formulated as follows:

If $\xi_1, \xi_2, \dots$ are independent identically distributed (Bernoulli) random variables taking the two values 0 and 1,

$$\mathbf{P}(\xi_n = 1) = p, \qquad \mathbf{P}(\xi_n = 0) = 1 - p,$$

then for any $\varepsilon > 0$

$$\mathbf{P}\left(\left|\frac{S_n}{n} - p\right| > \varepsilon\right) \to 0, \qquad n \to \infty, \tag{2}$$

where $S_n = \xi_1 + \dots + \xi_n$.

Poisson generalized this law of large numbers of Bernoulli to the case of differently distributed Bernoulli terms, giving it the following form:

If $\xi_1, \xi_2, \dots$ are independent (Bernoulli) random variables taking the two values 0 *and* 1,

$$\mathbf{P}(\xi_n = 1) = p_n, \qquad \mathbf{P}(\xi_n = 0) = 1 - p_n,$$

and $\sum_{n=1}^{\infty} p_n(1-p_n) = \infty$, then for any $\varepsilon > 0$

$$\mathbf{P}\left(\left|\frac{S_n}{n} - \frac{\mathbf{E}S_n}{n}\right| > \varepsilon\right) \to 0, \quad n \to \infty, \tag{3}$$

where $\mathbf{E}S_n = p_1 + \cdots + p_n$.

In 1867 Chebyshev investigated the question of the law of large numbers in the form (3) for an arbitrary sequence of independent random variables (not necessarily Bernoulli). The method Chebyshev developed, which is applicable for random variables with finite mathematical expectation and variance, ensures the law of large numbers (3) only under the condition

$$\frac{1}{n^2}\sum_{k=1}^{n} \mathbf{D}\xi_k \to 0, \quad n \to \infty. \tag{4}$$

This condition is usually called the *Markov condition.* Markov was the first to point out explicitly that the arguments of Chebyshev, who assumed that the expectations $\mathbf{E}\xi_k^2$ are bounded by the same constant, actually go through completely under the assumption (4).

In the paper "On sums of independent random variables" ([1928;4], [PS-4]), which was given to the printer on December 24, 1927, Kolmogorov found necessary and sufficient conditions for the validity of the (*generalized* in his terminology) law of large numbers for a "series scheme". Namely, suppose that for each $n \geq 1$ a sequence of independent random variables $\xi^n = (\xi_{n1}, \dots, \xi_{nn})$ is given. One says that the means

$$\Sigma_n = \frac{\xi_{n1} + \cdots + \xi_{nn}}{n}$$

are *stable* if there exists a sequence of numbers $A_1, A_2, \dots$ such that for any $\varepsilon > 0$

$$\mathbf{P}(|\Sigma_n - A_n| > \varepsilon) \to 0, \quad n \to \infty. \tag{5}$$

One also says that two systems of random variables $\xi^n = (\xi_{n1}, \dots, \xi_{nn})$ and $\overline{\xi}^n = (\overline{\xi}_{n1}, \dots, \overline{\xi}_{nn})$, $n \geq 1$, are *equivalent* if

$$\mathbf{P}(\Sigma_n \neq \overline{\Sigma}_n) \to 0, \quad n \to \infty,$$

where $\overline{\Sigma}_n = \frac{1}{n}(\overline{\xi}_{n1} + \cdots + \overline{\xi}_{nn})$.

Then Kolmogorov formulated the (generalized) law of large numbers as follows:

A necessary and sufficient condition for the stability of the means Σ_n, $n \geq 1$, is the existence of a system $\overline{\xi}^n = (\overline{\xi}_{n1}, \dots, \overline{\xi}_{nn})$ of independent random variables for each $n \geq 1$ that is equivalent to the system $\xi^n = (\xi_{n1}, \dots, \xi_{nn})$, $n \geq 1$, and satisfies

$$\frac{1}{n^2}\sum_{k=1}^{n} \mathbf{D}\overline{\xi}_{nk} \to 0, \quad n \to \infty.$$

From this Kolmogorov derived the following result, which summed up the long search for conditions implying the law of large numbers ([1928;4], [PS-4]):

Let $\xi_1, \xi_2, \ldots$ be a sequence of independent random variables, and let $S_n = \xi_1 + \cdots + \xi_n$. Then

$$\mathbf{P}\left(\left|\frac{S_n}{n} - \frac{\mathbf{E}S_n}{n}\right| > \varepsilon\right) \to 0$$

for any $\varepsilon > 0$ if and only if there exists for each $n \geq 1$ a sequence of independent random variables $\overline{\xi}_{n1}, \ldots, \overline{\xi}_{nn}$ such that as $n \to \infty$

$$\sum_{k=1}^{n} \mathbf{P}(\xi_k \neq \overline{\xi}_{nk}) \to 0,$$

$$\frac{1}{n}\sum_{k=1}^{n} [\mathbf{E}\xi_k - \mathbf{E}\overline{\xi}_{nk}] \to 0,$$

$$\frac{1}{n^2}\sum_{k=1}^{n} \mathbf{D}\overline{\xi}_{nk} \to 0.$$

In the case when the variables $\xi_1, \ldots, \xi_n$ are *independent and identically distributed*, Kolmogorov obtained the following result:

The means $\Sigma_n = \frac{1}{n}(\xi_1 + \cdots + \xi_n)$ are stable if and only if

$$n\mathbf{P}(|\xi_1| > n) \to 0, \quad n \to \infty. \tag{6}$$

If $\mathbf{E}|\xi_1| < \infty$, then the condition (6) holds, and the preceding assertion leads to the well-known criterion obtained earlier by Khinchin:

If $\xi_1, \xi_2, \ldots$ is a sequence of independent identically distributed random variables with $\mathbf{E}|\xi_1| < \infty$, then the law of large numbers is valid:

$$\mathbf{P}\left(\left|\frac{S_n}{n} - \mathbf{E}\xi_1\right| > \varepsilon\right) \to 0.$$

Somewhat later (in 1930: [1930;1], [PS-8]) Andreĭ Nikolaevich obtained another famous result of his, now a part of any more or less complete course in probability theory, a condition for the *strong* law of large numbers to be valid:

If $\xi_1, \xi_2, \ldots$ is a sequence of independent random variables with finite second moments and if

$$\sum_{n=1}^{\infty} \frac{\mathbf{D}\xi_n}{n^2} < \infty,$$

then

$$\frac{S_n - \mathbf{E}S_n}{n} \to 0 \quad (\mathbf{P}\text{-a.s.}).$$

In 1933 Andreĭ Nikolaevich himself observed in [1933;2] that for the case of *identically* distributed variables it is possible to get from the preceding assertion the following definitive result:

If $\xi_1, \xi_2, \ldots$ is a sequence of independent identically distributed random variables and $\mathbf{E}|\xi_1| < \infty$, then

$$\frac{S_n}{n} \to \mathbf{E}\xi_1 \quad (\mathbf{P}\text{-}a.s.);$$

but if $\mathbf{E}|\xi_1| = \infty$, then $\frac{1}{n}S_n$ diverges $\mathbf{P}$-a.s.

We recall that the strong law of large numbers was first formulated by E. Borel in 1909 for a Bernoulli scheme (in a number-theoretic interpretation):

If $\xi_1, \xi_2, \dots$ *is a sequence of independent Bernoulli random variables with*

$$\mathbf{P}(\xi_n = 1) = \frac{1}{2} \quad \textit{and} \quad \mathbf{P}(\xi_n = 0) = \frac{1}{2},$$

then as $n \to \infty$

$$\frac{S_n}{n} \to \frac{1}{2} \quad (\mathbf{P}\text{-}a.s.).$$

Cantelli later (1917) proved that:

If $\xi_1, \xi_2, \dots$ *is a sequence of independent random variables with finite fourth moment and* $\mathbf{E}|\xi_n - \mathbf{E}\xi_n|^4 \le c < \infty$ *for* $n \ge 1$, *then as* $n \to \infty$

$$\frac{S_n - \mathbf{E}S_n}{n} \to 0 \quad (\mathbf{P}\text{-}a.s.).$$

The term "strong law of large numbers" was introduced by Khinchin (1927, 1928), who gave a sufficient condition for it that is applicable also to *dependent* variables.

The above results of Kolmogorov about conditions for applicability of the strong law of large numbers to a sequence of independent random variables are distinguished by the completeness of the formulations and the transparency of the proofs. They are now present in almost all texts on probability theory, both as independent propositions and as illustrations of the effectiveness of using the "Kolmogorov inequality" (1).

At the end of 1927 Andreĭ Nikolaevich finished the paper [1929;4], [PS-5] on the *law of the iterated logarithm*—one of the remarkable theorems of probability theory and a refinement of the strong law of large numbers.

The history of the question can be presented briefly here as follows.

Making more precise the rate of convergence in the above Borel strong law of large numbers for a Bernoulli scheme, Hausdorff established in 1913 (*Grundzüge der Mengenlehre*, Viet, Leipzig, 1913) that $\mathbf{P}$-a.s.

$$S_n = o(n^{1/2+\varepsilon}), \quad \varepsilon > 0.$$

In 1914 Hardy and Littlewood ("Some problems of Diophantine approximations", *Acta Math.* **37** (1934), 155–239) showed that

$$S_n = O((n \log n)^{1/2}).$$

In 1922 Steinhaus ("Les probabilités dénombrables et leur rapport à la théorie de la mesure", *Fund. Math.* **4** (1922), 286–310) proved that

$$\limsup_n \frac{S_n}{\sqrt{8n \log n}} \le 1.$$

In 1923 Khinchin ("Über dyadische Brüche", *Math. Z.* **18** (1923), 109–116) found a formulation in which "iterated logarithm" first arose:

$$S_n = O((n \log\log n)^{1/2}).$$

Finally, in 1924 Khinchin was able ("Über einen Satz der Wahrscheinlichkeitsrechnung", *Fund. Math.* **6** (1924), 9–20) to obtain (for a Bernoulli scheme, and then also for a Poisson scheme) the result that is now commonly called the *law of the iterated logarithm*:

If $\xi_1, \xi_2, \dots$ are independent identically distributed Bernoulli random variables with $\mathbf{P}(\xi_n = 1) = \mathbf{P}(\xi_n = -1) = 1/2$, then

$$\limsup_n \frac{S_n}{\sqrt{2n \ln\ln n}} = 1 \quad (\mathbf{P}\text{-}a.s.),$$

which refines the strong law of large numbers stating that $\frac{S_n}{n} \to 0$ $\mathbf{P}$-a.s.

In 1929 Kolmogorov significantly broadened the area of applicability of the law of the iterated logarithm ([1929;4], [PS-5]):

Let $\xi_1, \xi_2, \dots$ be a sequence of independent random variables with zero mean, let $\sigma_n^2 = \mathbf{E}\xi_n^2$, and suppose that $B_n = \sum_{k=1}^n \sigma_n^2 \to \infty$ as $n \to \infty$. Assume that there is a sequence of constants M_n, $n \geq 1$, such that

$$|\xi_n| \leq M_n = o\left(\left(\frac{B_n}{\ln\ln B_n}\right)^{1/2}\right).$$

Then

$$\limsup_n \frac{S_n}{\sqrt{2B_n \ln\ln B_n}} = 1 \quad (\mathbf{P}\text{-}a.s.), \tag{7}$$

where $S_n = \xi_1 + \dots + \xi_n$.

Just as the "Kolmogorov inequality" was the central point in the proof of the theorems on convergence of series with probability 1 and the proof of the strong law of large numbers, and just as the idea of its derivation and the inequality itself became the source of further investigations by many authors, so it was with the proof of the law of the iterated logarithm: Kolmogorov created a new method that was included in the arsenal of fundamental tools of probability theory.

We outline the scheme Kolmogorov used to prove (7) (with slight modifications).

The assertion (7) is equivalent to the following statements:

(A) For any $\varepsilon > 0$ the function

$$\varphi^\varepsilon(n) = (1+\varepsilon)\sqrt{2B_n \ln\ln B_n}$$

is an *upper function* for S_n, $n \geq 1$, that is,

$$\mathbf{P}\{S_n > \varphi^\varepsilon(n) \text{ infinitely often}\} = 0. \tag{8}$$

(B) For any $\varepsilon > 0$ the function

$$\varphi_\varepsilon(n) = (1-\varepsilon)\sqrt{2B_n \ln\ln B_n}$$

is a *lower function* for S_n, $n \geq 1$, that is,

$$\mathbf{P}\{S_n > \varphi_\varepsilon(n) \text{ infinitely often}\} = 1. \tag{9}$$

Let $\{n_k\}$ be a nondecreasing sequence of integers such that $n_k \to \infty$ as $k \to \infty$. (This sequence is chosen below in a special way.) Then

$$\begin{aligned}&\{S_n > \varphi^\varepsilon(n) \text{ infinitely often}\}\\ &\qquad \subseteq \Big\{\max_{n_{k-1}<n\le n_k} S_n > \varphi^\varepsilon(n_{k-1}) \text{ infinitely often}\Big\}\\ &\qquad \subseteq \Big\{\max_{n\le n_k} S_n > \varphi^\varepsilon(n_{k-1}) \text{ infinitely often}\Big\}.\end{aligned}$$

Therefore, by the Borel–Cantelli lemma, to prove (8) it suffices to prove that (for a special choice of the subsequence $\{n_k\}$)

$$\sum_{k=1}^{\infty} \mathbf{P}\Big(\max_{n\le n_k} S_n > \varphi^\varepsilon(n_{k-1})\Big) < \infty. \tag{10}$$

The direct use of the "Kolmogorov inequality" (1) does not lead to the goal, and in the course of deriving sharp estimates for the probability $\mathbf{P}(\max_{n\le n_k} S_n > \varphi^\varepsilon(n_{k-1}))$ that will ensure convergence of the series in (10) Andreĭ Nikolaevich obtained the following:

(I) *The inequality*

$$\mathbf{P}\big(\max_{1\le k\le n} S_k \ge x\big) \le 2\mathbf{P}\Big(S_n \le x - \sqrt{2\textstyle\sum_{k=1}^n \mathbf{E}\eta_k^2}\Big) \tag{11}$$

($\eta_1, \dots, \eta_n$ *are independent random variables with* $\mathbf{E}\eta_k = 0$ *and* $\mathbf{E}\eta_k^2 < \infty$, *and* $S_k = \eta_1 + \dots + \eta_k$).

(II) *The exponential estimate*

$$\mathbf{P}(S_n \ge x) \le \begin{cases} \exp\left[-\frac{x^2}{2B_n}\left(1 - \frac{xM_n}{2B_n}\right)\right], & 0 \le xM_n \le B_n,\\ \exp\left[-\frac{x^2}{4M_n}\right], & xM_n \ge B_n,\end{cases} \tag{12}$$

for the problem of large deviations.

Choosing the sequences $\{n_k\}$ so that for some $\tau > 0$

$$B_{n_k-1} \le (1+\tau)^k < B_{n_k},$$

we deduce from (12) the estimate

$$\mathbf{P}(S_{n_k} \ge \varphi^\varepsilon(n_k)) \le [k\ln(1+\tau)^k]^{-(1+\varepsilon)^2(1-\mu)},$$

where μ is an arbitrary positive number which, if chosen sufficiently small, ensures that $(1+\varepsilon)^2(1-\mu) > 1$, and hence

$$\sum_{n=1}^{\infty} \mathbf{P}(S_{n_k} \ge \varphi^\varepsilon(n_k)) < \infty.$$

Together with (2) this implies the required inequality (10).

The proof that the functions $\varphi_\varepsilon(n)$ are lower functions for S_n, $n \ge 1$, that is, that (9) holds, is based on checking the condition for divergence of a series in the second part of the Borel–Cantelli lemma: *if the events* $A_1, A_2, \dots$ *are independent and* $\sum_{n=1}^\infty \mathbf{P}(A_n) = \infty$, *then* $\mathbf{P}(A_n$ *infinitely often*$) = 1$. To do this Kolmogorov first showed that for any $\varepsilon > 0$ there exist a subsequence $\{n_k\}$ and a $\gamma > 0$ such that

$$\begin{aligned}(13)\qquad &\mathbf{P}(S_{n_k} > \varphi_\varepsilon(n_k) \text{ infinitely often})\\ &\qquad \geq \mathbf{P}(S_{n_k} - S_{n_{k-1}} > (1-\gamma\varphi_0(n_k)) \text{ infinitely often}).\end{aligned}$$

Then after establishing that for some $\tau > 0$

$$\begin{aligned}(14)\qquad &\mathbf{P}(S_{n_k} - S_{n_{k-1}} > (1-\gamma\varphi_0(n_k))\\ &\qquad \geq \mathbf{P}\left(S_{n_k} > \left(1-\frac{\gamma}{2}\right)\varphi_0(n_k)\right) - \mathbf{P}\left(S_{n_k} > \frac{1}{3}\gamma\sqrt{\tau}\varphi_0(n_{k-1})\right),\end{aligned}$$

and after using the upper estimate already obtained for the second term on the right-hand side of this inequality, Kolmogorov came to the necessity of getting a "nice" lower estimate for the probability $\mathbf{P}(S_{n_k} > (1-\gamma/2)\varphi_0(n_k))$. He obtained such an estimate in the following form:

If $x > 0$, $xM_n/B_n \to 0$, and $x^2/B_n \to \infty$, then for any fixed μ and any sufficiently large n

$$\mathbf{P}(S_n \geq x) \leq \exp\left\{-\frac{x^2}{2B_n}(1+\mu)\right\}.$$

Thus,

$$\mathbf{P}\left(S_{n_k} > \left(1-\frac{\gamma}{2}\right)\varphi_0(n_k)\right) \geq (\ln B_{n_k})^{-(1+\mu)(1-\gamma/2)^2},$$

and by choosing $\mu > 0$ sufficiently small and τ sufficiently large, Kolmogorov arrived (for sufficiently large k) at the estimate

$$\mathbf{P}(S_{n_k} - S_{n_{k-1}} > (1-\gamma)\varphi_0(n_k)) \geq ck^{-\alpha},$$

where $c > 0$ and $\alpha < 1$. This estimate, the inequalities (13) and (14), and the second part of the Borel–Cantelli lemma imply that $\mathbf{P}(S_{n_k} > \varphi_\varepsilon(n_k)$ infinitely often$) = 1$ for any $\varepsilon > 0$. Thus, the functions $\varphi_\varepsilon(n)$ are lower functions for S_n, $n \geq 1$.

Kolmogorov's conditions for validity of the law of the iterated logarithm and the methods developed for proving it have served as a source of investigations by many authors. We mention some of them.

In 1937 Marcinkiewicz and Zygmund showed that in the Kolmogorov formulation of the law of the iterated logarithm the "small o" in the condition $M_n = o((B_n/\ln\ln B_n)^{1/2})$ cannot be replaced by "large O".

For the case of independent identically distributed variables $\xi_1, \xi_2, \ldots$ Hartman and Wintner in 1941 established the validity of the law of the iterated logarithm under only the assumption that $\mathbf{E}\xi_1 = 0$ and $\mathbf{E}\xi_1^2 < \infty$.

In 1965 Strassen obtained a "generalization" of the law of the iterated logarithm for independent identically distributed random variables $\xi_1, \xi_2, \ldots$ by showing that this law is simply characteristic of variables with finite second moments: *if* $\mathbf{E}\xi_1 = 0$ *and* $\limsup_n S_n/\sqrt{2n\ln\ln n} < \infty$ ($\mathbf{P}$-*a.s.*), *then* $\mathbf{E}\xi_1^2 < \infty$. Strassen also obtained an interesting *functional* variant of the law of the iterated logarithm.

Among the results of Feller (1943, 1946) connected with the law of the iterated logarithm there are assertions also for the case of variables with *infinite* second moment.

In the monograph *Asymptotic laws of probability theory* [XI-107] Khinchin established that for a standard Wiener process (Brownian motion) W_t, $t \geq 0$, the function $\varphi^\varepsilon(t) = (1+\varepsilon)\sqrt{2t\ln|\ln t|}$ is an upper function both as $t \to \infty$ (global form) and as $t \to 0$ (local form). And the function $\varphi_\varepsilon(t) = (1-\varepsilon)\sqrt{2t\ln|\ln t|}$ is a lower function.

The very concept of *upper* and *lower* functions was introduced by Khinchin, who directed attention to the problem of characterizing upper and lower functions. A fundamental step in this direction was made by I. G. Petrovskiĭ in the 1935 paper "On the first boundary value problem for the heat equation" [XI-79], where he used methods from the theory of differential equations, in contrast to the probabilistic methods of Khinchin and Kolmogorov (in the law of the iterated logarithm).

In that paper Petrovskiĭ showed that *a nondecreasing function $\varphi = \varphi(t)$ with $\varphi(0) = 0$ and $\varphi(t)/\sqrt{t} \uparrow \infty$ as $t \downarrow 0$ is (in a neighborhood of zero) an upper function for the Wiener process W_t, $t \geq 0$, if and only if*

$$\int_{0+}^{1} \frac{\varphi(t)}{t^{3/2}} \exp\left(-\frac{\varphi^2(t)}{2t}\right) dt < \infty.$$

Since the Wiener process is self-similar (if W_t, $t \geq 0$, is a Wiener process, then so is the process with $W_0' = 0$ and $W_t' = tW_{1/t}$ for $t > 0$), this result leads to the fact that *a nondecreasing function $\psi = \psi(t)$ is an upper function for W_t for large t if and only if*

$$\int_{1}^{\infty} \frac{\psi(t)}{t} \exp\left(-\frac{\psi^2(t)}{2}\right) dt < \infty.$$

(This result was formulated by Kolmogorov at the beginning of the 1930's, but he did not publish his proof.)

Petrovskiĭ's method goes as follows in modern terminology.

Consider the process $X_t = (t, W_t)$ in the domain

$$D = \{(t,w) : |w| \leq \varphi(t),\ 0 \leq t \leq 1\}.$$

We say that the point $x_0 = (0,0)$ is *regular* for the process X_t if $\mathbf{P}_{x_0}(\tau_D > 0) = 0$, where $\tau_D = \inf\{s > 0 : X_s \in \overline{D}\}$. Obviously, regularity of the point $x_0 = (0,0)$ means that the function $\varphi = \varphi(t)$ is a lower function, and *nonregularity* of it means that it is an upper function. Petrovskiĭ obtained a criterion for regularity (of the point $x_0 = (0,0)$) with the help of an explicit construction of barriers ("superharmonic functions for the process X_t") for the operator $\mathfrak{U} = \partial/\partial t + \frac{1}{2}\partial^2/\partial w^2$.

In May 1929 Andreĭ Nikolaevich finished his fourth year of graduate study, having eighteen mathematical publications in the years 1923–1928. The question arose of a position for him. Here is how Kolmogorov himself described the resolution of this question ([1986;2], p. 189).*

> Then came the question of where to continue my work. That year there was one vacancy at the Institute of Mathematics and Mechanics—the post of a senior researcher. Along with me, one of the older generation of mathematicians had a claim to the post; the director of the Institute, Dmitriĭ Fedorovich Egorov, was well aware of my scientific achievements, but he still thought it his duty to follow the guide-lines of age when appointing a scientific colleague. For me there was only one other attractive possibility. The Ukrainian Institute of Mathematics was opened in Kharkov in

**Editorial note.* This article is reprinted in the present book (pp. 145–162) from the *Russian Mathematical Surveys*; in this and further quotations the translation in that journal is used.

> 1928 and the Director was Sergeĭ Natanovich Bernstein, who was then at the height of his international fame and of his standing in the USSR. A special building had already been built for the Institute but the staff still had to be found. Bernstein sent me a proposal suggesting that I cooperate with him at the Institute where, according to his plan, I should only begin work there after a year's experience abroad, and for this he took steps to get me a Rockefeller scholarship. However, Aleksandrov was very much against Bernstein's plan and finally managed to persuade Egorov to give preference to my application.

Thus, in June 1929 Kolmogorov became a researcher at the Institute of Mathematics and Mechanics of Moscow University, with which he was associated for the rest of his life. (Looking ahead, we note that he became a Professor of Moscow University in March 1931 and was named Director of the Scientific Research Institute of Mathematics of Moscow University on December 1, 1933.)

In 1929 Andreĭ Nikolaevich, having had some boating experience, decided to organize a trip for three on the Volga. Recalling the companions he invited, which included P. S. Aleksandrov, Andreĭ Nikolaevich wrote the following [1986;2]:

> My personal contacts with Aleksandrov were also very limited at this time, although we met fairly often, for example at concerts in the Small Hall of the Conservatoire. We greeted each other but did not get into conversation. Apparently Aleksandrov's starched collars and general stiffness embarrassed me somewhat.
> ... To this day I cannot be quite sure how I came to suggest that Aleksandrov should be the third member. But he accepted straight away.

On June 16, 1929 they started from Yaroslavl along the Volga.

> For Aleksandrov a sailing holiday of this kind was something new, but he quickly undertook to be our quarter-master and even before leaving Moscow he was buying delicacies of all sorts. And it is from the day we set out, 16 June, that Aleksandrov and I date our friendship ...

In March 1981 Aleksandrov wrote something which he asked to publish on the eightieth birthday of Kolmogorov. The text [X-4]* included the following lines:

> My friendship with Kolmogorov occupies in my life quite an exceptional and unique place: this friendship had lasted for fifty years in 1979, and throughout this half-century it showed no sign of strain and was never accompanied by any quarrel. During this period we had no misunderstanding on questions in any way important to our outlook on life. Even when our views on any subject differed, we treated them with complete understanding and sympathy.

**Editorial note.* This article by Aleksandrov is also reprinted in the present book (pp. 141–143) from the *Russian Mathematical Surveys*, and the translation from there (with minor corrections) is used in quotations here.

In 1986 Andreĭ Nikolaevich remarked in "Memories of P. S. Aleksandrov" [1986;2] that their years of friendship "were the reason why all my life was on the whole full of happiness, and the basis of that happiness was the unceasing thoughtfulness on the part of Aleksandrov." In the same year 1986, at the meeting of the Moscow Mathematical Society (May 27) devoted to the memory of P. S. Aleksandrov (he died on November 16, 1982 at the age of 86), Andreĭ Nikolaevich said the following:

> I would probably have become a mathematician on my own, but my human qualities were formed to a significant degree under the influence of Pavel Sergeevich. He was really the most amazing person with regard to the richness and breadth of his views not only here but also in the whole world. His knowledge of music and painting, his sincere regard for people were extraordinary ...

In [1986;2] Andreĭ Nikolaevich gave a lively and humorous description of their 21-day trip on the Volga, which ended in Samara, and a subsequent trip to the Caucasus (Baku, Lake Sevan, Erivan (Yerevan), Tiflis (Tbilisi), etc.).

At Sevan Pavel Sergeevich worked "on various chapters of his joint monograph with Hopf, *Topology*." Kolmogorov wrote a paper on the theory of the integral and "was busy with ideas about the analytic description of Markov processes with continuous time, the end product of which later became the memoir "On analytic methods in probability theory" ([1931;1], [PS-9]).

The thirties (1930–1939)

A rapid broadening of Kolmogorov's creative activities in several directions of mathematics at once took place at the end of the 1920's and the beginning of the 1930's.

In 1929 Kolmogorov published his (not very well known to the broad circle of readers) paper "General measure theory and the calculus of probabilities" [1929;2], where he gave the first variant of his axiomatic construction of the foundations of probability theory, which was later to become the well-known "axiomatics of Kolmogorov" in *Grundbegriffe der Wahrscheinlichkeitsrechnung* ([1933;2]).

In this 1933 work he wrote of the necessity of constructing probability theory as "a very general and purely mathematical theory." He emphasized the imperative need to "single out in the exposition of probability theory those elements that are responsible for its intrinsic logical structure." He spoke of the fact that "the axiomatization of probability theory must be constructed on the basis of general measure theory and the theory of functions in purely metric spaces—a theory occupied with the study of those properties of functions that depend only on the measure of the sets on which the functions take a particular collection of values" (such as, for example, the orthogonality of two functions or the property of denseness of a system of orthogonal functions). He referred to "the space of elementary events in a given problem, and the probabilities of various sets of these events." He remarked that "the power of probability theory methods in connection with questions in pure mathematics is based to a significant degree on the use of the concept of independence of random variables." He directed attention to the fact that the concept of independence "has not yet been provided with an explicit purely mathematical formulation, although to provide one does not involve great difficulties."

E. Borel had expressed his thoughts on the importance of general measure theory in the construction of a foundation for probability theory as far back as 1909 ("Sur les probabilités dénombrables et leurs applications arithmétiques", *Rend. Circ. Math. Palermo* (1909), no. 26, 247–271); in 1923 these general ideas were developed in a paper of A. Łomnicki [XI-56].

Attempts to axiomatize probability theory had been made at the beginning of our century by G. Bohlmann [XI-11]. In 1917 Bernstein published a paper devoted to the construction of the foundations of probability theory. (In Bernstein's axiomatization the collection of events was regarded as a Boolean algebra and was based on a qualitative comparison of random events as being of greater or smaller probability.) Von Mises approached the foundations of probability theory from a very different point of view ([XI-64]–[XI-67]), in which the concept of the probability of a random event had to do with the result of a certain idealized experiment and the assumption of the existence of the limit of a frequency.

Four years after "General measure theory and the calculus of probabilities" Kolmogorov published (through Springer [1933;2], in German) the now classical monograph *Grundbegriffe der Wahrscheinlichkeitsrechnung*, in which the original idea of Borel took on a definitive form, and which became the basis for the whole subsequent development of probability theory, a model for its academic exposition, and the introduction to probability theory for many mathematicians.

"Reading Kolmogorov's *Grundbegriffe der Wahrscheinlichkeitsrechnung*," Itô wrote ([XI-36], p. xiii), "I was convinced that probability theory can be developed in terms of measure theory just as rigorously as the other areas of mathematics."

In the book *Quelques aspect de la pensée d'un mathématicien* (Blanchorol, Paris, 1970), Lévy describes his feelings after getting *Grundbegriffe*:

> *Dés* 1924, *Je m'ètais peu à peu habitué à l'idée qu'il ne fallait pas se borner à ce que j'appelais les vraies lois de probabilité. J'avais cherché à prolonger une vraie loi. Si arbitraire que ce fût, j'étais arrivé à l'idée d'une loi définie dans une certaine famille borelienne. Je ne songeais pas à me dire que c'était là la vraie base du calcul des probabilités; je n'avais pas l'idée de publier cette idée si simple. Puis, un jour, je reçus le mémoire d'A. Kolmogorov sur les fondements du calcul des probabilités. Je compris quelle occasion j'avais perdue. Mais c'était trop tard. Quand saurai-je distinguet ce qui, dans mes idées, merite d'être publié?*

In his autobiography (*Enigmas of chance. An autobiography*, Univ. of California Press, Berkeley, 1985), M. Kac describes the beginning of his mathematical life in collaboration with Hugo Steinhaus in 1935–1938:

> Our work began at a time when probability theory was emerging from a century of neglect and was slowly gaining acceptance as a respectable branch of pure mathematics. The turnabout came as a result of a book by the great Soviet mathematician A. N. Kolmogorov on foundations of probability theory, published in 1933.

In the Foreword to *Grundbegriffe* (the Russian edition was in 1936) Kolmogorov remarked that he would like to indicate the places "that go beyond the limits ... of the circle of ideas already sufficiently familiar to the specialists in general outline." He included among these places the following:

1) probability distributions in infinite-dimensional spaces;
2) differentiation and integration of expectations with respect to a parameter;
3) (especially) the theory of conditional mathematical expectations.

Here he noted that "all these new concepts and problems necessarily arise in considering completely concrete physical problems," referring to his joint paper with M. A. Leontovich, "On computing the mean Brownian area ([1933;4], [PS-14]), and to Leontovich's paper [XI-49]. Interestingly, in [1933;4], [PS-14] the purely physical part was written by Kolmogorov and the mathematical part by Leontovich.

The significance of all these new results (besides the axiomatics proper of probability theory) is now very clear after the lapse of more than fifty years since *Grundbegriffe* appeared ([1933;2], [1936;2], [1974;1]).

For instance, the fundamental theorem in *Grundbegriffe* (§ 4, Chapter III) on the unique construction of a probability measure on an infinite-dimensional space from a consistent collection of finite-dimensional distributions laid the foundation for the theory of random processes, which has become an extensive and independent area of probability theory with an enormous number of applied directions.

Further, using the Radon–Nikodým theorem (which, incidentally, appeared in modern form in the 1930 paper of Nikodým [XI-74]), Kolmogorov gave a definition of the concepts of the conditional probability $\mathbf{P}(A \mid \mathcal{G})$ of an event A with respect to a σ-subalgebra $\mathcal{G}$, the conditional probability $\mathbf{P}(A \mid \eta)$ of A with respect to a random element η, the conditional expectation $\mathbf{E}(\xi \mid \mathcal{G})$ of a random variable ξ with respect to a σ-subalgebra $\mathcal{G}$, and the conditional expectation $\mathbf{E}(\xi \mid \eta)$ of a random variable ξ with respect to a random element η—concepts now used in the basic arsenal of tools in modern probability theory.

In the summer of 1930 Andreĭ Nikolaevich finished and sent to the printer one of the most remarkable of his papers on probability theory: "On analytic methods in probability theory" ([1931;1], [PS-9]), which includes the foundations of the general theory of Markov random processes and which reveals the deep connections that both this theory and probability theory as a whole have with the theory of ordinary differential equations, partial differential equations, and mathematical physics.

Writing in "Analytic methods" (as [1931;1] is often called) that the topic of his investigation is the study of processes with continuous time, Kolmogorov especially emphasized this circumstance and the resulting essential novelty of the methods presented.

In the paper [X-2] dedicated to Kolmogorov on his fiftieth birthday, Aleksandrov and Khinchin have the following to say about "Analytic methods":

> In the whole of probability theory in the twentieth century it would be hard to find another investigation that has been so fundamental for the further development of the science and its applications as this paper of Andreĭ Nikolaevich. In our day it has led to the development of an extensive area of study in probability: the theory of random processes, which in its scope and in the number of its applications can rival the "classical" parts of probability theory. The differential "Kolmogorov equations" that govern Markov processes and that have been mathematically grounded rigorously and in all their breadth, contained as special cases all the equations (Smoluchowski, Chapman, Fokker–Planck, and so on) that up to then had

been derived and applied by physicists for isolated reasons by rule-of-thumb methods, without sufficient justification and without any clear explanation of the premises on which they were based. An enormous number of investigations worldwide have been based and continue to be based on these equations of Kolmogorov; they have proved to be fundamental both for the subsequent development of the theory and for the mathematical treatment of the most diverse applied problems.

The main object of investigation in "Analytic methods" is the transition probability $P(s,x;t,A)$ that an event A is realized at a time $t > s$ when the state at the time s is x. In addition to the naturally arising boundary conditions it is assumed that the transition probability satisfies the fundamental equation (expressing the Markov property)

$$P(s,x;t,A) = \int P(s,x;u,dy)P(u,y;t,A), \quad 0 \le s < u < t, \tag{15}$$

which is now called the *Kolmogorov–Chapman equation* (S. Chapman found such an equation in [XI-111]).

As is now known, under very broad conditions the equation (15) allows one to construct a Markov process $X = (X_t)_{t\ge 0}$ for which the conditional probability $\mathbf{P}(X_t \in A \mid X_s = x)$ coincides with $P(s,x;t,A)$. In "Analytic methods" Kolmogorov did not work directly with the realization $X = (X_t)_{t\ge 0}$ but occupied himself with the question of deriving differential equations for the transition probability by starting from (15), thereby creating in full generality and breadth a new analytic method—a method based on differential equations for investigating the probabilistic properties of continuous-time random processes whose evolution is subject to the Markov laws (15).

First of all, Kolmogorov introduced in "Analytic methods" the concept of "differential characteristics", considering firstly processes with a discrete state space $E = \{\dots, i, j, \dots\}$ and secondly continuous diffusion processes with values in the number line $E = R$.

In the first case these differential characteristics for $P_{ij}(s,t) = P(s,i;t,\{j\})$ are the limits (assumed to exist)

$$\begin{aligned} A_{ii}(t) &= \lim_{\Delta\downarrow 0} \frac{p_{ii}(t,t+\Delta)-1}{\Delta}, \\ A_{ij}(t) &= \lim_{\Delta\downarrow 0} \frac{p_{ij}(t,t+\Delta)}{\Delta}, \quad i \ne j. \end{aligned} \tag{16}$$

In the second case these characteristics for $F(s,x;t,y) = P(s,x;t,(-\infty,y])$, assumed to have a density $f(s,x;t,y) = \partial F(s,x;t,y)/\partial y$, are the limits

$$\begin{aligned} A(s,x) &= \lim_{\Delta\downarrow 0} \frac{1}{\Delta} \int_{-\infty}^{\infty} (y-x) f(s,x;s+\Delta,y)\,dy, \\ B^2(s,x) &= \lim_{\Delta\downarrow 0} \frac{1}{\Delta} \int_{-\infty}^{\infty} (y-x)^2 f(s,x;s+\Delta,y)\,dy, \end{aligned} \tag{17}$$

for the existence of which Kolmogorov determined conditions, and which have the following real meaning: $A(s,x)$ is the *mean rate of change of the parameter* x *in the course of an infinitesimally small time interval*; $B^2(s,x)$ is the *differential variance*.

In each of these cases Kolmogorov then derived his famous:
backward, or *first, differential equations* (with respect to s, x)

$$-\frac{\partial p_{ij}(s,t)}{\partial s} = \sum_{k \in E} A_{ik} p_{kj}(s,t), \tag{18}$$

$$-\frac{\partial}{\partial s} f(s,x;t,y) = A(s,x) \frac{\partial}{\partial x} f(s,x;t,y) + \frac{B^2(s,x)}{2} \frac{\partial^2}{\partial x^2} f(s,x;t,y); \tag{19}$$

forward, or *second, differential equations* (with respect to t, y)

$$\frac{\partial p_{ij}(s,t)}{\partial t} = \sum_{k \in E} p_{ij}(s,t) A_{kj}(t), \tag{20}$$

$$\frac{\partial f(s,x;t,y)}{\partial t} = -\frac{\partial}{\partial y}[A(t,y) f(s,x;t,y)] + \frac{1}{2} \frac{\partial^2}{\partial y^2}[B^2(t,y) f(s,x;t,y)]. \tag{21}$$

(The equation (21) was used by Fokker and Planck in physics papers on diffusion theory.)

Kolmogorov posed the question of the existence and uniqueness of solutions of these equations along with questions about the differentiability of the transition probabilities. These problems were subsequently addressed in papers of Kolmogorov himself (for example, [1951;1]), Feller, Gikhman, and many others.

The differential characteristics introduced by Kolmogorov later found their embodiment in a general setting in the framework of the semigroup approach to the theory of Markov processes. In this theory the *infinitesimal operator* of the corresponding semigroup of operators associated with a Markov process was introduced as a differential characteristic of the process. Necessary and sufficient conditions on the infinitesimal operator were obtained for it to determine the transition function uniquely (Feller, Dynkin, and others).

The Markov process under study in "Analytic methods" was not considered from the point of view of its sample path properties. Kolmogorov worked only with the transition probabilities and their differential characteristics. A powerful method for explicitly constructing Markov diffusion processes was the method of stochastic differential equations, developed by Itô ([XI-31]–[XI-35]; see also [XI-36]) and Gikhman ([XI-17], [XI-18]) in the 1940's and 1950's. (Finite-difference stochastic differential equations had been considered as far back as 1934 by Bernstein [XI-8].)

The essence of Itô's method is to start from the simplest process — a Wiener process $W = (W_t)_{t \geq 0}$ with $\mathbf{E}\Delta W_t = 0$ and $\mathbf{E}(\Delta W_t)^2 = \Delta t$ and construct processes $X = (X_t)_{t \geq 0}$ as solutions of the *stochastic differential equation*

$$dX_t = A(t, X_t)dt + B(t, X_t)dW_t. \tag{22}$$

Starting out from the intuitive notion that such a process $X = (X_t)_{t \geq 0}$ leaving a point x at time t behaves (locally) like a Wiener process with instantaneous drift $A(t,x)$ and instantaneous diffusion $B^2(t,x)$, and from the meaning of the Kolmogorov "differential characteristics" $A(t,x)$ and $B(t,x)$, Itô gave a constructive way of producing Markov diffusion processes $X = (X_t)_{t \geq 0}$ whose transition probabilities $p(s,x;t,\Gamma) = \mathbf{P}(X_t \in \Gamma \mid X_s = x)$ satisfy the Kolmogorov equations.

With this goal Itô first gave a precise meaning to the concept of a "stochastic differential equation", for which he had to define the "stochastic integral of a nonanticipative function with respect to a Wiener process", which has now become a standard tool; second, using the method of successive approximations, he showed that Lipschitz conditions and linear growth of the coefficients $A(t,x)$ and $B(t,x)$ with respect to x ensure the existence and uniqueness of a solution to (22); third, he established that this solution is a Markov process; fourth, using his famous "Itô's formula" for change of variables, which says that if $f = f(t,x)$ is continuous and has continuous derivatives $\partial f/\partial t$, $\partial f/\partial x$, and $\partial^2 f/\partial x^2$, then

$$df(t,X_t) = \left[\frac{\partial f}{\partial t}(t,X_t) + A(t,X_t)\frac{\partial f}{\partial x}(t,X_t) + \frac{B^2(t,X_t)}{2}\frac{\partial^2 f}{\partial x^2}(t,X_t)\right] dt + B(t,X_t)\frac{\partial f}{\partial x}(t,X_t)\, dW_t,$$

he showed that the backward Kolmogorov equations hold for the transition density of the Markov process.

Later, in the 60's and 70's (especially after work of D. Struik and S. R. S. Varadhan [XI-101]), the *martingale approach* began to be developed. Under very weak assumptions on $A(t,x)$ and $B(t,x)$ it led to a proof of the existence and uniqueness of a so-called *weak solution* of the stochastic differential equation (22), and that enabled substantial progress to be made in the solution of the problem, posed in "Analytic methods", of the existence of a Markov diffusion process having "differential characteristics" $A(t,x)$ and $B(t,x)$ with hardly any essential restrictions on them ([XI-98], [XI-21], [XI-28]).

Besides methods for studying transition probabilities with the help of differential equations, "Analytic methods" (and also another 1931 paper, "A certain generalization of the Laplace–Lyapunov theorem" ([1931;4], [PS-12])) gave a fundamentally new proof of the Laplace–Lyapunov–Lindeberg theorem, based on the idea that the sums S_n, $n \geq 1$, of independent random variables $\xi_1, \dots, \xi_n, \dots$ (with zero mean) form a Markov process that for a suitable normalization becomes a Markov diffusion process. In this way Kolmogorov gave a method for constructing *asymptotic expansions* for the probabilities $P_n(x) = \mathbf{P}(S_n/\sqrt{\mathbf{D}S_n} \leq x)$, and posed the question of the probability of all the inequalities of the form $a(t_k) < S_k < b(t_k)$ being valid, $k = 1, \dots, n$. In essence, this problem is (in a contemporary interpretation) a typical boundary problem of the "invariance principle", and consists in the following.

Suppose that $\xi_{n1}, \dots, \xi_{nn}$ is a sequence of independent random variables for each $n \geq 1$, $S_{nk} = \sum_{i \leq k} \xi_{ni}$, $\mathbf{E}\xi_{ni} = 0$, $\sum_{i \leq n} \mathbf{D}\xi_{ni} = 1$, and $\sum_{i \leq n} \mathbf{E}|\xi_{ni}|^3 = L_n \to 0$ as $n \to \infty$. The question is to determine when

$$R_n \equiv |\mathbf{P}(a(\mathbf{D}S_{nk}) \leq S_{nk} \leq b(\mathbf{D}S_{nk}),\ 1 \leq k \leq n) - \mathbf{P}(a(t) \leq W_t \leq b(t),\ 0 \leq t \leq 1)|$$

converges to 0 as $n \to \infty$ for sufficiently smooth bounds $a(t)$ and $b(t)$, $0 \leq t \leq 1$, and what the rate of convergence is, where $W = (W_t)_{t \geq 0}$ is a standard Wiener process. Kolmogorov found a boundary problem to which the search for the probability $P = \mathbf{P}(a(t) \leq W_t \leq b(t),\ 0 \leq t \leq 1)$ reduces, and he gave for the probabilities $P_n = \mathbf{P}(a(\mathbf{D}S_{nk}) \leq S_{nk} \leq b(\mathbf{D}S_{nk}),\ 1 \leq k \leq n)$ an asymptotic expansion with the probability P as its first term. Later Yu. V. Prokhorov [XI-87] obtained the estimate

$$R_n \equiv |P_n - P| = o(L_n^{1/4}(\ln L_n)^2)$$

for sufficiently smooth functions $a(t)$ and $b(t)$, $a(0) < 0 < b(0)$, while Skorokhod [XI-98], using the "method of a single probability space" for the case of identically distributed bounded terms, obtained for R_n the estimate $O((\ln n)/\sqrt{n})$, subsequently reduced to $O(1/\sqrt{n})$ by Nagaev [XI-71] and Sakhanenko [XI-94], who got rid of the $\ln n$ and removed the assumption of boundedness.

From June 1930 to March 1931 Kolmogorov was on a ten-month trip to Germany and France. Together with Aleksandrov they first spent three days in Berlin and then traveled to Göttingen. About this period Andreĭ Nikolaevich wrote the following in his "Memories of P. S. Aleksandrov" [1986;2]:

> At that time Göttingen was regarded as the leading mathematical centre of Germany and as a worthy rival to Paris in France and Princeton in the USA. Göttingen had attained this position with only a very limited number of permanent staff. There were four full professors—Hilbert, Courant, Landau, and apparently Bernstein (when he reached the age of 68 Hilbert had to retire, and Hermann Weyl had already been invited to succeed him). There were many of Courant's young collaborators there as Assistenten (Friedrichs, Rellich, Hans Levi, and others). Although she was not a full Professor, Emmy Noether was already being considered the head of the school of modern general algebra. Her students, van der Waerden and Deuring, were already there as Assistenten.
> ... Most of the corpus of Göttingen mathematicians were grouped around Hilbert, Courant, Landau, and Emmy Noether. It was a very friendly group and Aleksandrov was regarded as a full member of the group ...
>
> I too had various scientific contacts in Göttingen. First of all with Courant and his students working on limit theorems, where diffusion processes turned out to be the limits of discrete random processes; then with H. Weyl in intuitionistic logic; and lastly with Landau in function theory.

After Göttingen, Andreĭ Nikolaevich went to Munich to visit Carathéodory; he recalled [1986;2]: "he liked my work on measure theory and he insisted on having it published as quickly as possible," although Carathéodory reacted rather coldly to the work on the generalization of the concept of the integral.

Having an invitation from Fréchet to visit and work with him on the coast of the Mediterranean in Sanary not far from Toulon (on probability theory with Kolmogorov, and on set-theoretic topology with Aleksandrov), they arrived in Sanary after a short journey (to the Bavarian Alps, Ulm, and Freiburg in Germany, Lake d'Annecy and Marseilles in France).

"At that time Fréchet was studying Markov chains with discrete time and different types and sets of states. We discussed with him all the Markov problems over a wide range. This rather monotonous life continued for about a month, interrupted by occasional short outings," recalled Andreĭ Nikolaevich [1986;2], remarking further that, finding himself in Paris, he was naturally interested in "seeing how my work was regarded there and in getting some advice about continuing my work from the leaders of the older generation of mathematicians—Borel and Lebesgue. But my contact with them, alas, was limited to a small number of official visits.

However, Borel's help was essential if I was to get my French visa extended. Permission was granted immediately on receipt of a letter signed Emile Borel, Ancien Ministre de la Marine.

In mathematical matters I gained much from my contacts with P. Lévy. Time and again he invited me to his home, where we had lengthy serious scientific discussions. For lack of time I did not manage to form any interesting relations with the representatives of the younger generation."

In March 1931 Andreĭ Nikolaevich became a Professor at Moscow University, and from December 1, 1933 he began as Director of the Scientific Research Institute of Mathematics at Moscow University, a post he held until January 15, 1939 (and also for a short period in 1951–1953).

In 1930 and 1932 Andreĭ Nikolaevich published two papers on geometry: "On the topological and group-theoretic foundations of geometry" ([1930;2], [MM-15]) and "On the foundations of projective geometry" ([1932;5], [MM-20]).

The first of them gives the foundation of classical geometries of constant curvature for an n-dimensional space, based on topology and group theory. The second gives a new construction of projective geometry on the basis of Pontryagin's theorem that the only connected locally compact topological division rings with a countable base are the real numbers, the complex numbers, and the noncommutative division ring of quaterions. This theorem led to the possibility of giving a direct construction of both real and complex projective geometry.

Kolmogorov's now classical work on topology belongs to the extraordinarily productive 1930's. In this area his merit was primarily his introduction in algebraic topology (simultaneously with and independently of J. Alexander) of the concept of the ∇-operator and the construction with its help of cohomology groups (also called the upper Betti groups or ∇-groups) ([1936;6], [MM-29]), which provided a powerful and convenient tool for investigating various questions in topology, in particular, problems connected with continuous mappings. Further, Kolmogorov ([1936;8], [MM-30]) and Alexander defined a multiplication operation in cohomology groups that thereby turned them into rings (cohomology rings), and this played an exceptionally important role in subsequent investigations. A third outstanding contribution of Kolmogorov to topology (the papers [1936;12]–[1936;15], [MM-32]–[MM-35]) was a proof of "the duality law concerning closed sets in arbitrary locally compact completely regular topological spaces satisfying the condition of acyclicity in the dimensions mentioned in the statement of the result" ([XI-1], p. 187). (For details on these results see the commentaries of G. S. Chogoshvili about Kolmogorov's work in homology theory ([MM], English pp. 467–475).)

The list of remarkable publications of Kolmogorov on topology should include his 1937 paper "On open mappings" ([1937;1], [MM-36]), in which he constructed a masterly example of a continuous open mapping (that is, it carries open sets into open sets) of a one-dimensional continuum onto a two-dimensional continuum. Commenting on this paper ([MM], English p. 476), Andreĭ Nikolaevich wrote: "The problem of the possibility of raising dimension under open mappings was intensely interesting to Aleksandrov. For some time we worked together to find a proof of the impossibility of raising dimension. In these searches it gradually became clear why we had not succeeded. This analysis of our failures led in the end to a counterexample," which served as a stimulus for further investigations by Soviet topologists in the area of open mappings (L. V. Keldysh, B. A. Pasynkov, and others).

Also among the topological work of Andreĭ Nikolaevich in the 1930's was his paper "On normability of a general linear topological space" ([1934;2], [MM-23]), where, in particular, definitions were first given for a *linear topological space* and for the boundedness and convexity of a subset of such a space, and where a necessary and sufficient condition was given for a linear topological space to be normable.

In 1935 and 1936 two papers of Kolmogorov on the theory of approximation appeared ("On the order of the remainder term of Fourier series of differentiable functions" ([1935;4], [MM-27]) and "On best approximation of functions of a given function class" ([1936;1], [MM-28])). As was often the case with work of Kolmogorov, they led to the creation of a whole new direction in the theory of approximation of functions.

In the first of these papers Kolmogorov considered the class $\mathbf{F}^{(p)}$ of all periodic functions $f = f(x)$ that are continuous together with their derivatives of order $p-1$, which satisfy the Lipschitz condition $|f^{(p-1)}(x) - f^{(p-1)}(y)| \leq |x-y|$, while the pth derivatives satisfy $\sup_x |f^{(p)}(x)| \leq 1$. Letting

$$R_n(f,x) = f(x) - \left[\frac{1}{2}a_0 + \sum_{k=1}^{n}(a_k \cos kx + b_k \sin kx)\right]$$

be the remainder term of the Fourier series of a function $f = f(x)$, Kolmogorov considered the question of the value

$$C_n^{(p)} = \sup_{f \in \mathbf{F}^{(p)}} |R_n(f,x)|.$$

As far back as 1910 Lebesgue showed that $C_n^{(1)}$ has order $(\log n)/n$. In his paper Kolmogorov showed that in the general case

$$C_n^{(p)} = \frac{4}{\pi^2}\frac{\log n}{n^p} + O\left(\frac{1}{n^p}\right),$$

and in the case of odd p the following *precise* formula holds:

$$C_n^{(p)} = \frac{1}{\pi}\int_0^{2\pi}\left|\sum_{k=n+1}^{\infty}\frac{\sin kx}{k^p}\right|dx.$$

This work of Kolmogorov was developed by many authors and in varied directions: other approximating expressions were taken instead of partial Fourier sums, other function classes were taken instead of the class $\mathbf{F}^{(p)}$, and so on. Considerable progress in approximation theory was made in the 1940's by a student of Kolmogorov, Academician Sergeĭ Mikhaĭlovich Nikol′skiĭ, as well as by their students and successors. In [XI-75] Nikol′skiĭ told of the systematic visits of Aleksandrov and Kolmogorov in the 1930's to Denpropetrovsk, where they gave lectures and led research seminars, which, in particular, promoted the beginning of research activity in the theory of approximation of functions in this city.

Of exceptional importance and fundamental significance was the second of Kolmogorov's papers mentioned above: "On best approximation of functions of a given function class" ([1936;1], [MM-28]), where he introduced new characteristics, later called "Kolmogorov widths", of approximation properties of function classes. These characteristics have attracted considerable attention, especially beginning in the 1960's.

The statement of the problem formulated by Kolmogorov in [1936;1], [MM-28] was as follows.

Assume that a certain distance $\rho(f,g)$ has been introduced for the functions f and g under consideration, and consider the problem of approximating f by linear combinations $\varphi_c = c_1\varphi_1 + \cdots + c_n\varphi_n$ with fixed functions $\varphi_1, \ldots, \varphi_n$. Chebyshev had treated the problem of choosing the coefficients $c = (c_1, \ldots, c_n)$ so as to make the distance $\rho(f,g)$ as small as possible for a *fixed* function f. Kolmogorov now posed a new problem.

Let $F = \{f\}$ be a class of functions, and let

$$D_n(F) = \inf_{(\varphi_1,\ldots,\varphi_n)} \sup_{f\in F} \inf_{c=(c_1,\ldots,c_n)} \rho(f,\varphi_c).$$

For a given n and F it is required to find $D_n(F)$ and clear up the question of the existence of optimal functions $\varphi_1, \ldots, \varphi_n$ and their uniqueness (to within a linear transformation).

For the case $\rho(f,g) = \left[\int_0^1 (f-g)^2 dx\right]^{1/2}$ and the class F_1 of differentiable functions $f = f(x)$ with $\int_0^1 (f'(x))^2 dx \le 1$ Kolmogorov showed that

$$D_n(F_1) = \frac{1}{\pi n}, \qquad n = 1, 2, \ldots,$$

and that the functions $1, \sqrt{2}\cos \pi k x$, $k = 0, \ldots, n-1$, are optimal functions $\varphi_1, \ldots, \varphi_n$.

For the class F_p^*, $p \ge 1$, consisting of all p-times differentiable functions $f = f(x)$ satisfying the conditions

$$\int_0^1 (f^{(p)}(x))^2 dx \le 1,$$
$$f(0) = f(1), f'(0) = f'(1), \ldots, f^{(p-1)}(0) = f^{(p-1)}(1)$$

Kolmogorov found that

$$D_{2m-1}(F_p^*) = D_{2m}(F_p^*) = \frac{1}{(2\pi m)^p}, \qquad m = 1, 2, \ldots,$$

and established that for $n = 2m+1$ the functions $1, \sqrt{2}\sin 2\pi k x, \sqrt{2}\cos 2\pi k x$, $k = 1, \ldots, m$, can be taken as optimal.

In §5 of [1936;1], [MM-28] Andreĭ Nikolaevich gave his proofs an intuitive geometric interpretation that served as grounds for calling the infimum $D_n(F)$ the nth *width* of the set F.

While carrying out investigations in diverse areas of mathematics in the 1930's, Andreĭ Nikolaevich obtained in probability theory (commonly regarded as one of his main areas of expertise) a number of other results of fundamental significance in addition to those already mentioned.

At the end of the 1920's and beginning of the 1930's there appeared several papers of Bruno de Finetti devoted to the study (in a modern interpretation) of probabilistic properties of random processes $X = (X_t)_{t\ge 0}$ with homogeneous independent increments, which is equivalent to the study of the structure of distributions of so-called *infinitely divisible laws*.

Random variables $\xi = \xi(\omega)$ having infinitely divisible laws are characterized by the condition that for any finite $n \ge 1$ their distribution coincides with the distribution of a sum $\xi_{n1} + \cdots + \xi_{nn}$ of independent identically distributed random variables $\xi_{n1}, \ldots, \xi_{nn}$. (The importance of this class of infinitely divisible distributions is that under very general conditions they occur as limits of normalized sums of independent random variables.)

De Finetti proposed [XI-102] a certain sufficiently general formula for the characteristic function $f(t) = \mathbf{E}e^{it\xi}$ of an infinitely divisible random variable ξ. Namely, combining the normal type and the Poisson type (with jumps of different sizes), he determined the following formula for $f(t)$:

$$f(t) = \exp\Big\{iat - \frac{\sigma^2}{2}t^2 + c\int(e^{iut} - 1)dF(u)\Big\}, \tag{23}$$

where $F(u)$ is the distribution function of the sizes of the jumps. However, this formula (23) did not encompass the general case, but described only a certain subclass of infinitely divisible distributions.

In 1932 Kolmogorov gave an exhaustive answer to de Finetti's problem for the case of variables ξ with *finite second moment*, $\mathbf{E}\xi^2 < \infty$ ([1934;2], [1934;3], [PS-13]):

A function $f = f(t)$ is the characteristic function of an infinitely divisible law of a random variable ξ with $\mathbf{E}\xi^2 < \infty$ if and only if it can be represented in the form

$$f(t) = \exp\Big\{iat + \int_{-\infty}^{\infty}\frac{e^{itx} - 1 - itx}{x^2}\,dK(x)\Big\}, \tag{24}$$

where $a \in R$, and $K = K(x)$ is a nondecreasing bounded function; the integrand is assumed to be equal to $-t^2/2$ when $x = 0$.

The general case, including also the possibility of *infinite variance*, was investigated by a very different method in 1934 by Lévy [XI-46]. In 1937 Khinchin [XI-108] showed that Lévy's result can be obtained also by Kolmogorov's method.

The formula now commonly called the "Lévy–Khinchin formula" and giving the characteristic function $f = f(t)$ of an infinitely divisible distribution has the form

$$f(t) = \exp\Big\{i\alpha t - \frac{t^2\sigma^2}{2} + \int_{-\infty}^{\infty}\Big(e^{itx} - 1 - \frac{itx}{1+x^2}\Big)\frac{1+x^2}{x^2}\Lambda(dx)\Big\}, \tag{25}$$

where $\alpha \in R$, $\sigma^2 \geq 0$, and Λ is a finite measure on $(R, \mathcal{B}(R))$ with $\Lambda(\{0\}) = 0$.

Other forms of notation are also used, for example,

$$f(t) = \exp\Big\{i\alpha t - \frac{t^2\sigma^2}{2} + \int_{-\infty}^{\infty}(e^{itx} - 1 - ith(x))F(dx)\Big\},$$

where F is a nonnegative measure such that $\int_{-\infty}^{\infty}\min(1, x^2)F(dx) < \infty$, and $h = h(x)$ is a bounded Borel function with compact support and behaving like x in a neighborhood of zero.

Kolmogorov's remarkable paper "On the empirical determination of a distribution law" ([1933;5], [PS-15]) appeared in 1933, and has become classical and one of the central papers in all the statistics of nonparametric goodness-of-fit testing.

The formulation of the main result in this paper is simple and beautiful. Let $\xi = (\xi_1, \xi_2, \ldots)$ be a sequence of independent identically distributed random variables with *continuous* distribution function $F(x) = \mathbf{P}(\xi_1 \le x)$, and let

$$F_n(x;\xi) = \frac{1}{n}\sum_{k=1}^{n} I(\xi_k \le x)$$

be the empirical distribution function. *Then*

$$\lim_n \mathbf{P}\{\sqrt{n}\sup_n |F_n(x;\xi) - F(x)| \le \lambda\} = \mathcal{K}(\lambda), \tag{26}$$

where $\mathcal{K}(\lambda) = \sum_{k=-\infty}^{\infty}(-1)^k e^{-2\lambda^2 k^2}$.

To appreciate the significance of this paper we should recall that to test the hypothesis that a distribution $F = F(x)$ is true from the results of observing $\xi_1, \dots, \xi_n$, Cramér (1928) and von Mises (1931) considered the "omega-square" statistic

$$\omega_n^2 = \int_{-\infty}^{\infty} [F_n(x;\xi) - F(x)]^2 dx.$$

However, no precise assertions about the asymptotic behavior of ω_n^2 were given. (For the history of the question and the contemporary state of this whole circle of problems see the commentary of É. V. Khmaladze [PS], English pp. 574–583, about Kolmogorov's paper.)

It clearly follows from the result (26) of Kolmogorov that $D_n = \sup_x |F(x;\xi) - F(x)| \to 0$ *in probability.* It is interesting to note that the very same (fourth) volume of the journal *Giorn. Inst. Ital. Attuari* for 1933 containing Kolmogorov's paper contains two well-known mathematical statistics papers by V. S. Glivenko ([XI-22]) and F. Cantelli ([XI-38]), both entitled "Sulla determinazione empirica delle leggi di probabilita", and both including a proof of the convergence $D_n \to 0$ now *with probability* 1 (Glivenko for continuous functions $F = F(x)$ and Cantelli in the general case). Later, N. V. Smirnov in 1944 [XI-99] and A. Dvoretsky, J. Kiefer, and J. Wolfowitz in 1956 [XI-23] showed that the Kolmogorov statistic D_n satisfies the inequality

$$\mathbf{P}(D_n > \alpha) \le ce^{-2nd^2}, \quad d > 0,\ n \ge 1,$$

which, by the Borel–Cantelli lemma, naturally implies also that $D_n \to 0$ with probability 1.

In 1936 and 1937 Kolmogorov undertook a broad study ([1936;7], [1937;3], [PS-23]) of questions on the asymptotic behavior of the transition probabilities from one state into another in an unboundedly increasing number of steps for the case of a Markov chain with a *countable* set of possible states. His classification of Markov chains according to the *arithmetic properties* of the transition probabilities $p_{ij}^{(n)}$ from a state i to a state j in n steps (nonessential and essential states, indecomposable classes, cyclic subclasses, etc.), along with his classification according to the *asymptotic properties* of the probabilities $p_{ii}^{(n)}$ (recurrence, nonrecurrence, zero and positive states, etc.) remain to this day a brilliant example of investigating the behavior of such a complex (in spite of the discrete time) stochastic object as a Markov chain with countable set of states.

The exceptional breadth of Kolmogorov's scientific interests manifested itself also in his more "practical" papers, where probabilistic considerations were applied to problems in biology, genetics, physics, geology, and so on.

For example, in the paper "On the solution of a biological problem" ([1938;14], [PS-25]) concerning a simple model of a branching random process, he found an asymptotic expression for the probability that descendants die out as the number of generations increases.

In [1940;7], [PS-26] Kolmogorov wrote that in a "discussion of questions of genetics in the fall of 1939 much attention was given to the problem of verifying the consistency of Mendel's laws." In this paper, entitled "On a new confirmation of Mendel's laws" (published in *Doklady*), Kolmogorov analyzed some statistical material and conclusions published by N. I. Ermolaeva (in the journal *Yarovizatsiya* **2 (23)** (1939), 79–86) and wrote:

"This material, contrary to the opinion of Ermolaeva herself, is a striking new confirmation of Mendel's laws" (leading in the simplest case to a splitting in the ratio 3 : 1).

In the article "On the statistical theory of crystallization of metals" ([1937;5], [PS-22]) Kolmogorov gave under fairly general assumptions a rigorous solution of the problem of the rate of flow of a crystallization process, observing that "the study of a crystal growth process when the formation of crystallization centers is random is of essential significance in metallurgy" and that "taking into account the encounters of grains of crystalline substance that arise near individual crystallization centers presents well-known difficulties." The formulas Kolmogorov gave here for the probability of inclusion of a given point in a crystallizing mass (the formula (3) in [1937;5]) and also the number of crystallization centers (the formulas (6) and (6a) in [1937;5]) are even now fundamental in the general theory of crystallization of metals.

In 1933 Kolmogorov and M. A. Leontovich published the paper "On computing the mean Brownian area" in a physics journal ([1933;4], [PS-14]). In this paper they solved a problem S. I. Vavilov posed to them about the expectation $\mathbf{E}S_t$ of the area S_t swept out in time t by a disk of radius ρ whose center is moving in the plane like a Brownian particle.

If we confine ourselves to the principal term, then their formula for $\mathbf{E}S_t$ has the form

$$\mathbf{E}S_t \sim \frac{4\pi Dt}{\ln(1.26Dt\rho^{-2})} \qquad (Dt\rho^{-2} \gg 1),$$

where D is the diffusion coefficient.

The way this problem was solved is worth mentioning: it is very closely connected with the methods developed by Kolmogorov in his "Analytic methods".

Namely, let $P_L(x, y; t)$ be the probability that a Brownian particle starting at the point (x, y) intersects the given boundary Γ of a domain G containing (x, y) at least once in a time t, with the stipulation that the first intersection takes place on a specified part L of the whole boundary Γ. Then $P_L(x, y; t)$ satisfies the "first Kolmogorov equation" (19) with the following conditions: $P_L(x, y; 0) = 0$ if $(x, y) \in G$, $P_L(x, y; t) \to 1$ for any $t > 0$ as (x, y) tends to some point of the part L of Γ, and $P_L(x, y; t) \to 0$ for $t > 0$ as (x, y) tends to a point in $\Gamma \setminus L$. These conditions uniquely determine the function $P_L(x, y; t)$, and that gives the possibility of finding it. (The same method was proposed simultaneously by Pontryagin, Andronov, and Witt in the paper "On a statistical treatment of dynamical systems" [XI-81].)

A small note of Kolmogorov in 1934, "Random motions" ([1934;10], [PS-19]), has the subtitle "On the theory of Brownian motion" and is devoted to the question of a general description of Brownian motion with inertia, a topic then of lively interest to physicists. For the investigation of the motion of a Brownian particle the theory of Einstein and Smoluchowski [XI-118] did not take into account its energy, and thus the particle did not have finite velocity. In 1930, L. Ornstein and G. Uhlenbeck investigated refinements of the theory of Brownian motion from the point of view of inertia. In this "refined" theory the sample paths of the motion of a particle were differentiable (true, with infinite acceleration).

Kolmogorov looked at this circle of problems in a general setting, assuming that the state of the system under study is described by $2n$ coordinates $q = (q_1, \dots, q_n)$, $\dot{q} = (\dot{q}_1, \dots, \dot{q}_n)$ for which a probability density $G(t, q, \dot{q}; t', q', \dot{q}')$, $t < t'$, is defined, and (in the spirit and on the basis of "Analytic methods") showed that this density satisfies the corresponding forward equation (equation of Fokker–Planck type; this is equation (9) in [1934;10]), which is a degenerate equation of parabolic type. After this work the mathematical theory was developed by N. S. Piskunov, one of Kolmogorov's students.

The cycle of Kolmogorov's work on the theory of Brownian motion includes also his paper [1937;7], [PS-24], finished in 1936, which touches upon general questions in reversibility of statistical laws of nature.

The essence of the problem is as follows. In a thermodynamic understanding, Brownian motion is irreversible, in the sense that for a large number of diffusing particles there is a "smoothing" of the probability distribution describing the state of the particles as time increases. Conversely, as time decreases, the "heterogeneity" of this distribution increases. Schrödinger was apparently the first to point out that a diffusion process will nevertheless have a certain reversibility if in considering the behavior of the process on the time interval $[t_0, t_1]$ the probability distribution is fixed at both the initial time t_0 and the final time t_1.

In the paper [1937;7] Kolmogorov gave necessary and sufficient conditions for statistical reversibility (in the sense that the "usual" transition probability densities coincide with the "reverse" ones) in the very general situation of an n-dimensional Markov diffusion process.

The investigations of Andreĭ Nikolaevich in the 1930's on the theory of Brownian motion and diffusion processes include also his 1937 joint paper "An investigation of a diffusion equation connected with an increasing amount of substance and its application to a certain biological problem" ([1937;4], [MM-38]), written with Petrovskiĭ and Piskunov. This was the first paper establishing the existence of wave solutions of parabolic equations and the convergence to them of solutions of a Cauchy problem as $t \to \infty$.

The diffusion equation treated in this paper has the form $(k > 0)$

$$\frac{\partial u}{\partial t} - k\frac{\partial^2 u}{\partial x^2} = F(x), \tag{27}$$

where $F = F(u)$ is a sufficiently smooth function defined on $[0, 1]$ and satisfying the conditions $F(0) = F(1) = 0$, $F(u) > 0$ for $0 < u < 1$, $F'(0) = \alpha > 0$, and $F'(u) < \alpha$ for $0 < u \leq 1$.

A solution $u = u(x, t)$ is sought that satisfies the initial condition $u(x, 0) = f(x)$, where

$$f(x) = \begin{cases} 1, & x > 0, \\ 0, & x < 0. \end{cases} \tag{28}$$

It turned out that (27) has a solution of traveling wave type, that is, a solution of the form

$$u(x,t) = W(x - ct) \tag{29}$$

for all $c \geq \sqrt{2k\alpha}$. Further, the problem (27), (28) has a solution that converges with respect to shape and velocity to a solution of traveling wave type as $t \to \infty$. In the commentaries of G. I. Barenblatt ([MM], English pp. 481–487) and A. I. Vol′pert ([XI-79], pp. 333–358) there is a fairly thorough description of the state of investigations in mathematical physics (in particular, the theory of combustion), whose beginnings stem from this paper of Kolmogorov, Petrovskiĭ, and Piskunov.

The very short 1935 note by Kolmogorov, "The Laplace transformation in linear spaces" ([1935;3], [MM-26]), was the first paper in which the *characteristic function* of a probability measure on a Banach space was defined; in other words, the concept of the characteristic function was generalized to the infinite-dimensional case. In this note he also gave a definition of normal distributions and of nth-order moment forms, and he mentioned the possibilities for generalizing the central limit theorem to linear spaces and the importance of the concepts introduced for constructing a nonlinear quantum theory. He wrote:

> If we want to make a nonlinear quantum theory, we will have to consider the distributions themselves, or their characteristic functions, or, finally, the whole collection of moments.

L. LeCam (1947) and É. Mourier (1950, 1953) later came to the concept of the characteristic function. The contemporary state of problems involving probability distributions in Banach spaces is reflected in the monograph *Probability distributions in Banach spaces* by N. N. Vakhaniya, V. I. Tarieladze, and S. A. Chobanyan ("Nauka", Moscow, 1985; English transl., Reidel, Dordrecht, 1987).

We dwell further on some publications of Kolmogorov in the astonishingly fruitful 1930's: 5 in 1931, 6 in 1932, 9 in 1933, 9 in 1934, 4 in 1935, 17 in 1936, 9 in 1937, 16 in 1938, and 5 in 1939.

His paper "A simplified proof of the Birkhoff–Khinchin ergodic theorem" appeared in 1937 (in German) and 1938 (in Russian) ([1937;6], [1938;12], [MM-39]). In the same year he finished a joint paper with I. M. Gel′fand, "On rings of continuous functions on topological spaces" ([1939;4], [MM-41]), which explained that in a fairly general situation the purely algebraic structure of the ring of continuous functions on a topological space with a sufficiently "nice" topology determines (up to a homeomorphism) the topological space itself.

In 1939 he published the paper "On inequalities between upper bounds of the successive derivatives of an arbitrary function on an infinite interval" ([1939;3], [MM-40]), the sources of which go back to work of Landau and Hadamard. The statement of the problem and a formulation of the result obtained by Kolmogorov are as follows.

Consider functions $f = f(x)$ on the real line, and let $M_k(f) = \sup_x |f^{(k)}(x)|$, $k = 0, \ldots, n$. Then ([MM-40], English p. 277):

To a triple of positive numbers M_0, M_k, M_n, $0 < k < n$, there corresponds a function $f = f(x)$ such that

$$M_0 = M_0(f), \quad M_k = M_k(f), \quad M_n = M_n(f)$$

The "Komarovka house".

if and only if

$$M_k \leq C_{nk} M_0^{\frac{n-k}{n}} M_n^{\frac{k}{n}},$$

where the C_{nk} *are constants that can be determined.*

For a history of the problem, a discussion of this result, and subsequent investigations in this direction see [MM], English pp. 442–447 (commentary of V. M. Tikhomirov and G. G. Magaril-Il′yaev).

In 1935 an important event occurred in the lives of Andreĭ Nikolaevich and Pavel Sergeevich Aleksandrov—the acquisition of part of an old manor-house in the village of Komarovka near Bolshevo. Andreĭ Nikolaevich recalls:

> On returning from a trip along the Volga and to the Caucasus in the fall of 1929 Pavel Sergeevich and I decided to live together somewhere near Moscow. That same fall we leased a three-room half-house in the village of Klyazma, where we settled together with my aunt Vera Yakovlevna. She was then already 66 years old, but she was still able to manage our household.
>
> In 1931 we moved to a country house in the same village owned by Pavel Sergeevich's brother, the well-known surgeon Mikhail Sergeevich Aleksandrov. Here we occupied the whole house in the winter, but in the summer Pavel Sergeevich and I moved into a garret in order to make space for the owners. On the balcony adjoining the garret we liked to sleep out under the stars until late in the fall. In the mornings we ran out to bathe in the river Ucha right up to the time when it froze over. In the winter we incorporated a barefoot run wearing only shorts in our morning exercise.
>
> But in 1935 the need clearly arose to have our own comfortable home. The problem unexpectedly resolved itself. While skiing around the area in the early spring of that year, we saw a fairly large white house on the edge of a field and said in unison, "How

> nice it would be if that house were for sale." Miraculously, it turned out that the house, which belonged to Anna Sergeevna Shtekker, a sister of Konstantin Sergeevich Stanislavskiĭ, was indeed for sale. It was not just the white house that was for sale, but the whole property containing also a log house and a barn that could be converted into a house. Anna Sergeevna's son, Georgiĭ Andreevich Shtekker, had skillfully put together a group of buyers, and as the result of this transaction Pavel Sergeevich and I became the owners (together with the artist Vladimir Ivanovich Kozlinskiĭ) of a large part of the white house.

In "Memories of P. S. Aleksandrov" [1986;2] Andreĭ Nikolaevich wrote:

> This "house in Komarovka" satisfied all our needs: there was room for a large library and we could put up our guests in separate rooms for several days and even for longer periods.
>
> By the end of the 30's we were both well settled in. As a rule, of the seven days of the week, four were spent in Komarovka, one of which was devoted entirely to physical recreation—skiing, rowing, long excursions on foot (our long ski tours covered on average about 30 kilometers, rising to 50; on su‘ March days we went out on skis wearing nothing but shorts, for as much as four hours at a stretch). On the other days, morning exercise was compulsory, supplemented in winter by a 10 kilometer ski run. We were never walruses, bathing every day all the year round; we bathed at any time we felt like it. Especially did we love swimming in the river just as it began to melt, even when there were still snow drifts on the banks. When the frost was not too severe the morning run of about one kilometer was done barefoot and wearing only shorts. I swam only short distances in icy water but Aleksandrov swam much further. It was I however who skied naked for considerably longer distances.
>
> One of our favorite ways of arranging ski-runs was this: we invited young mathematicians to, say, Kalistov, and from there we set out in the direction of Komarovka. Some who did not get as far as Komarovka caught a bus and set off for home. Those who got to Komarovka were offered a shower and then if one felt like it a romp in the snow and then dinner. In the golden years of the Komarovka house the number of guests at the dinner table after skiing could be as many as fifteen.

The sketch made by Andreĭ Nikolaevich gives a good idea of the daily routine at Komarovka.

As Kolmogorov wrote in [1986;2]:

> There were two cases in which this arrangement could be altered; a) when scientific research became exciting and demanded an unlimited length of time; b) on sunny days in March when skiing was the only occupation.

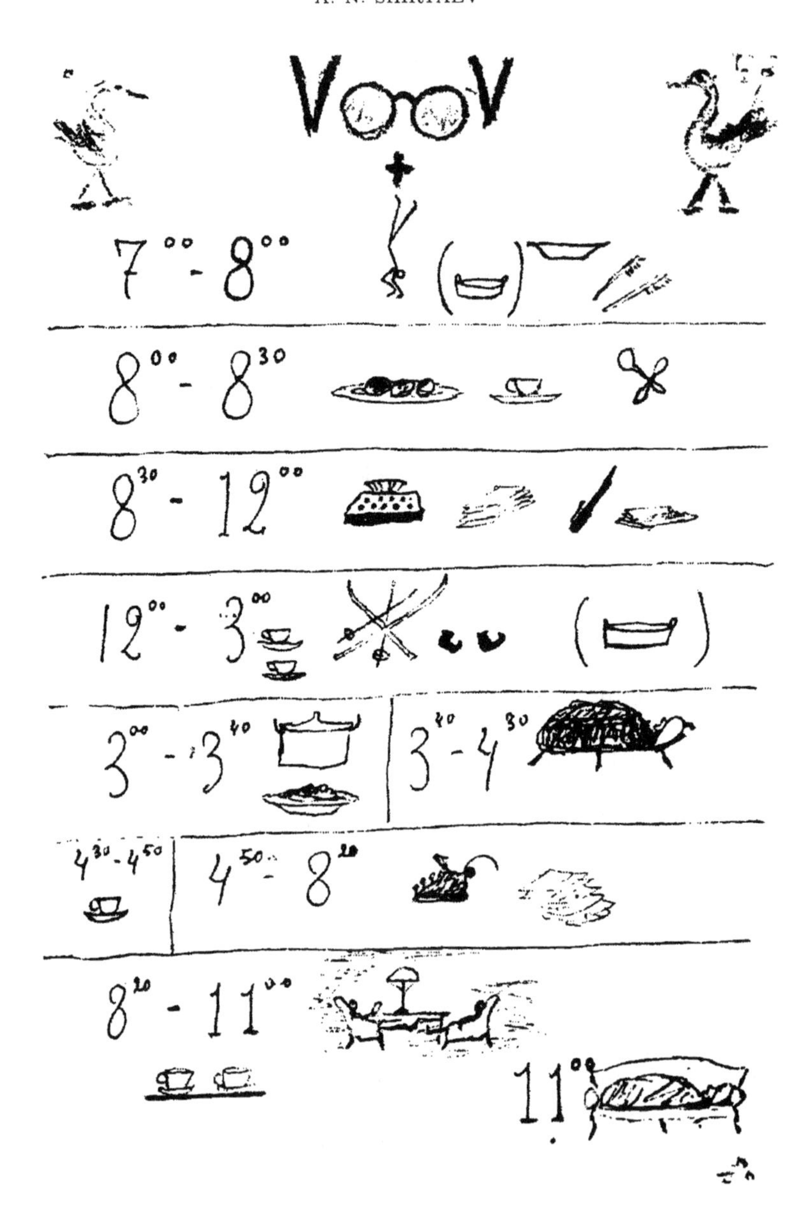

Daily routine at Komarovka, sketched by Kolmogorov.

(The newspaper "Mayak" of the Pushkin Region of the Moscow Oblast dated March 8, 1989 contains the following:[3] "405 years ago, in the property registers of the Moscow State of 1584, there is the first mention of 'the village of Komarovka on the Klyazma River; indeed, near that village a mill grinds with one set of millstones, plowed tracts of average soil 42 and one eighth *chet'i* [unit of area] in the field, and two more, and also 300 shocks of hay', the patrimony of the Troitse-Sergiev Monastery until 1764.

[3]This text from the newspaper was kindly presented to me by neighbors of the "Komarovka house": G. S. and B. V. Zakhoder.

"In 1852 Komarovka is registered in the second police district of the Moscow region; in it are 6 farmsteads and 69 people live there.")

There is more below about certain other aspects of the exceptionally energetic and fruitful activities of Andreĭ Nikolaevich in the 1930's (in particular, those involving questions in the teaching of mathematics). Here we mention the following facts from his biography.

In 1930, by a decision of the State Academic Council of the Narkompros of the RSFSR, A. N. Kolmogorov was approved for the academic rank of Professor on the "mathematics" faculty (PR no. 014075), and in 1935, by a decision of the qualification commission of the Narkompros of the USSR, he was awarded (without a defense) the academic degree of Doctor of the Physical and Mathematical Sciences (DT no. 000038).

At the Second All-Union Mathematical Congress in Leningrad in 1934 it was decided to create a new mathematics journal called "Uspekhi Matematicheskikh Nauk" (briefly, UMN). From the time it was founded (1936) to 1944 it was printed first as isolated issues, but from 1946 it began to appear as a periodical. Beginning in 1936 until the year of his death, Kolmogorov was a permanent member of the editorial board of UMN, and was its editor-in-chief from 1946 to 1955 and from 1982 to 1987.

From December 1, 1933 to April 15, 1939 Kolmogorov occupied the post of Director of the Scientific Research Institute (SRI) of Mathematics of Moscow University.

In 1939 (January 29) he was elected a member (Academician) of the Academy of Sciences of the USSR. From 1939 to 1942 he was Academician Secretary of the Physical and Mathematical Sciences Section and a member of the Presidium of the Academy. From 1938 to 1958 he was head of the Probability Theory Section of the Steklov Institute of Mathematics.

The forties (1940–1949)

The end of the 1930's and beginning of the 1940's saw Kolmogorov's work on the *theory of random processes with stationary increments* and the related *theory of isotropic turbulence*, work that was exceptional in its conceptual significance, the fundamentality of its ideas, and the most diverse possibilities for applications.

In [PS], English p. 521, Andreĭ Nikolaevich remarked that his "interest in the spectral theory of stationary random processes arose in connection with work of Khinchin and E. E. Slutskiĭ," who had studied such processes as far back as the beginning of the 1930's. In the same place he referred to his report "The statistical theory of oscillations with continuous spectrum" delivered at the General Meeting of the Academy of Sciences in 1947 ([1947;6], [PS-34]), in which he underscored, in particular, the fundamental importance of working with the Stieltjes integral with the goal of obtaining a general representation of stationary oscillatory processes, including almost periodic oscillations and oscillations with continuous spectrum.

The 1940 papers "Curves in Hilbert space invariant under a one-parameter group of motions" ([1940;2], [MM-42]) and "Wiener's spiral and some other interesting curves in Hilbert space" ([1940;3], [MM-43]) involve (the "L^2-theory" of) random processes $\xi = (\xi_t)_{-\infty<t<\infty}$ with *stationary increments* and their various subclasses (including wide-sense stationary processes, Wiener processes, and so on)

A. G. Kurosh, A. A. Glagolev, A. N. Kolmogorov,
I. G. Petrovskiĭ, and L. S. Pontryagin (1940).

from the point of view of the structure of their correlation function

$$B_\xi(\tau_1, \tau_2) = \mathbf{E}[\xi_{t+\tau_1} - \xi_t]\overline{[\xi_{t+\tau_2} - \xi_t]}$$

(see Theorem 2 in [1940;2], [MM-42]) and of the possibility of a spectral representation of the process $\xi = (\xi_t)_{-\infty<t<\infty}$ itself, which (Theorem 3 in the same place) Kolmogorov gave in the form

$$\xi_t = \int_{-\infty}^{\infty} (e^{i\lambda t} - 1)\Phi(d\lambda) + x_0 + x_1 t. \tag{30}$$

If the process $\xi = (\xi_t)_{-\infty<t<\infty}$ under consideration (which is continuous in the mean square) happens not only to have stationary increments but to be simply *stationary*, then $x_0 = \int_{-\infty}^{\infty} \Phi(d\lambda)$, $x_1 = 0$, and it is easy to get from (30) a spectral representation for $\xi = (\xi_t)$ as a Lebesgue–Stieltjes integral

$$\xi_t = \int_{-\infty}^{\infty} e^{i\lambda t}\Phi(d\lambda) \tag{31}$$

with respect to a random measure $\Phi(\Delta)$ with orthogonal values ($\mathbf{E}\Phi(\Delta_1)\overline{\Phi(\Delta_2)} = 0$ for $\Delta_1 \cap \Delta_2 = \varnothing$). This representation was discovered independently by Cramér (1942, [XI-43]) and Maruyama (1949, [XI-60]). See also the papers of Loève [XI-57] and of Blanc-Lapierre and Fortet [XI-10] devoted to general "harmonizable" random processes.

The paper "Wiener's spiral and some other interesting curves in Hilbert space" is closely related to [1940;2], [MM-42] and is devoted to some special cases of processes $\xi = (\xi_t)_{-\infty<t<\infty}$ with stationary independent increments. In essence, this article takes up processes ξ having the property of *self-similarity*, which means that for any $k \neq 0$ there exists a similarity transformation A_k such that for all t

$$\xi_{kt} = A_k \xi_t.$$

It turned out that the "structure" function $B_\xi(\tau_1, \tau_2)$ for such processes can be represented in the form

$$B_\xi(\tau_1, \tau_2) = c[|\tau_1|^\gamma + |\tau_2|^\gamma - |\tau_1 - \tau_2|^\gamma], \tag{32}$$

where c and γ are real constants satisfying the inequalities $c \geq 0$ and $0 \leq \gamma \leq 2$.

We remark that many publications in recent years with applications, for example, in statistical physics have investigated such random processes ξ having self-similarity properties (see [XI-97]).

After these papers on processes with stationary increments came Kolmogorov's classical papers on stationary (in the wide sense) random processes. Here, as in [1940;2] and [1940;3], he made broad use of Hilbert space techniques, which was reflected, for example, in the title of the 1941 paper "Stationary sequences in Hilbert space" ([1941;1], [PS-27]).

In this paper he introduced new concepts (subordination of one stationary sequence to another, regularity, singularity, and minimality) that served as sources of numerous subsequent investigations for vector processes with continuous time. (See [XI-90], [XI-91] for details.)

Kolmogorov's concept of *subordination* of a stationary sequence $\eta = (\eta_n)$, $n = 0, \pm1, \dots$, to another stationary sequence $\xi = (\xi_n)$, $n = 0, \pm1, \dots$, means that these sequences are jointly stationary, and that the smallest closed linear subspace $H(\xi)$ containing the elements ξ_n, $n = 0, \pm1, \dots$, contains all the elements η_n, $n = 0, \pm1, \dots$. An unexpected result discovered here by Kolmogorov is that subordination can be expressed in purely spectral terms. Namely, a sequence η is subordinate to ξ if and only if there exists a function $\varphi(\lambda) \in L^2(F_{\xi\xi})$ for which the spectral functions $F_{\eta\eta}(\lambda)$ and $F_{\xi\eta}(\lambda)$ in the representations of the covariance functions

$$B_{\eta\eta}(n) = \int_{-\pi}^{\pi} e^{i\lambda n} dF_{\eta\eta}(\lambda), \quad B_{\xi\eta}(n) = \int_{-\pi}^{\pi} e^{i\lambda n} dF_{\xi\eta}(\lambda)$$

are such that $F_{\eta\eta}(\lambda) = \int_{-\infty}^{\lambda} |\varphi(u)|^2 dF_{\xi\xi}(\lambda)$ and $F_{\xi\eta}(\lambda) = \int_{-\infty}^{\lambda} \varphi(u)\, dF_{\xi\xi}(\lambda)$, with $F_{\xi\xi}(\lambda)$ in $B_{\xi\xi}(\eta) = \int_{-\pi}^{\pi} e^{i\lambda n} dF_{\xi\xi}(\lambda)$ $(B_{\xi\eta}(n) = \mathbf{E}\xi_{n+k}\overline{\eta}_k)$.

The property of being *singular* (one also says deterministic) of a sequence $\xi = (\xi_n)$ with $\mathbf{E}\xi_n = 0$ for $n = 0, \pm1, \dots$ means that the space $H(\xi)$ coincides with $H_{-\infty}(\xi) = \bigcap_n H_n(\xi)$, where $H_n(\xi)$ is the closed linear manifold generated by the variables ξ_k, $k \leq n$.

On the other hand, the property of *regularity* (being purely nondeterministic) means that the space $H_{-\infty}(\xi)$ is trivial.

Appealing to results on boundary properties of functions analytic in the disk, Kolmogorov obtained the following widely known result:

A nondegenerate stationary sequence ξ is regular if and only if the spectral function $F_{\xi\xi}(\lambda)$ has a density $f_{\xi\xi}(\lambda)$ such that

$$\int_{-\pi}^{\pi} \ln f_{\xi\xi}(\lambda)\, d\lambda > -\infty. \tag{33}$$

Kolmogorov's definition of *minimality* for a sequence ξ means that the smallest closed linear subspace $\widehat{H}(\xi)$ containing all the ξ_n with $n \neq 0$ does not coincide with the smallest closed linear subspace $H(\xi)$ containing all the ξ_n, $n = 0, \pm1, \dots$.

He showed that:

A stationary sequence ξ is minimal if and only if there exists a spectral density $f_{\xi\xi}(\lambda)$ such that $f_{\xi\xi}(\lambda) > 0$ almost everywhere with respect to Lebesgue measure, and

$$\int_{-\pi}^{\pi} \frac{d\lambda}{f_{\xi\xi}(\lambda)} < \infty. \tag{34}$$

In addition, if all these conditions are satisfied, then

$$d_\xi \equiv \inf_{\xi_0 \in \hat{H}(\xi)} \mathbf{E}[\xi_0 - \widehat{\xi}_0]^2 = (2\pi)^2 \left[\int_{-\pi}^{\pi} \frac{d\lambda}{f_{\xi\xi}(\lambda)}\right]^{-1}. \tag{35}$$

Very closely related to the paper [1941;1] is Andreĭ Nikolaevich's continuation of it in "Interpolation and extrapolation of stationary random sequences" ([1941;2], [PS-28]). He wrote in the introduction to the latter:

> Spectral conditions are established for being able to extrapolate stationary random sequences from a sufficiently large number of terms with an arbitrary specified accuracy.

In this paper Kolmogorov not only gave a rigorous statement of the problems of *extrapolation* and *interpolation* of (real) random sequences, but also obtained the first results on the magnitude of errors in these problems.

His results and those of Wiener [XI-14] in this area created in the theory of random processes a whole direction of study with broad applications both in science and in technology.

Denoting by

$$\sigma_E^2(n,m) = \inf_{(a_1,\dots,a_n)} \mathbf{E}[\xi_{t+m} - \widetilde{\xi}_{[t-1,\dots,t-n]}]^2$$

the minimal error in predicting the value ξ_{t+m}, $m \geq 0$, from the values $\widetilde{\xi}_{[t-1,\dots,t-n]}$ of the form $a_1\xi_1 + \cdots + a_n\xi_n$, $n > 0$, and letting $\sigma_E^2(m) = \lim_{n\to\infty} \sigma_E^2(n,m)$, Kolmogorov obtained an *explicit formula* for $\sigma_E^2(m)$, expressed in spectral terms. He showed that if $\int_0^\pi \log f_{\xi\xi}(\lambda)\,d\lambda = -\infty$, then the prediction error $\sigma_E^2(m)$ is $= 0$ for all $m \geq 0$. In the case of regular sequences, when the integral $\int_0^\pi \log f_{\xi\xi}(\lambda)\,d\lambda$ is finite, Kolmogorov gave an explicit formula for the error $\sigma_E^2(m)$ (Theorem 2 in [1941;2]).

In the interpolation problem Kolmogorov introduced the minimal error

$$\sigma_I^2(n) = \inf \mathbf{E}[\xi_t - \widehat{\xi}_{[t-n,t+n]}]^2$$

in interpolating the value of ξ_t from the values of the form

$$\widehat{\xi}_{[t-n,t+n]} = \sum_{k=\pm1,\dots,\pm n} a_k \xi_{t+k}$$

and for $\sigma_I^2 = \lim_{n\to\infty} \sigma_I^2(n)$ he established that if the integral $R = \frac{1}{\pi}\int_0^\pi \frac{d\lambda}{f_{\xi\xi}(\lambda)}$ is equal to $+\infty$, then $\sigma_I^2 = 0$, while if $R < \infty$, then $\sigma_I^2 = R^{-1}$ (Theorem 3 in [1941;2]).

Above we mentioned a report of Kolmogorov [1947;2] at the 1947 General Meeting of the Academy of Sciences, characterized by A. M. Yaglom ([PS], English pp. 545–551) as "the first popular survey of the spectral theory of stationary random processes, one of the most important areas of the mathematical theory of random functions, developed only shortly before (with the active participation of

Kolmogorov himself) but little known outside a narrow circle of specialists." (In this commentary of Yaglom the reader can find detailed historical and bibliographical references related to the spectral theory of stationary processes; see also [XI-27], [XI-41], [XI-90], [XI-91], [XI-121].)

It is difficult to overestimate the (purely *physical*, in essence) papers of Kolmogorov on the *theory of turbulence*, which served as a foundation for the whole subsequent development of both representations and the theory of the local structure of turbulent motion and its applications.

Commenting ([MM], English p. 487) on his work in turbulence at the beginning of the 1940's, Kolmogorov wrote:

> My interest in the study of turbulence of liquid and gas flows began at the end of the 1930's. It was clear to me from the start that the main mathematical tool for investigation had to be the theory of random functions of many variables (random fields), which had just been conceived at that time. Moreover, I soon realized that one could hardly hope to create a self-contained, pure theory. In the absence of such a theory it was necessary to use hypotheses obtained by processing experimental data. It was also important to get talented co-workers able to work in a mixed setting, combining development of the theory with experiment.
>
> In the last respect I was lucky: A. M. Obukhov, who had transferred from Saratov University to Moscow University, began his undergraduate thesis under me in 1939, and then became my graduate student. Almost at the same time M. D. Millionshchikov began to work for me as a graduate student of the Moscow Institute of Aviation. Later A. S. Monin and A. M. Yaglom also became my students.
>
> In 1946 O. Yu. Shmidt proposed that I head the Turbulence Laboratory at the Institute of Theoretical Geophysics of the Academy of Sciences (in 1949 the leadership of this laboratory passed to Obukhov). I did not do experimental work directly, but expended my energy on computational and graphical processing of the data obtained by other researchers.

The presence of chaotic fluctuations in the velocity $u(x,t)$, the pressure $p(x,t)$, and other hydrodynamic characteristics in liquid and gas flows (which is called turbulence) makes it less realistic to study individual fields of turbulent motions. This is what makes a statistical description of flows interesting, as was already clear to Reynolds, the founder of the theory of turbulence, at the end of the last century. However, his proposed averaging with respect to a given integral of time or space turned out not to be very convenient because of the difficulty in obtaining sufficiently simple and reliable equations for the average field.

Kolmogorov understood the averaging in a probabilistic sense—the sense of averaging over an ensemble. He thus proposed regarding fields of hydrodynamic characteristics as random functions of spatial and time coordinates, which is what is now commonly done.

His deep physical intuition enabled Kolmogorov to discover the general qualitative and quantitative laws that determine the probabilistic-statistical regime of fine-scale fluctuations in the locally isotropic developed turbulence with a sufficiently large Reynolds number on the basis of two similarity hypotheses he formulated in the article "The local structure of turbulence in an incompressible viscous fluid for very large Reynolds numbers", dated December 28, 1940. These hypotheses made it possible to derive fundamental quantitative relations having the character of new laws of nature, first and foremost Kolmogorov's famous **two-thirds law** saying that *the mean square difference of the velocities at two points a distance* r *apart (not too small nor too large) is proportional to* $r^{2/3}$.

The so-called longitudinal and transversal structure functions $B_{dd}(r)$, $B_{nn}(r)$ of the velocity field, introduced by Kolmogorov in [1941;4], [MM-45], have been experimentally measured many times, and on a significant interval of values of r the two-thirds law has been confirmed ($B_{dd}(r) \sim r^{2/3}$), as has the relation $B_{nn}(r) \sim \frac{4}{3}B_{dd}(r)$. (For details see [1941;4], [MM-45], and [MM], English pp. 488–504.)

A report on the development following from his papers [1941;4] and [1941;6] on turbulence theory was given by Kolmogorov at the 1961 International Colloquium on the Mechanics of Turbulence in Marseilles ([1962;6], [MM-58]), relating to work of Obukhov [XI-76]. In the report Kolmogorov proposed replacing his two similarity hypotheses in [XI-76] by two refined hypotheses relating now to the normalized differences of the velocities, and supplementing them by a third hypothesis postulating the logarithmic normality of the probability distribution of the energy dissipation rate ε_r averaged over a sphere of radius r, and the linearity of the dependence of the variance of $\log \varepsilon_r$ on $\log(L/r)$, where L is the characteristic scale of length in the flow under consideration.

These three new hypotheses led to a refinement of the two-thirds law: to a new formula $B_{dd}(r) \sim r^{2/3}(L/r)^{-k}$ that takes into account remarks of L. D. Landau on the necessity of considering the changeability of the dissipation of energy as the ratio L/r increases. (For details see [MM], English pp. 393, 496.)

Summarizing Andreĭ Nikolaevich's role in the theory of turbulence, we present the concluding lines of Obukhov's paper "Kolmogorov flow and its laboratory simulation" [XI-77]:

> Kolmogorov's personal contribution to the study of turbulence and his ideas relating to the general theory of dynamical systems are fundamental reference points in the evolution of investigations in the most complicated phenomenon of nature—turbulence in connection with diverse areas of knowledge.

Kolmogorov responded vigorously to appeals to him not only as a universal mathematician, but also as an applied mathematician possessing an astounding ability to penetrate to the essence of a problem, to discover the principal and determining factor, to bring clarity to the situation.

A good example of this is Andreĭ Nikolaevich's articles on the theory of gunnery written during the Second World War (1941–1945).

In "Determination of the center of scattering and measure of accuracy from a limited number of observations" [1942;1] submitted September 15, 1941 Kolmogorov noted that "I was asked to give my opinion about the differences between the existing methods for estimating the measure of accuracy from experimental data", and he presented a critical comparison of various approaches, modestly commenting that the paper mainly could be only of methodological interest.

Together with colleagues from the Mathematical Institute, the Mechanics and Mathematics Department of Moscow University, the Naval Artillery Research Institute, and other organizations, Kolmogorov performed a great deal of theoretical and computational work on the effectiveness of gunnery systems. The nature of these extensive investigations can be judged from his two publications "The number of hits in several shots and general principles for estimating the effectiveness of a gunnery system" [1945;1] and "Artificial scattering in the case of hitting with a single shot and scattering in a single measurement" [1945;2].

In [1945;1] he considered the number μ of hits in a group of n shots ($\mu = 0, \dots, n$). Letting $P_m = \mathbf{P}(\mu = m)$ and $R_m = \mathbf{P}(\mu \geq m)$, and letting $\mathbf{E}\mu$ be the expectation of the number of hits, Kolmogorov posed the question of defining the concept of an "effectiveness index" of a gunnery system. He noted that the widespread "arguments in the literature about the comparative merits and deficiencies of an estimate 'according to the mathematical expectation' and 'according to the probability' often do not have sufficient clarity," and he posed the question as to whether the collection of probabilities $P_0, \dots, P_n$ characterizing the gunnery system from the point of view of the probability distributions of the number of hits "can be replaced by some single variable

$$W = f(P_0, \dots, P_m)$$

depending on the P_k, which can can be taken as the index of effectiveness of the gunfire. After a discussion of this question (§ 1 in [1945;1]), Kolmogorov derived some precise formulas for the probabilities P_k, and then gave practically convenient formulas for computing approximations of them.

The next circle of questions treated in this paper involves, first, a classification of factors affecting the result of the gunnery in order to choose a rational gunnery system, and, second, the following problem on artificial scattering.

Let $p_i = p_i(\alpha_i, \beta_i)$ be the probability of a hit with the ith round; it depends on the bearing α_i and the sight β_i. Let $(\overline{\alpha}_i, \overline{\beta}_i)$ be the combination (unique, as a rule) of values of α_i and β_i maximizing the probability of a hit:

$$\max p_i(\alpha_i, \beta_i) = p_i(\overline{\alpha}_i, \overline{\beta}_i),$$

and let $\overline{\alpha} = (\overline{\alpha}_1, \dots, \overline{\alpha}_n)$ and $\overline{\beta} = (\overline{\beta}_1, \dots, \overline{\beta}_n)$.

The problem is whether

$$\max E = W(\overline{\alpha}, \overline{\beta}),$$

that is, whether maximal effectiveness of the gunnery will be achieved by firing so that the probability of hitting is as large as possible for each separate round.

In the paper it is remarked that in the two special cases when

$$W = \mathbf{E}\mu$$

and when W has the form ($c_i \geq 0$)

$$W = c_1 P_1 + \dots + c_n P_n,$$

and the events B_i, $i = 1, \dots, n$, consisting of hitting the target with the ith round are independent, then the property $\max W = W(\overline{\alpha}, \overline{\beta})$ is valid, and hence in these cases the most rational gunnery system is that in which the probability of hitting the target is maximized for each individual round.

However, with different criteria for the effectiveness W of gunfire that is no longer true in general, and this leads to the conclusion that to achieve the greatest possible effectiveness of gunfire on the whole one should intentionally choose settings for individual rounds that deviate from the settings that give the greatest probability of hitting the target for each round. This kind of gunnery is called gunfire with artificial scattering, and a typical situation when it is useful is when "it is most essential to make at least a small number of hits, which may be considerably fewer than the total number n of rounds fired."

In [1945;2] Kolmogorov considered the "artificial scattering" situation in which just a single hit and scattering in one dimension suffice for achieving the set purpose, for example, in the bombardment of a long narrow strip (a bridge) perpendicular to the plane of the gunfire.

In 1949 Andreĭ Nikolaevich wrote the paper "Solution of a probability problem related to the mechanism of stratification" ([1949;3], [PS-37]). In a commentary on this paper A. B. Vistelius wrote ([PS], English pp. 591–597):

> At the time the paper was published the geological sciences made practically no use of such concepts as random variable, probability distribution function, and a sequence of values of a random variable. It was a period in which the foundation was laid for the creation of a scientific basis for several geological disciplines by introducing in them ideas of a stochastic nature for the quantities studied by these sciences. This fundamental restructuring, which subsequently gave rise to mathematical geology, was strongly promoted not only by the paper cited, but also by the personal advice and views of Kolmogorov in the period from 1945 to 1950.

In 1948 the Russian translation of Cramér's book "Mathematical methods of statistics" appeared. Giving a great amount of attention to the organization of mathematical statistics education and the evolution of research in mathematical statistics in Russia, Kolmogorov wrote a detailed Foreword and also edited the translation itself. In the Foreword he notes:

> The systematic courses in mathematical statistics that have been taught up to the present time are structured on a theoretical foundation that certainly does not correspond to contemporary needs ... Investigations in special problems of mathematical statistics have grown beyond that level of exposition of mathematical and probability-theoretic premises ... Cramér's book is an attempt at a systematic presentation of the fundamental questions in mathematical statistics from a completely modern viewpoint.

To promote research in statistics in the USSR, Andreĭ Nikolaevich gave the two reports "Fundamental problems of theoretical statistics" [1949;6] and "The real meaning of the results of analysis of variance" [1949;7] in 1948 at the Second All-Union Conference on Mathematical Statistics in Tashkent. In March 1950 he published the long paper "Unbiased estimators" ([1950;1], [PS-38]). Besides a systematic treatment of properties of unbiased estimators and of various methods for constructing them with the help of sufficient statistics, he showed the importance in applications of unbiased estimators for problems of statistical control and rejection in industrial mass production.

Kolmogorov's paper "Unbiased estimators" together with the subsequent paper "Statistical acceptance sampling when the admissible number of defective items is zero" [1951;4] stimulated a broad formulation of the theory and practice of probabilistic methods of sampling. (See the commentaries of Yu. K. Belyaev and Ya. P. Lumel'skiĭ in [PS], English pp. 585–587.)

Kolmogorov is rightfully regarded as one of the founders of the modern theory of *branching* random processes. (The term itself was introduced by Andreĭ Nikolaevich in a Moscow University seminar he led in 1946–1947.)

Although isolated problems connected with simple models of branching processes had been treated even earlier (Fisher [XI-103], Steffenson [XI-100], Leontovich [XI-50], and Kolmogorov ([1938;14], [PS-25])), it was Kolmogorov's articles "Branching stochastic processes" (with N. A. Dmitriev) ([1947;4], [PS-32]) and "Computation of the final probabilities for branching stochastic processes" (with B. A. Sevast′yanov) ([1947;5], [PS-33]) that started the stormy development of the now-independent new area of probability theory called the theory of branching random processes.

The papers [1947;4] and [1947;5] take up models of Markov branching processes with several types of particles for both the case of discrete time and the case of continuous time. Later more complicated models were considered that took into account the dependence of reproduction on the age of the particles, their locations in space, the dependence on energy, and so on. In addition to the contemporary state of the theory of branching random processes, the books [XI-95], [XI-106], and [XI-6] and the commentary of Sevast′yanov in [PS] (English pp. 538–539) contain a wealth of material on diverse applications to biology, chemistry, physics, engineering, and so on.

An exceptionally important role in the development of probability theory was played by the 1949 book of Gnedenko and Kolmogorov entitled "Limit distributions for sums of independent random variables" [1949;1], which is devoted to the theory of limit theorems—a theory in which the concepts of infinitely divisible and stable laws are central.

In the Foreword to the book of Gnedenko and Kolmogorov the problem of limit theorems leading to these laws is described as follows.

If $\xi_1, \xi_2, \dots$ is a sequence of independent identically distributed random variables and $S_n = \xi_1 + \dots + \xi_n$, then it is natural to pose the general question of conditions under which the limit relation

$$\mathbf{P}\left(\frac{S_n - A_n}{B_n} \le x\right) \to V(x), \quad n \to \infty$$

holds for this or that choice of constants A_n and B_n, and of what limit distributions $V(x)$ can appear.

This problem was completely solved by Khinchin, who established that all the conceivable laws $V(x)$ are the so-called "stable" distributions, of which it was said in [1949;1] that "the circle of actual applied problems in which they will play an essential role will in time prove to be fairly large." (As we now know, this has indeed happened.)

Passing to problems connected with infinitely divisible distributions, the authors of the book especially emphasize the importance of considering a *series scheme* of random variables

$$\xi^n = (\xi_{n1}, \dots, \xi_{nn})$$

that are independent within each series, since in the framework of this scheme "there can be quite meaningful and practically interesting limit theorems involving sums of independent terms and leading to distribution laws that differ essentially from normal laws" ([1949;1], English p. 8).

Posing the problem of conditions under which a limit relation of the form

$$\mathbf{P}\left(\frac{S_n - A_n}{B_n} \leq x\right) \to V(x), \quad n \to \infty,$$

is possible for $S_n = \xi_{n1} + \cdots + \xi_{nn}$ and the problem of the form of the limit laws $V(x)$, Gnedenko and Kolmogorov confined themselves to requiring "limiting negligibility" in the variability of each separate term (in the form $\sup_{1\leq k\leq n} \mathbf{P}(|\xi_{nk} - a_{nk}| \geq \varepsilon B_n) \to 0$ as $n \to \infty$, where the a_{nk} are constants), and gave an exemplary presentation of the whole circle of problems that more than one generation of probabilists has learned from.

In the eighth chapter of the book the authors dwell on results relating to the rate of convergence of the probabilities $V_n(x) = \mathbf{P}((S_n - A_n)/B_n \leq x)$ to $V(x)$, noting that Chebyshev had already emphasized the importance of asymptotic expansions for $V_n(x)$ and had given such an asymptotic expansion in powers of $n^{-1/2}$ for the case of the central limit theorem (though without a precise justification).

In 1951 (December 14) Gnedenko and Kolmogorov were awarded the Chebyshev Prize of the Academy of Sciences for the book *Limit distributions for sums of independent random variables.*

At the end of the 1940's Andreĭ Nikolaevich began the major task of heading the mathematics section of the 2nd edition of the "Great Soviet Encyclopedia" (GSE). He not only prepared the glossary, selected authors, and edited articles, but also he himself wrote articles about many diverse areas of mathematics. From 1949 to 1958 he wrote 88 articles for the GSE (5 in 1949, 19 in 1950, 7 in 1951, 25 in 1952, 14 in 1953, and 9 in 1954, and so on).

Among these articles a special role is played by his well-known article "Mathematics" [1954;6], in which he "traced the historical development of mathematics in a condensed form and on a theoretical basis, indicated the key points of this development, and proposed an original scheme for dividing it into periods."

The present author has heard repeatedly from representatives of the older generation about a report at a meeting of the Moscow Mathematical Society in 1944 (December 11) devoted to problems in probability theory. Recently two pages of his notes for this report were found in the home archives of Kolmogorov (hand-written by him and typed by him). I present them here in entirety:

A. N. Kolmogorov
Problems in probability theory

This report will contain a characterization of the contemporary state of probability theory, together with an attempt to outline the prospects for its development in the coming years. Besides a general characterization of the main directions of work that seem especially urgent to the speaker (indicated by Roman numerals), examples are given of precisely formulated individual problems that merit the attention of researchers (indicated by Arabic numerals).

Kolmogorov and Aleksandrov at Komarovka.

I. *Axiomatics and problems of applicability.*
   1. *A logical justification of mathematical statistics, that is, of methods for testing hypotheses, estimating parameters, controlling and regulating mass production from sample observations.*
   2. *Construction of a general theory of observation, together with the effect of the observer on the system being observed, with the purpose of clarifying the logical foundations of quantum physics.*
   3. *Clarification of the logical nature of probability-theoretic analogies in number theory.*

II. *Limit theorems.*
   4. *Refinement of the basic limit theorem.*
   5. *Limit theorems for distributions in function spaces as a universal source of special limit theorems of classical type.*

III. *Infinite-dimensional probability distributions.*
   6. *Distributions invariant under various transformation groups for scalar, vector, and tensor functions. The problem is of interest from the standpoint of the statistical mechanics of continua, in particular, the statistical theory of turbulence.*
   7. *Distributions invariant under various groups of motions for configurations of particles in space. The problem is of interest from the standpoint of the statistical theory of crystals and crystallization.*

IV. *Classical random processes.*
   8. *General solution of the Smoluchowski equation.*
   9. *Estimation of a stationary process from a bounded segment of it (estimation of parameters and questions of prediction).*
   10. *Nonlinear spectral analysis of random processes with continuous spectrum.*

11. *Statistical properties of dynamical systems "in the general case".*

12. *Interrelations between reversible and irreversible processes.*

The speaker is not undertaking to list all essential problems in related sciences that require probability theory for their solution. Mentioned above are only those applied questions that are inseparably linked to the basic lines of development of probability theory itself; and moreover, of course, only those for which the formulation of the corresponding probability-theoretic problems is clear for the speaker. In particular, the general theory of "classical random processes" is broadly represented, since here, as in the case of the general theory of "dynamical systems", an independent purely mathematical branch of research has already taken shape to a sufficient degree. On the other hand, questions in quantum physics appear only in problem 2, *since in this area it seems difficult to project a long independent line of purely mathematical investigations.*

Then followed Andreĭ Nikolaevich's signature and the date: November 18, 1944.

After reading this text, one cannot but be surprised by the fact that half a century ago Kolmogorov precisely defined problems that have been the subjects of numerous investigations right up to the present day!

In August 1943 Andreĭ Nikolaevich got a large "warehouse" notebook in which he began his diary. Here are some excerpts from it:

Sunday, August 1, 1943
New moon.

... *Why this notebook now? There are two reasons*:

1) *The idea of a diary as a means of discipline has long attracted me. To write down what I have done, what I want to change in my life, what I should do, and then to check my performance; the idea is not new but is just as useful at* 40 *as at* 16.

2) *By the age of 40 I was becoming more sensitive to how life is passing by ... and going away, of the extent to which what has been lived through has its own value in comparison with what is to come* (*at* 16, *even at* 30, *everything still seems in preparation for a more significant future*). *Therefore, the need arose to fix the present at the very moment it passes from nonexistence as not yet having been, to nonexistence as already past ...*

Tuesday, September 14
morning.

The frost on the neighboring roofs melts under the sun's rays. It seems we will finally have a clear sunny day.

Yesterday I added to my report a little admonition about applied mathematics:

At each given moment there is only a fine layer between the "trivial" and the impossible. Mathematical discoveries are made in this layer. Therefore, in most cases a made-to-order applied problem either is trivial to solve or cannot be solved at all ... It is another matter if the applications are selected (*or adapted!*) *to match a new mathematical apparatus of interest to a particular mathematician ...*

Chebotarev naturally remarked that no consistent development of mathematics "by layers" happens at all, but that the researcher randomly "gnaws his way" into some point of an unknown realm remote from anything known previously.

Anna Dmitrievna and Andreĭ Nikolaevich in Komarovka (1940's).

In the evening, however, Anat. Iv.[4] *and I discussed the question of how long one would have to wait for this or that mathematical discovery if a particular mathematician had not done it.* Pro domo mea *I expressed the opinion that without me an everywhere divergent Fourier–Lebesgue series would not have been constructed over the past* 20 *years. As a more significant and more long-term example, A. I. proposed the Hilbert theorem on a finite system of invariants.*

A. I. was talking about the general problem of a finite system of invariants for an arbitrary group of linear transformations (according to Weyl).

To cancel out my backwardness in the whole area of linear algebra and continuous groups I had to read Weyl's book: A. I. and Lev Semenovich have drawn from it the whole of their knowledge.

On December 1 Andreĭ Nikolaevich presented[5] a "schedule for becoming a great man if one has enough desire and diligence for that" (see the table).

In 1941 (March 13) Kolmogorov and Khinchin were awarded the Stalin (State) Prize of 2nd Degree for a cycle of papers on the theory of random processes.

Kolmogorov was awarded the Order of the Red Banner of Labor in 1940, the Order of Lenin in 1944 and 1945, and a medal "for valorous work in the Great Patriotic War" in 1945.

In the fall of 1942 Andreĭ Nikolaevich married Anna Dmitrievna Egorova, a friend from his school years.

The fifties (1950–1959)

The general theory of Hamiltonian systems, information theory, the ergodic theory of dynamical systems, ε-entropy, superposition of functions, Hilbert's thirteenth problem, etc.—these are areas of mathematics in which Andreĭ Nikolaevich Kolmogorov worked in the 1950's and which are certainly associated with his name. Many scientific branches and schools grew out of his papers in these areas.

[4] A. I. Maslov. [A. N. Sh.]

[5] The text is shortened somewhat. [A. N. Sh.]

1944–1953	Minor analysis course	Investigations in linear algebra, group representations, and multidimensional differential geometry	Random process and dynamical systems. Homogeneous fields of random variables and turbulence. Foundations of probability theory and mathematical statistics. Theory of observation and experiment	Z	Z	Investigations on the foundations of mathematics	Work of a direct pedagogical nature only	Algebra and elements of analysis for secondary school	Z
1954–1963	Major analysis course	?	?	Investigations in mathematical physics	Z	Investigations in logic	Z	Under favorable conditions, university and academic work	Secondary school geometry and trigonometry
1964–1973	Second edition of a short course in analysis	Z	?	?	Course in mathematical physics	Course in logic. Investigations in the history of science	?	Logic for secondary school	Preparation of a complete collection of his mathematical papers on his 70th birthday
1974–1983	?	Z	Z	Z	?	“A history of the forms of human thought”	Z	“Mathematical diversions”	Writing the memoirs of his past life

Kolmogorov's work in the theory of dynamical systems during this period consisted of two cycles. The first was directly connected with problems in classical mechanics ([1953;3], [1954;1], [1954;2], or [MM-51], [MM-52], [MM-53]), and the second with problems in information theory.

In his own commentary on his research on classical mechanics Andreĭ Nikolaevich wrote ([MM], English pp. 503–504):

> My work on classical mechanics arose under the influence of von Neumann's work [XI-72] on the spectral theory of dynamical systems and especially under the influence of the classical work of Bogolyubov and Krylov in 1937 (see [12] cited in [1954;2]).
>
> I became extremely interested in the question of what ergodic sets can be (in the Bogolyubov–Krylov sense) in the dynamical systems of classical mechanics, and which of the types of these sets can fill a set of positive measure (this question remains unsolved to the present day). To put together some concrete information we organized a seminar devoted to the study of individual examples. My reflections on these and related topics met with a broad response from young Moscow mathematicians.

At the International Congress of Mathematicians in Amsterdam in 1954 Kolmogorov presented the report "The general theory of dynamical systems and classical mechanics" at the concluding session [1954;2]. Its theme, in the terminology of Poincaré, had to do with the "fundamental problem of dynamics"—to investigate the behavior of quasiperiodic motions of Hamiltonian systems under a small perturbation of the Hamiltonian function (a "small" change in the Hamiltonian function $W(p)$ is understood as a change into a function $W(p) + \theta S(q, p, \theta)$ with a small parameter θ).

A notable result in the theory developed here by Kolmogorov is that *quasiperiodic motions are preserved for most initial conditions and in the general position case*, that is, when $\det(\partial^2 W/\partial p^2) \neq 0$.

Kolmogorov's theory and its subsequent development made it possible to solve a number of problems that had awaited solution for a long time. For example, his theory implies the stability of a nonsymmetric heavy rigid body rotating rapidly about a fixed point, the stability of the motion of an asteroid of negligible mass in the planar restricted three-body problem, and the preservation of most magnetic surfaces under small changes of the magnetic field in toroidal systems.

In his report [1954;2] Kolmogorov, talking about the method of proof, remarked that it was based on a reworking of "the idea, widely discussed in the literature on celestial mechanics, of the possibility of avoiding the appearance of abnormally 'small denominators' when computing perturbations of orbits." The following example of a small denominator is well known: $2\omega_1 - 5\omega_2 = 0.007$, where $\omega_1 = 299.1''$ and $\omega_2 = 120.5''$ are the frequencies of the motions of Jupiter and Saturn. The presence of such small denominators leads to large mutual perturbations in the motions of these planets because of the fact that expressions of the form $m\omega_1 + n\omega_2$ appear as denominators in perturbation theory series having the form

$$\sum_{m,n\neq 0} a_{mn} \frac{\exp[i(m\omega_1 + n\omega_2)]}{m\omega_1 + n\omega_2}.$$

The method Kolmogorov proposed and used to overcome the difficulties connected with small denominators was later developed by his student V. I. Arnol′d and by J. Moser and is now known as "KAM theory" (the Kolmogorov–Arnol′d–Moser theory).

In 1965 Kolmogorov and Arnol′d were awarded the Lenin Prize for their work in the theory of perturbations of Hamiltonian systems.

The second cycle of Kolmogorov's work on the theory of dynamical systems has to do with applications of ideas in information theory to the study of ergodic properties of these systems (see p. 56 below).

Andreĭ Nikolaevich turned to problems in information theory proper at the beginning of the 1950's under the influence of Shannon's work, observing that (see his Foreword in the Russian translation [1963;9]):

> The significance of Shannon's work for pure mathematics did not receive sufficient recognition at first. I recall that as far back as the ICM in Amsterdam (1954) my American colleagues, specialists in probability theory, regarded my interest in Shannon's work as somewhat exaggerated, since this was more technology than mathematics. Now such opinions need hardly be refuted.
>
> It is true that Shannon left to his successors the rigorous "justification" of his ideas in some difficult cases. However, his mathematical intuition was amazingly precise.

From this it becomes clear that mathematical work was needed to give information theory a stable mathematical foundation.

In this respect the first work was done by Khinchin in the papers [XI-109] and [XI-110] devoted to proving basic theorems in information theory for the discrete case, and by Gel′fand, Kolmogorov, and Yaglom in the papers "On a general definition of amount of information" ([1956;8], [IA-2]) and "The amount of information and entropy for continuous distributions" ([1958;2], [IA-4]), in which they now considered the general case, established general properties of the "amount of information", obtained explicit formulas for the information characteristics in the Gaussian case, and formulated coding theorems for reproducing messages with a specified accuracy.

At a session of the Academy of Sciences in 1956 on scientific problems in the automation of production, Andreĭ Nikolaevich made a long plenary report, "The theory of information transmission" ([1957;1], [IA-3]), in which he expounded the basic ideas of information theory and explained the limits of its applicability.

As remarked by R. L. Dobrushin in his commentary ([IA], English pp. 222–225), these papers "began a tradition of presenting results in information theory on a level of mathematical rigor which since that time has invariably been present in the work of those experts in information theory who regard themselves as mathematicians as well as those who regard themselves as engineers."

Thinking about Shannon's ideas in information theory led Andreĭ Nikolaevich to a completely unexpected synthesis of these ideas with those he had presented in his papers of the 1930's on the theory of approximations and the theory of algorithms.

Let us dwell on this somewhat more closely.

As a *measure of the indeterminacy of discrete messages* ξ taking discrete values $x_1, x_2, \dots$ with probabilities $p_1, p_2, \dots$ Shannon used the concept of "entropy" $H(\xi)$, defined as

$$H(\xi) = -\sum_i p_i \log p_i.$$

He also defined the concept of the *information* $I(\xi, \eta)$ contained in the object ξ with respect to η, setting

$$I(\xi, \eta) = \sum_{i,j} p_{ij} \log \frac{p_{ij}}{p_i q_j}$$

in the case of discrete random variables ξ and η, with

$$p_{ij} = \mathbf{P}(\xi = x_i, \eta = y_j), \quad p_i = \mathbf{P}(\xi = x_i), \quad q_j = \mathbf{P}(\eta = y_j),$$

or

$$I(\xi, \eta) = \iint \log \frac{p(x,y)}{p(x)q(y)} p(x,y)\, dx\, dy$$

in the case of variables ξ and η having joint distribution density $p(x, y)$ and one-dimensional densities $p(x)$ and $q(y)$, respectively.

For the case of *continuous* messages all the natural analogues of the Shannon entropy lead to an entropy value equal to infinity. In this connection Kolmogorov repeatedly stressed that for the case of arbitrary messages the fundamental concept should be *not the entropy, but the amount of information* $I(\xi, \eta)$ of one object ξ with respect to another η.

Starting out from this concept, Kolmogorov gave a general definition of the *ε-entropy* $H_\varepsilon(\xi)$ of a random object ξ as the quantity

$$H_\varepsilon(\xi) = \inf I(\xi, \eta),$$

where (for a fixed distribution P_ξ of the object ξ) the infimum is taken over all pairs (ξ, η) of random variables satisfying the following restriction: their joint distribution $P_{\xi\eta}$ belongs to some given class W_ε depending on the parameter ε. (For example, $W_\varepsilon = \{(\xi, \eta) : \mathbf{E}\rho(\xi, \eta) \leq \varepsilon\}$, where ρ is some metric in the space of values of the objects under consideration.)

The quantities $H_\varepsilon(\xi)$ were considered by Shannon under the name "rate of message creation". In [1957;1], [IA-3] Kolmogorov wrote:

> Although choosing a new name for this quantity does not change the essence of the matter, I decided to propose a renaming that underscores a broader interest of the concept and its deep analogy with ordinary exact entropy ... I wanted especially to emphasize the interest in investigating the asymptotic behavior of the ε-entropy as $\varepsilon \to 0$. The cases studied previously ... are only very special cases of the laws that can be encountered here. To understand the perspectives that are opening up it may be of interest to look at my note [1956;7], presented in different terms.

The note [1956;7] is "On some asymptotic characteristics of totally bounded metric spaces". In this paper Kolmogorov introduced the notion of the ε-entropy of a *nonrandom* object C which is a subset of a metric space (X, ρ). It is denoted by $\mathcal{H}_\varepsilon(C)$ and defined to be the logarithm to base 2 of the minimal number $N_\varepsilon(C)$ of elements of a covering of C by sets of diameter at most 2ε.

Along with the ε-entropy $\mathcal{H}_\varepsilon(C)$, later called the *absolute* ε-entropy, Kolmogorov introduced also the *relative* ε-entropy $\mathcal{H}_\varepsilon(C, X)$, defined to be the logarithm to the base 2 of the minimal number $N_\varepsilon(C, X)$ of elements in an ε-net in X for the set C.

By intention, $\mathcal{H}_\varepsilon(C)$ and $\mathcal{H}_\varepsilon(C, X)$ are constructed according to the same scheme as for the so-called "Kolmogorov widths" he introduced back in 1936. For example, the logarithm to base 2 of the inverse function of the "width"

$$\mathcal{E}_N(C, X) = \inf_{A \in \Sigma_N} \sup_{x \in C} \inf_{y \in A} \|x - y\|,$$

where Σ_N is the collection of N-point approximating sets, coincides with the ε-entropy $\mathcal{H}_\varepsilon(C, X)$.

As almost always with new concepts introduced by Kolmogorov, the ε-entropy way of estimating the "metric massiveness" of function classes and spaces served as a basis for the creation of new directions of research in approximation theory. (About this see Tikhomirov's commentary in [IA], English pp. 231–239.)

In 1958 Andreĭ Nikolaevich published "A new metric invariant of transitive dynamical systems and of automorphisms of Lebesgue spaces" ([1958;6], [IA-5]). (A somewhat reworked variant of this paper was published later, [1985;3].) In this paper ideas from information theory led Kolmogorov to introduce entropy characteristics in the theory of dynamical systems (this is the "second cycle" referred to on p. 54).

A dynamical system is understood in [1958;6] to be a one-parameter group $\{S^t\}$ of measure-preserving transformations of a probability space $(X, \mathfrak{X}, \mu)$. By analogy with the theory of stationary random processes, Kolmogorov introduced the concept of a quasiregular dynamical system, or, in modern terms, a *K-system*. The importance of this concept for an ergodic system revealed itself a few years later when Sinaĭ explained that many classical dynamical systems not having anything in common with probability theory are K-systems.

For quasiregular dynamical systems Kolmogorov introduced the concept of entropy (a somewhat modernized refined variant of the corresponding definition was given in [1985;3]). In a paper published a little later Sinaĭ proposed a definition of entropy that was applicable to any dynamical system (for details see Sinaĭ's commentary in [IA], English pp. 247–250).

For the case of discrete time ($t = 1, 2, \ldots$; $S^2 = S$) the now commonly accepted definition of the *Kolmogorov–Sinaĭ entropy* is that given below.

Let $A = \{A_1, \ldots, A_N\}$ be a finite partition of X, that is, $\bigcup_i A_i = X$ and $A_i \cap A_j = \varnothing$ for $i \neq j$. The entropy associated with this partition is

$$H(A) = -\sum_{i=1}^{N} \mu(A_i) \log \mu(A_i).$$

Let $\mu_{i_1,\ldots,i_r} = \mu(A_{i_1} \cap SA_{i_2} \cap \cdots \cap S^{r-1}A_{i_r})$, and let

$$H_r(A) = -\sum_{(i_1,\ldots,i_r)} \mu_{(i_1,\ldots,i_r)} \log \mu_{(i_1,\ldots,i_r)}.$$

The Kolmogorov–Sinaĭ entropy of the dynamical system determined by a measure-preserving transformation S of a probability space $(X, \mathfrak{X}, \mu)$ is defined to be the quantity

$$H(S) = \sup_A \lim_{r\to\infty} \frac{H_r(A)}{r}.$$

The concept of the entropy of a dynamical system has played a prominent role in ergodic theory and first of all in the solution of the problem of metric classification of dynamical systems, that is, the problem of finding a (complete) set of invariants that imply a metric isomorphism between dynamical systems.

We recall that the first example of a metric invariant was the *spectrum* of the dynamical system. In the class of ergodic dynamical systems with *pure point spectrum* this exhausts the complete system of metric invariants (von Neumann [XI-73], Halmos and von Neumann [XI-105]). But for dynamical systems with *continuous spectrum*, in particular, for the most important subclass of them—systems with countably multiple Lebesgue spectrum (say, for Bernoulli automorphisms)—there were no approaches to the problem of metric classification before Kolmogorov's paper.

The entropy of a dynamical system turned out to be a fundamentally *new invariant* of a metric isomorphism of dynamical systems, independent of their spectra, as follows from the fact that the entropy can take any admissible values on the class of systems with countably multiple Lebesgue spectrum. Thus, the new invariant permitted a "splitting" of dynamical systems with countably multiple Lebesgue spectrum into a continuum of invariant subclasses with different values of the entropy and hence not metrically isomorphic.

Further, all the K-systems (which, from the point of view of the theory of random processes, correspond to processes with the very weak condition of relaxing the dependence between the values of the process on time intervals that are far apart) have countably multiple Lebesgue spectrum and positive entropy. Since no metric invariants distinguishing K-systems were known besides entropy, the natural question arose as to whether K-systems with the same values of entropy were metrically isomorphic.

The first examples of a nontrivial isomorphism of Bernoulli automorphisms were given by L. D. Meshalkin, a student of Kolmogorov. Sinaĭ showed that Bernoulli automorphisms with the same entropy are weakly isomorphic, that is, each of them can be realized as a factor of the other. A complete solution of the isomorphism problem was obtained by the American mathematician D. Ornstein, who showed that *Bernoulli automorphisms with equal entropy are metrically isomorphic.* After some time it was explained that the entropy is not a complete system of metric invariants in the class of all K-systems. (Ornstein and Shields showed that the number of nonisomorphic types of K-systems with the same entropy is uncountable.) For details see Sinaĭ's commentary on Kolmogorov's ergodic theory papers ([IA], English pp. 247–250), along with the monographs [XI-40] and [XI-78], where it is shown that the entropy theory of dynamical systems begun by Kolmogorov's paper [1958;6], [IA-5] is now a whole extensive direction in ergodic theory.

The entropy characteristics of "metric massiveness" ($\mathcal{H}_\varepsilon(C), \mathcal{H}_\varepsilon(C,X), \dots$) introduced by Kolmogorov made it possible for him to give a lucid interpretation of the results of A. G. Vitushkin on the nonrepresentability in general of an r-smooth function of n variables as a superposition of l-smooth functions of m variables when $n/r > m/l$. These investigations led Kolmogorov directly to Hilbert's thirteenth problem, which was to prove that there is a *continuous* function of three variables that *cannot* be expressed as a superposition of continuous functions of two variables.

In 1955 Andreĭ Nikolaevich started a student seminar devoted to the theory of approximate representation of functions of several variables, including problems of approximate nomography. As he recalled ([MM], English pp. 518–519) about the work of this seminar: "I stated Hilbert's thirteenth problem already in the introductory lecture as a long-term plan that would almost certainly not be achieved."

In Hilbert's formulation, his "thirteenth problem" consisted in showing that a solution $f = f(x, y, z)$ of the 7th-degree equation

$$f^7 + xf^3 + yf^2 + zf + 1 = 0,$$

to which a general 7th-degree algebraic equation can be reduced, cannot be represented as a superposition of continuous functions of two variables. (See the collection *Hilbert's problems* [XI-82].)

Kolmogorov's 1956 paper "On the representation of continuous functions of several variables as superpositions of continuous functions of fewer variables" ([1956;6], [MM-55]) begins with these words:

Theorem 3 *below has the following surprising corollary: any continuous function of arbitrarily many variables is representable as a finite superposition of continuous functions of at most three variables. For an arbitrary function of four variables such a representation has the form*

$$f(x_1, x_2, x_3, x_4) = \sum_{r=1}^{4} h^r[x_4, g_1^r(x_1, x_2, x_3), g_2^r(x_1, x_2, x_3)].$$

In 1957 Arnol′d showed that each continuous function of *three variables* can be represented as a superposition of continuous functions of *two variables.* Hilbert's conjecture was thereby *refuted.*

Finally, in the same year 1957 Kolmogorov made the final step, showing that *each continuous function $f(x_1, \ldots, x_n)$ of n variables is representable as a superposition of continuous functions of a single variable and the addition operation:*

$$f(x_1, \ldots, x_n) = \sum_{q=1}^{2n+1} \chi_q \left[\sum_{p=1}^{n} \varphi^{pq}(x_p) \right],$$

where the "inner" functions φ^{pq} are universal, and only the "outer" functions χ_q depend on the function $f(x_1, \ldots, x_n)$ being decomposed ([1957;2], [MM-56]).

Arnol′d recalls ([MM], English p. 519) that "Kolmogorov called this result his most technically difficult achievement".

In 1953 Kolmogorov published the paper "Some recent work in the area of limit theorems in probability theory" ([1953;2], [PS-41]), and wrote in the introduction:

> In the mid 1940's some people thought that the question of limit theorems of classical type (that is, the question of the limit behavior of the distributions of sums of a large number of independent terms or terms connected in a Markov chain) was mainly a closed subject. But in reality there was a significant revival of work in these very "classical directions" at the end of the 1940's, due to the fact that the accuracy of the estimates obtained for the remainder terms in limit theorems was far from adequate, and to the fact that

> in a number of old problems that had yielded to solution only under complicated and restrictive conditions it was possible to obtain a very simple and complete solution.

This paper of Kolmogorov was important in that it analyzed and presented various estimates for the closeness of probability distributions and various forms of convergence for them. In the same paper Andreĭ Nikolaevich proposed a new formulation of the problem of approximating the distributions of sums $S_n = \xi_{n1}+\cdots+\xi_{nn}$ of independent random variables satisfying conditions of limiting negligibility. (Under this assumption the limit distribution, if it exists, is infinitely divisible.)

The essence of this new formulation consists in the following.

Investigations of convergence to specific infinitely divisible laws did not reveal completely the mechanism behind the behavior of distributions of sums of independent random variables. The cardinal change in Kolmogorov's statement of the problem was that he suggested approximating the distributions of the sums S_n not by an *individual* distribution (and then studying the rate of this approximation), but rather using *whole families* of infinitely divisible distributions close to the S_n for the approximation. Moreover, in this paper he expressed the idea of possibly getting uniform theorems not only for a fixed sequence $(\xi_{n1},\dots,\xi_{nn})$, but also for whole classes of random variables.

In 1955 Yu. V. Prokhorov proved [XI-86] that for the distance $\rho(F,G) = \sup_x |F(x) - G(x)|$ between distribution functions $F = F(x)$ and $G = G(x)$ it is possible for any distribution F to construct a sequence of infinitely divisible distributions D_n such that

$$\rho(F^{*n}, D_n) \to 0, \quad n \to \infty,$$

where $F^{*n} = F * \cdots * F$ is the n-convolution of F.

In other words,

$$\rho(F^{*n}, \mathcal{D}) \equiv \inf_{D\in\mathcal{D}} \rho(F^{*n}, D) \to 0, \quad n \to \infty, \tag{36}$$

where $\mathcal{D}$ is the class of all infinitely divisible distributions.

In the paper "Two uniform limit theorems for sums of independent terms" ([1956;2], [PS-43]), dated November 12, 1956, Kolmogorov took a fundamental step forward: he showed that the convergence in (36) *is uniform over the class $\mathcal{F}$ of all distributions*, that is,

$$\psi(n) \equiv \sup_{F\in\mathcal{F}} \inf_{D\in\mathcal{D}} \rho(F^{*n}, D) \to 0, \quad n \to \infty, \tag{37}$$

and that

$$\psi(n) \le cn^{-1/5}, \tag{38}$$

where c is a constant.

The same paper gave a corresponding theorem also for the case of terms with different distributions.

These results of Kolmogorov provided a powerful stimulus to subsequent investigations about the correct order of decrease of the function $\psi(n)$ as $n \to \infty$.

In 1960 Yu. V. Prokhorov [XI-88] showed that $\psi(n) \le cn^{-1/3}(1+\ln n)^2$. In 1963 F. M. Kagan [XI-37] established the estimate $\psi(n) \le cn^{-1/3}(1+\ln n)$. In the 1963 paper "Approximation of distributions of sums of independent terms by infinitely divisible distributions" ([1963;1], [PS-51]) Kolmogorov himself obtained the estimate $\psi(n) \leq cn^{-1/3}$.

In [XI-62] and [XI-63] (1960 and 1961) Meshalkin established the *lower* estimate

$$\psi(n) \geq cn^{-2/3}(1+\ln n)^{-7/2}.$$

Finally, in 1980–1983 Arak and Zaĭtsev obtained in [XI-2], [XI-4], [XI-5], and [XI-30] the *definitive* result

$$c_1 n^{-2/3} \leq \psi(n) \leq c_2 n^{-2/3}, \tag{39}$$

where c_1 and c_2 are constants.

An exposition of this and related results, along with a thorough historical and bibliographical description of the problems under discussion can be found in the book [XI-3] by Arak and Zaĭtsev.

The result (39) is interesting in part because it implies that the convolution F^{*n} has an infinitely divisible approximation whose order in the central limit theorem (in the uniform metric) is considerably better than the $n^{-1/2}$ given by the classical Berry–Esseen estimate.

In the paper [1956;2] Kolmogorov gave a number of inequalities for Lévy's *concentration function* $Q(l,\xi) = \sup_x \mathbf{P}(x \leq \xi \leq x+l)$ of a random variable ξ. This function is a convenient characteristic of the "dispersion" of a random variable, especially for a quantitative description of the degree of increase in the "dispersion" when independent random variables are summed. Developing results of Lévy on properties of the concentration function, Kolmogorov obtained an estimate of $Q(l,S_n)$ for the sum $S_n = \xi_1 + \cdots + \xi_n$ in terms of the concentration functions of the individual independent terms. In 1961 Rogozin [XI-89] strengthened the Kolmogorov inequality, giving it the form

$$Q(l,S_n) \leq cl\Big\{\sum_{i=1}^{n} l_i^2[1-Q(l_i,\xi_i)]\Big\}^{-1/2}. \tag{40}$$

Later Miroshnikov and Rogozin obtained in [XI-68] and [XI-69] the inequality

$$Q(l,S_n) \leq cl\Big\{\sum_{i=1}^{n} \mathbf{E}[\min(|\widetilde{\xi}_i|, l_i/2)]^2 Q_i^{-2}(l_i)\Big\}^{-1/2}, \tag{41}$$

where $\widetilde{\xi}_i$ is the symmetrization of the random variable ξ_i, that is, $\widetilde{\xi}_i = \xi_i - \xi_i'$, where ξ_i and ξ_i' are independent and identically distributed random variables. (The estimate (40) follows from (41) for $2l \geq l_i$, $i = 1,\ldots,n$.)

In probability theory the 1950's were the years of creation of a new area of the theory of random processes—*the theory of functional limit theorems* (in particular, "the invariance principle") for random processes—and papers of Kolmogorov, Yu. V. Prokhorov, and Skorokhod played a prominent role in its formation.

As far back as 1931, in the article "A generalization of the Laplace–Lyapunov theorem" ([1931;4], [PS-12]), Kolmogorov considered a problem that in modern terms could be called a problem in the theory of "functional limit theorems". The program report of Andreĭ Nikolaevich on December 11, 1944 (mentioned at the end of the preceding section "The forties") contained "Limit theorems for distributions in function spaces as a universal source of special limit theorems of classical type" as one of the problems in probability theory. In the forties and fifties a number of individual results were published relating to the "invariance principle" (Erdős and Kac [XI-119] in 1947; Doob [XI-26] in 1949; Donsker [XI-24], [XI-25] in 1951–1952; Gikhman [XI-19] in 1953; Fortet and Mourier [XI-104] in 1954; Maruyama [XI-61] in 1955, and so on).

At the November 30 meeting of the Moscow Mathematical Society in 1948, Kolmogorov gave the report "Measures and probability distributions in a function space", in which he proposed the idea of regarding *the distributions of a random process as measures on the Borel algebra of some function space* and indicated that in this setting it is natural to understand convergence of probability distributions of random processes as *weak convergence* of the corresponding measures on the function space. In 1953 Kolmogorov's student Yu. V. Prokhorov stated the following important result [XI-85]: *for a family of probability measures on an arbitrary metric space to be relatively compact it suffices that the family be dense* (the necessity is true for complete separable metric spaces). He then constructed a general theory of weak convergence of probability distributions of random processes ([XI-87], 1956).

In 1955–1956 Skorokhod considered random processes with values in the space D (of functions continuous from the right and having limits from the left) and introduced in it a *metric* turning it into a Hausdorff topological space. In the paper "On Skorokhod convergence" ([1956;1], [PS-42]) Kolmogorov gave a more convenient metric (equivalent to Skorokhod's) in which D is *separable.* In the same paper Kolmogorov showed that it follows from general topological considerations that there is a metric on D turning it into a *complete* separable metric space, and he posed the problem of the simplest possible explicit construction of such a metric. (Shortly thereafter such a metric was constructed by Prokhorov [XI-87].)

The main ideas and results of the general theory of weak convergence of probability measures on metric spaces were presented in the joint report of Kolmogorov and Prokhorov ([1956;3], [PS-44]) at a 1956 conference on probability theory and mathematical statistics in Berlin.

Kolmogorov's plans relating to the concept of "algorithm" belong to the 1950's. The purpose, on the one hand, was to give the broadest possible mathematical definition of this concept, and, on the other hand, to see that this general definition did not reduce to an extension of the already formulated concept of "computable function". These ideas were expounded by Kolmogorov in his report "On the concept of an algorithm" ([1953;1], [IA-1]) at a meeting of the Moscow Mathematical Society (March 17, 1953) and developed by his student V. A. Uspenskiĭ in his diploma thesis "A general definition of algorithmic computability and algorithmic reducibility", completed in the first half of 1952. The 1958 paper "On the definition of algorithm" ([1958;9], [IA-6]) by Kolmogorov and Uspenskiĭ and the commentary of Uspenskiĭ and Semenov ([IA], English pp. 251–260) give detailed analyses of the development of these ideas of Kolmogorov in the theory of algorithms, and present the contemporary state of this circle of problems.

At the end of the 1940's and beginning of the 1950's Andreĭ Nikolaevich initiated and immediately offered a new course, "Analysis III", in the Mechanics and Mathematics Department of Moscow University. It was intended to replace individual courses in the theory of functions of a real variable, measure theory, integral equations, and the calculus of variations by a single course. In 1954 and 1960 the two separate volumes of the book *Elements of the theory of functions and functional analysis* ([1954;12], [1960;8]) appeared, taken from lecture notes for the new course taught by Kolmogorov and S. V. Fomin.

A. N. Kolmogorov.

The second edition came out in 1968 as a single book, went through several editions (the 5th edition in 1981; the 6th in 1989) and has been translated into many languages.

As always, Andreĭ Nikolaevich devoted much of his powers and energy to scientific organization and pedagogical activities. In 1951–1953 he was Director of the Scientific Research Institute of Mechanics and Mathematics at Moscow University, from December 25, 1954 to February 1, 1958 he was Dean of the Mechanics and Mathematics Department of Moscow University, from 1939 to 1966 he headed the Probability Theory Branch of this department, and from 1939 to 1958 he headed the Probability Theory Department of the Steklov Institute of Mathematics. During the spring semester of 1958 he worked as a Professor at the University of Paris.

On April 28, 1953 Kolmogorov was elected an honorary member of the Moscow Mathematical Society, having been a member since February 1, 1930.

The sixties (1960–1969)

This period includes the fundamental papers of Kolmogorov on **the logical foundations of information theory and probability theory and their restructuring on an algorithmic basis.** It can be said that while Kolmogorov's efforts in the 1950's had been directed toward the use of ideas from information theory in diverse areas of mathematics, he now created new areas of mathematics: *algorithmic information theory and algorithmic probability theory.*

Analyzing various approaches to the introduction of the fundamental concepts of information theory, Kolmogorov singled out the following three:

I. *The purely combinatorial approach.*

II. *The probabilistic approach.*
III. *The algorithmic approach.*

In a commentary to his work in information theory and some of its applications ([IA], English pp. 219–221) Kolmogorov described the meaning of the first approach:

> In a combinatorial approach the amount of information communicated by designating a specific element in a set of N objects is taken to be the logarithm of N to base 2 (R. Hartley, 1928). For example, there are
>
> $$C(m_1, \ldots, m_s) = \frac{n!}{m_1! \cdots m_s!}$$
>
> different words in an s-element alphabet that contain m_i occurrences of the ith letter $(m_1 + \cdots + m_s = n)$. Therefore, the amount of information of interest to us is
>
> $$H = \log_2 C(m_1, \ldots, m_s).$$
>
> As $n, m_1, \ldots, m_s$ tend to infinity the asymptotic formula
>
> $$H \sim \left(\sum_i \frac{m_i}{n} \log_2 \frac{m_i}{n} \right) \tag{1}$$
>
> comes into effect. The reader has probably already noticed the similarity between this formula and the formula
>
> $$H = n \sum_i p_i \log_2 p_i \tag{2}$$
>
> from probabilistic information theory.
>
> If our word was formed according to the well-known scheme by independent trials, then the asymptotic formula (1) is an obvious consequence of the formula (2) and the law of large numbers, but the circle of practical applications of (1) is considerably broader (see, for example, works on information transmission on nonstationary channels). It seems to me an important problem in general to get rid of extraneous probabilistic assumptions wherever possible. I have insisted repeatedly in my lectures on the independent value of the purely combinatorial approach to information theory problems.
>
> My work and that of my collaborators on the ε-entropy and the ε-capacity of compact classes of functions is based on the purely combinatorial approach to the concept of entropy. Here the ε-entropy $H_\varepsilon(K)$ is the amount of information necessary for distinguishing some individual function in the class of functions, while the ε-capacity $C_\varepsilon(K)$ is the amount of information that can be coded by elements of K under the condition that elements of K no closer than ε to each other can be reliably distinguished.

The probabilistic approach to the fundamental concepts of information theory was described above (in "The fifties").

Kolmogorov in the mountains (1961).

As for the algorithmic approach, Kolmogorov's plan was to give definitions of the concepts of entropy (in other words, complexity) and amount of information based on the concepts of algorithm and computable function. This is how Kolmogorov outlined the sources and essence of the algorithmic approach in his report on April 24, 1963 to the Probability Theory Section of the Moscow Mathematical Society:

One often has to work with very long sequences of symbols. Some of them, for example, sequences of digits in a five-place table of logarithms, admit a simple logical definition and can correspondingly be obtained by computations (perhaps cumbersome) according to a simple program. Others, however, presumably do not admit any sufficiently simple "regular" construction: for example, a sufficiently long segment of a "table of random numbers".

The question arises of ways of constructing a precise mathematical theory permitting one to catch the difference mentioned above.

Following the conventions in information theory, we confine ourselves to binary sequences, that is, sequences of the form $x = (x_1, \dots, x_n)$, where $x_i = 0$ or 1. We denote the set of such sequences of length n by D^n, and we let $E = D^1 \cup D^2 \cup \cdots$ be the set of all binary sequences.

Kolmogorov stressed that there are many possible ways of introducing for such sequences x a measure $K(x)$ of "complexity" that corresponds qualitatively to the stated plan, though it is apparently difficult to avoid a certain arbitrariness. In [IA] (English p. 221) Kolmogorov wrote:

> The basic discovery made by me and R. Solomonoff simultaneously was that with the help of the theory of algorithms it is possible to limit this arbitrariness, defining the "complexity" almost invariantly (replacing one method of description by another leads only to the addition of a bounded summand).

We identify each sequence $x = (x_1, \dots, x_n)$ with the natural number whose binary representation is (uniquely) determined by the tuple $\langle 1, x_1, \dots, x_n \rangle$. This identification permits us to speak of partial recursive (computable) functions f defined on E and with values in E. For each such function f Kolmogorov defined

$$K_f(x) = \begin{cases} \min\{n : x \in f(D^n)\} & \text{for } x \in f(E), \\ +\infty & \text{for } x \notin f(E) \end{cases}$$

and called this quantity the *complexity* of the object x for the "method of assignment f."

Kolmogorov next introduced the class $\mathcal{F}_0$ of functions f_0 called *optimal* and having the property that for any other function f there is a constant C depending on f_0 and f such that for all $x \in E$

$$K_{f_0}(x) \leq K_f(x) + C.$$

The fundamental result discovered independently by Kolmogorov and Solomonoff consists here in the fact that this class $\mathcal{F}_0$ of computable functions f_0 is *nonempty.* This basic fact justifies calling any function $K(x) = K_{f_0}(x)$ with $f_0 \in \mathcal{F}_0$ the "complexity" (or "measure of complexity") of the sequence x. For any two functions f_1 and f_2 in $\mathcal{F}_0$, $|K_{f_1}(x) - K_{f_2}(x)| \leq C(f_1, f_2)$ for any $x \in E$. Thus, in this sense all optimal functions in $\mathcal{F}_0$ are equivalent, and hence ([IA-13], English p. 212): "from the asymptotic point of view the complexity $K(x)$ of an element x under the restriction of effective methods of assignment does not depend on the random features of the optimal method chosen."

Along with the concept of the "complexity" $K(x)$, also called the "simple Kolmogorov entropy" of the individual object x, Kolmogorov also introduced the *conditional entropy* $K(y|x)$ of an object y for a known x, and the *information* $\mathfrak{I}(y|x)$ about y contained in x.

The development of a new direction, algorithmic information theory, began with all these concepts introduced and investigated by Kolmogorov in the papers "Three approaches to the definition of the amount of information" ([1965;1], [IA-10]), "On the logical foundations of information theory and probability theory" ([1969;1], [IA-12]), "Combinatorial foundations of information theory and the calculus of probabilities" ([1983;4], [IA-13]), as well as in papers of his students and other authors (for details see the commentary of A. Shen in [IA], English pp. 226–230).

The ideas connected with the introduction of the concept of complexity led Kolmogorov to new approaches to defining which concrete sequences $x = (x_1, \dots, x_n)$ consisting of, say, 0's and 1's should naturally be regarded as *random* and which should not be interpreted as such.

For example, if we flip a fair coin honestly $2n$ times and use 0 or 1 for an expression depending on which side of the coin lands up, then for sufficiently large n outcomes of type $(0, \dots, 0)$ or $(0, 1, 0, 1, \dots, 0, 1)$ are hardly perceived as "random", but from the probability theory point of view each such sequence, like any other, has the very same probability 2^{-2n}. Thus, classical probability theory does not give an answer to the question of how to distinguish "random" sequences from "nonrandom" ones, nor what meaning should be invested in the very concept of "randomness" of an *individual* sequence.

The idea itself of distinguishing, in a set of *infinite* (say, binary) sequences $x = (x_1, x_2, \dots)$, a subset whose elements could be called "random" goes back to von Mises ([XI-64], [XI-65]), who employed the German word "Kollektiv" for this.

Kolmogorov and C. R. Rao (India, 1962).

According to von Mises' scheme, for a sequence $x = (x_1, x_2, \dots)$ to be "random" it is necessary first of all that the limit $\lim_{n\to\infty} S_n/n$ exist, where $S_n = x_1 + \cdots + x_n$ is the number of 1's in $(x_1, \dots, x_n)$. The example of the sequence $(0, 1, 0, 1, \dots)$, for which this limit exists, shows that the condition is necessary, but it certainly cannot be regarded as a sufficient condition for "randomness", since such a sequence can hardly be perceived as "random".

It was for this reason that von Mises added another requirement, whose meaning is that the average frequency of 1's must be preserved if instead of the whole sequence one looks at an infinite subsequence of it obtained by means of any *admissible selection rule* (which is not precisely defined by von Mises). In 1940 Church [XI-112] presented a possible definition of an "admissible selection rule", thereby giving a formal definition of randomness for a sequence $x = (x_1, x_2, \dots)$ and refining the plan of von Mises.

The significance of Kolmogorov's paper entitled "On tables of random numbers" ([1963;4], [IA-9]), which, as he notes in the preface to [IA-9], reflects a certain stage in his "attempts to make sense of von Mises' frequency interpretation of probability," is first of all that it proposes a more general selection scheme which extends the class of admissible selection rules by the statement that "the order of the terms in the subsequence does not have to coincide with their order in the original sequence." (D. Loveland arrived at the same scheme independently in 1966 ([XI-54], [XI-55]).) The class of sequences arising according to this selection scheme is called the *von Mises–Kolmogorov–Loveland class of random sequences*, and they all are random in the sense of von Mises and Church. However, as shown by Loveland, the converse is not true.

In "On tables of random numbers" (see also especially "From the author" in the Russian translation of the paper ([IA], English p. 176)) Kolmogorov, in referring to his broader (compared to von Mises) definition of an algorithm for forming random selections, underscores that "the main difference from von Mises is the strictly *finite* character of all the concepts and the introduction of a quantitative estimate of stability of the frequencies." By this Andreĭ Nikolaevich wanted to emphasize that degree of randomness can also be defined not only for an infinite sequence $x = (x_1, x_2, \dots)$, but also for a (sufficiently long) finite sequence $x = (x_1, \dots, x_n)$.

It is natural to call the approach to this problem worked out by Kolmogorov in [1963;4] and based on the introduction of a *finite system of selection rules* the *frequency* approach. Somewhat later Andreĭ Nikolaevich and his students Martin-Löf and L. A. Levin developed an approach based on the concept of *complexity*, in which a sequence $x = (x_1, \dots, x_n)$ is said to be random if its complexity is maximal. (Since $K(x = (x_1, \dots, x_n)) \leq n + c$, it is natural to call the sequence x random if $K(x_1, \dots, x_n) \geq n - c$.) For details see [IA], English p. 242.

Andreĭ Nikolaevich repeatedly turned to the circle of problems on "randomness and complexity" in succeeding years. At the 1970 International Congress of Mathematicians in Nice he gave the report "Combinatorial foundations of information theory and the calculus of probabilities" (unfortunately, published only in 1983 ([1983;4], [IA-13])). At the Fourth Soviet-Japanese Symposium on Probability Theory in Tbilisi in 1982 he delivered the report "On the logical foundations of probability theory" [PS-53], in which he made clear the idea that "randomness" is the absence of "regularity" and explained how the concept of complexity of a finite object permits this idea to be given a precise meaning.

In 1986 Kolmogorov and Uspenskiĭ delivered the forum report "Algorithms and randomness" [1987;3] at the First World Congress of the Bernoulli Society. The full text was published in 1987 in the journal *Theory of Probability and its Applications* and contains a most exhaustive exposition of the ideas and results of Andreĭ Nikolaevich and his students and successors on the algorithmic approach to the definition of the concept of randomness. (See also the long survey article [XI-51].)

Among Kolmogorov's investigations in the 1960's a special place is occupied by his papers in linguistics and philology dealing with analysis of the statistics of speech and with poetics.[6] Andreĭ Nikolaevich's plans in this direction, on the one hand, are closely connected to the probabilistic and algorithmic approaches to information theory, and, on the other hand, reflect his lasting interest in the analysis of the regularities peculiar to the form and language of works of fiction. (According to the diaries Kolmogorov became interested in poetics as far back as the 1940's.)

The global idea expressed by Andreĭ Nikolaevich and explaining the direction of these investigations is that the "entropy of speech" (that is, the measure of the amount of information transmitted by speech) can be broken up into two components:

1) information external to speech (involving meaning, semantics);
2) information proper to speech (linguistic information).

---

[6]The description given of these investigations and the list of references (including that in Section V) are based on material prepared by A. V. Prokhorov at the request of the author.

The first component characterizes the diversity enabling us to transmit various semantic information.

The second, called the "residual entropy" by Kolmogorov, characterizes the diversity of possible ways to express the same or equivalent semantic information. In other words, it can be said that this component characterizes the "flexibility" of speech, the "flexibility" of expression. The presence of the residual entropy makes it possible to give speech a special artistic, in particular, phonetic, expressiveness when transmitting intended semantic information.

In the light of this general idea concrete problems were formulated and solved on computing the total "entropy of speech" and residual entropy. Besides Kolmogorov, this work was done by A. V. Prokhorov, N. G. Rychkova-Khimchenko, N. D. Svetlova, A. P. Savchuk, and others.

In 1960–1961 Kolmogorov worked out a new method of consistent estimation of the entropy of speech. This method, an improvement of a method of Shannon [XI-113] for determining entropy with the help of experiments on guessing (predicting) continuations was free of the fundamental indeterminacy of the Shannon method. Namely, when the guesser used an optimal strategy, Kolmogorov's method led to a consistent sharp estimate of the entropy instead of the upper and lower Shannon estimates. The practice of carrying out experiments on guessing continuations permitted only an upper estimate to be obtained for the true entropy. However, this estimate proved to be sharper than Shannon's estimate. (If an optimal strategy is not used, then all the estimates can be excessive, and the lower estimates are unreliable; see [XI-93], [XI-80], [XI-92], [XI-120].)

In 1962 Kolmogorov proposed a purely combinatorial approach to the definition of a measure of the amount of information transmitted by speech. In this approach the "combinatorial entropy of speech" was defined to be the limit

$$H_k = \lim_{n\to\infty} \frac{N_n}{n},$$

where N_n is the number of texts formed from n letters and made up of words in a specific dictionary and subject to specific grammatical rules. As was supposed, the estimates obtained for the entropy of speech under this approach turned out to be larger on the average than the upper estimates found by the method of guessing continuations for concrete prose texts ([1961;1], [IA-10], [XI-120]).

From the point of view of defining the "flexibility" of speech and estimating the "residual entropy", Kolmogorov concentrated his efforts on the study of poetry. The basis for this was that in poetry there are precisely formulated laws that can be established independently of the content and not directly connected with it. The possibility of giving poetry a suitable phonetic expressiveness is based on the existence of a large number of possible variants of the same or equivalent content. In contemporary poetics an enormous amount of work has accumulated that involves studies of the metrics and rhythmics of verse by statistical methods. At Kolmogorov's initiative there has been much research in revising and improving results obtained by well-known researchers of verse like A. Belyĭ, B. Tomashevskiĭ, G. Shengeli, K. Taranovsky, R. Jacobson, and others. The main results obtained in this direction by Kolmogorov and his students and collaborators can be formulated as follows.

I. *Clarification of the laws of metrics.* Given: general and particular definitions of meter; an idea of meter form and phonetic meter form; a rigorous formally logical definition of the classical meters ([1963;6], [1968;10], 1968;11], [1984;3]); a description of and differentiation between the nonclassical Russian meters (first and foremost dolniks, logaoedic meters, free verse, and pure accented verse ([1962;2] [1963;5], [1963;6], [1964;1], [1964;2], [1965;6]).

II. *Classification and statistics of rhythmic variations of meter.* Formulated and verified here is the general fundamental position that the phonetic structure of speech is subordinate to simple statistical regularities that can be calculated with the help of probability theory. (These regularities are realized under the pressure of having to transmit semantic information, unless a systematically conducted artistic tendency opposes this pressure.) A general method is indicated for constructing theoretical models of different meters; the hypothesis "imitation of randomness" is formulated (see [1963];7], [XI-84], [XI-83], [1958;1]).

III. *Analysis of the "residual" entropy and estimation of it.* In the paper "Statistics and probability theory in the investigation of Russian versification" (*Abstracts and annotations of a symposium on the complex study of artistic creativity,* "Nauka", Leningrad, 1963) Kolmogorov and A. V. Prokhorov obtained an estimate of the residual entropy and computed the "expenditure of entropy" on individual modes of phonetic expressiveness of verse.

The work carried out under the guidance of Kolmogorov led to a large flow of research in the area of mathematical methods for studying the language in works of fiction. During the 1960's there were two seminars on these topics in the Mechanics and Mathematics Department at Moscow University, and linguists and philologists took part for the first time (A. Zaliznyak, V. Ivanov, M. Gasparov, and V. Rozentsveĭg). Many papers on poetics were written under the direct influence of Andreĭ Nikolaevich's ideas (by Ivanov, Gasparov, Krasnoperova, and others). This work drew the constant attention and interest of Academician V. Zhirmunskiĭ, Professor S. Bondi, Professor K. Taranovsky of Harvard University, and R. Jacobson.

Andreĭ Nikolaevich's organizational activities in science during the 1960's were marked by two major events: the creation of the Statistical Methods Laboratory of the Mechanics and Mathematics Department of Moscow University, and the Physics and Mathematics Boarding School No. 18 affiliated with the university.

The Statistical Methods Laboratory under the Probability Theory Branch, conceived by Andreĭ Nikolaevich, was opened in 1960 with the object of unifying the efforts and intensifying the work on applications of probability-theoretic and statistical methods.[7] Kolmogorov determined the following basic directions for the Laboratory: optimal control theory and statistical decision making (led by Girsanov), the theory of reliability (Gnedenko, Belyaev), the design of experiments (Nalimov, V. V. Fedorov), statistics and linguistics (A. V. Prokhorov), statistics in medicine (Meshalkin), statistics in geology (A. M. Shurygin), nonlinear spectral analysis of random processes (Kolmogorov, Shiryaev, Zhurbenko). A number of colleagues from academic institutions (L. N. Bol′shev, Shiryaev, and others) took part in the work of the Laboratory as volunteers.

[7]The material about the Statistical Methods Laboratory was prepared by I. G. Zhurbenko at the request of the author. About the "Laboratory" see also the paper of V. V. Nalimov in this collection (Russian pp. 501–518; not among the articles translated).

Kolmogorov during an expedition on the
research vessel "Dmitriĭ Mendeleev" (1971).

An all-union seminar on the turbulence of fluids and gases was organized within the Laboratory and became widely known. Participants were Kolmogorov, Millionshchikov, Obukhov, S. A. Khristianovich, L. I. Sedov, Monin, A. M. Yaglom, and others. It was on the basis of this seminar that Andreĭ Nikolaevich got the idea of taking part first-hand in two voyages aboard the research vessel "Dmitriĭ Mendeleev" (in 1969 and 1971) for studying ocean turbulence.

As the scientific leader of these voyages, Andreĭ Nikolaevich worked together with Monin, V. Pak, Zhurbenko, M. V. Kozlov, and others on the development and direct application of highly robust (with respect to noise in neighboring frequencies) methods of spectral analysis.[8] He stressed that these methods were especially important because of the possibility of applying them to the spectral analysis of nonstationary processes. Remarking that it is very laborious to compute the statistical parameters of a random space-time turbulent field, Kolmogorov saw a particular role of his on these voyages as being (in his words) "to refine the procedures of computation effectively and directly on the ship, to determine the necessary duration of the realizations, the discretization step, and so on," and, by analyzing the computational results, "to efficiently plan further measurements, estimate the extent to which they are representative, and judge the suitability and quality of the test samples of the measuring apparatus."

On April 25, 1971 Andreĭ Nikolaevich received the following poem as a birthday salutation from the participants of the voyage:

[8]For details see Zhurbenko's paper (Russian pp. 445–450; not among the articles translated).

April 25, 1971
Research vessel "Dmitriĭ Mendeleev"
Pacific Ocean

TO ACADEMICIAN ANDREĬ NIKOLAEVICH KOLMOGOROV

Your birthday you now meet
While on unknown paths,
But you already knock at the threshold
Of treasures hidden in the blue sea!

You are greeted by the latitudes
On the blue lace of the longitudes,
The closed company of constellations
Recognizes you as its star!

Neptune, lord of the ocean
Sends from the depths a seal
To you, as a permanent license
To study his domain!

Even without that the steep waves
Are prepared to give you the key,
Faithful to your laws
In their changable fate!

To you the white "Mendeleev"
From the heart sends a sea-greeting,
Of all known anniversaries
Yours will be remembered for many years!

You, breathing in the blue expanse
Are like a youth in the prime of his powers,
His mind great and his aquiline gaze
turned to the oceans!

The ship is ever white for you,
Awaits your newest voyages,
At your service is the Mendeleev
Ever ready to go forward!

In the Statistical Methods Laboratory Andreĭ Nikolaevich, influenced by J. Neyman and E. Scott, developed work connected with a statistical analysis of results of an active effect on atmospheric phenomena, including artificial stimulation of rain.

After studying Neyman's methods, Andreĭ Nikolaevich concluded that in the given circle of problems parametric methods are too "delicate" (nonrobust) with respect to the relatively "coarse" atmospheric data. In this connection he created his own nonparametric methods (conceptually close to Fisher's ideas), which met with approval in a number of experiments. Together with Neyman, Kolmogorov was persistent among practicians in defending the idea of randomization of an experiment and the necessity of obtaining "clean" data; in every way he supported Neyman's thesis that "statistics is powerless against bad initial data."

From the moment of creation of the Statistical Methods Laboratory (that is, from 1960) Andreĭ Nikolaevich was its effective head (on a voluntary basis), while heading the Probability Theory Branch of the Mechanics and Mathematics Department at Moscow University. On May 14, 1963 he was confirmed as research director of the laboratory. On January 17, 1966 he passed his position as head of the Probability Theory Branch, which he had held since 1934, to Boris Vladimirovich Gnedenko, while remaining a professor there. At the same time he was named head (without extra pay) of the Statistical Methods Laboratory. On April 1, 1967 the laboratory became the Interdepartmental Statistical Methods Laboratory of Moscow University, and Andreĭ Nikolaevich was its head, devoting to it much effort, time, and energy.

A specialized library of probability and statistics literature was created for the Laboratory. Kolmogorov gave a substantial part of the money from his 1963 International Prize from the Balzan Foundation to acquire foreign journals and books.

On February 1, 1976 Kolmogorov became head of the new Mathematical Statistics Branch, created at his initiative in the Mechanics and Mathematics Department. In the same year the Interdepartmental Laboratory was restructured into three laboratories (under the Probability Theory Branch and the newly created Mathematical Statistics Branch): "Probability Theory", "Mathematical Statistics", and "Computational Tools in Probabilistic and Statistical Research". The work in these laboratories was in three fundamental areas: educational work with students and graduate students, fundamental research in probability theory and mathematical statistics, and applied research.

In 1963 four boarding schools specializing in physics and mathematics were opened by Moscow, Leningrad, Novosibirsk, and Kiev Universities by a decision of the Council of Ministers of the USSR.

The creation of Boarding School No. 18—a new type of school—affiliated with Moscow University was inseparably (now even officially) connected with Kolmogorov's name, as reflected in the fact that it is simply called the "Kolmogorov Boarding School" or "Kolmogorov School".

In answer to the question of how he imagined the first stages of familiarizing a future scholar with science, Andreĭ Nikolaevich said:

> Tracing the biographies of well-known scientific scholars, we find in most cases at the start of their paths a schoolteacher fascinated by science who gave individual attention to the able student, a first academic advisor who pointed out a suitable topic for independent study, often deliberately adapted to the resources of just that student. We frequently note also one or more close friends, peers who support one another. I think that these subtle human interrelations forming the scholar to be will keep their significance in the future.
>
> Now that our country requires many capable and well-trained researchers in the most diverse areas of science and engineering, we certainly need a broad system of organizational measures in which

Kolmogorov with his students.

> there is a place for a choice of optional studies with older pupils, specialized schools, various forms of work outside school (groups of pupils at institutes of higher learning, olympiads, and so on), a broad acquaintance of young people with the special nature of the work of universities and engineering institutes teaching the new technology (like the Moscow Institute of Physics Engineering), a suitable arrangement of competitive enter exams in this kind of institutes, and the broad involvement in scientific work of the students of those institutes in which the training of future scientific researchers is only an auxiliary task. However, all these practical measures will not give the expected result unless they are backed up by an individual concern about the development of each youth — each potential future scholar—of which I spoke at the beginning.

Following these principles, Andreĭ Nikolaevich gave himself wholly to the boarding school from its very inception, regarding direct personal involvement with students, and later also work in a broader design connected with improving the secondary school mathematical curriculum, as necessary and important to the country and as his own civic responsibility for mathematical education.

Both in the very organization and in the first days and years of existence of the boarding school, Andreĭ Nikolaevich contributed much for its formation by his *personal* participation. In the course of fifteen years (!) he not only lectured and went through exercises for the students, but also wrote summaries of his lectures for them, talked to them about music, art, and literature, and hiked with them.

We mentioned above that Andreĭ Nikolaevich had worked in a secondary school as far back as 1922–1925, teaching mathematics and physics at the Potylikha experimental-model school. There is an entry about the work of this period in

A. N. Kolmogorov.

his service record under no. 1: "Before beginning work at Moscow State University (1929)—total length of job, 3 years. Basis—copy of certificate no. 87, March 1925." At the school Andreĭ Nikolaevich had been secretary of the school council and a tutor at the boarding house, of which he was always very proud. Forty years later, as chairman of the council of trustees of the boarding school, he not only supervised the work, but also concerned himself directly with the needs and problems of the students, many of whom were from small towns and villages (the boarding school did not accept residents of university towns).

For more details about the work of the school see [1974;10] and [1981;2]. A documentary film about the school was made in 1970, with the title "Ask, boys!".

In 1964 Kolmogorov headed the mathematics section of the Committee of the Academy of Sciences and the Academy of Pedagogical Sciences of the USSR for determining the content of secondary school education. For grades 6 to 8 and 9 to 10 this section put out new programs in mathematics in 1968 that were a basis for further improving the content of mathematical education and a basis for writing textbooks. Andreĭ Nikolaevich himself took a direct part in the preparation of the school texts *Algebra and the foundations of analysis*: *grades* 9 *and* 10, and *Geometry for grades* 6 *to* 8 (see, for example, [1983;1] and [1981;3]).

We complete this brief description of Andreĭ Nikolaevich's activities in the 1960's connected with secondary school by giving one of his annual personal reports.

Report on the work of a member of the Academy of Pedagogical Sciences of the USSR for the year 1969.

1. The first three articles "Scientific bases for school mathematics" were written for the periodical *Matematika v shkole* (two published in 1969, the third to be published in 1970).
2. Editing of a trial geometry textbook for the sixth grade (authors: Nagibin, Semenovich, and Cherkasov).
3. Heading the mathematics section of the Committee on the Content of Mathematics Education during the first half of the year, I reviewed many textbooks for this committee.
4. Directing the teaching of mathematics at the boarding school of Moscow State University. Lectures for the ninth-grade students (during the first half-year) and for the tenth-grade students (during the second half-year).
5. Preparing material for *Kvant*, a periodical for the older students, to begin publishing in 1970.

At the end of the 1950's and the beginning of the 1960's Andreĭ Nikolaevich posed some problems for his students V. P. Leonov and Shiryaev involving questions in nonlinear analysis of random processes. The problems led to the creation of a technique for computing semi-invariants of nonlinear transformations [XI-47], and to the development of the theory of spectral analysis of higher-order moments of stationary random processes ([XI-48], [XI-114], [XI-29]). Under the guidance of Kolmogorov and with his participation, research was performed on conditions for the validity of various ergodicity and mixing properties of processes, and on conditions for the validity of some fundamental limit theorems for random processes ([XI-15], [XI-16], [XI-48], [XI-96], [XI-115]).

In [1962;1], [PS-50] Kolmogorov, Sinaĭ, and Arato constructed a mathematical theory that enabled them to estimate the parameters of the movement of the Earth's axis of rotation and that was based on the idea that the fine structure of the movements can be described well by a model for a complex stationary Gaussian process.

Under Andreĭ Nikolaevich's leadership work was begun in 1964 at the Statistical Methods Laboratory on studying the nature of the 11-year periodicity of solar activity, with the object of confirming a conjecture of E. E. Slutskiĭ (that we have *the exact periodicity with preservation of phase over hundreds of periods masked by perturbations, and not a self-oscillatory process with wandering phase*).

At the sixth All-Union Conference on Probability Theory and Mathematical Statistics in 1960 (Vilnius) Kolmogorov and Shiryaev gave a report entitled "Applications of Markov processes to the detection of disorder in an industrial process", which served as the starting point of a broad development and refinement of *statistical sequential analysis* ([1988;2], [XI-116]), *methods of optimal nonlinear filtering* [XI-52], and *martingale theory* ([XI-53], [XI-28]).

On April 6, 1961 Andreĭ Nikolaevich gave a report at Moscow State University with the title "Automata and life" [1961;1]. It made an unforgettable impression on the listeners because of its startling power, depth, and extraordinary thoughts.

A. N. Kolmogorov (1968).

In 1967 Kolmogorov and J. M. Barzdin published a paper "On the realization of networks in three-dimensional space" ([1967;1], [IA-11]), which arose in connection with an attempt to explain the fact that the (human) brain is structured so that its main mass is occupied by nerve fibers (axons), while the nerve cells together with their branches (neurons) are located only on its surface. The constructions given in this paper confirmed that this very structure of the nerve network is optimal.

Andreĭ Nikolaevich turned 60 on April 25, 1963. On this day the Presidium of the Supreme Soviet of the USSR issued a decree: **On awarding the title of Hero of Socialist Labor to Academician A. N. Kolmogorov:**

> For outstanding services in the area of mathematics and on the occasion of his sixtieth birthday Andreĭ Nikolaevich Kolmogorov has been awarded the title of Hero of the Socialist Labor, and has been given the Order of Lenin and the Golden "Hammer and Sickle" Medal.

In recognition of his outstanding services Kolmogorov was awarded the Balzan Foundation International Prize in Mathematics in 1963.[9]

---

[9]See footnote 11 on p. 84.

From December 1, 1964 to December 13, 1966 Kolmogorov was the President of the Moscow Mathematical Society.

In 1965 Kolmogorov and Arnol′d were awarded the Lenin Prize for their work on the theory of perturbations of Hamiltonian systems.

In 1966 Kolmogorov was elected a member of the Academy of Pedagogical Sciences of the USSR.

The seventies and eighties (1970–October 20, 1987)

Improving the teaching of mathematics in secondary schools continued to be a principal direction of activity for Andreĭ Nikolaevich in the 1970's and 1980's. He worked at the boarding school and in 1970 created, together with Academician I. K. Kikoin, a popular science periodical of physics and mathematics, *Kvant*, for the pupils. Kolmogorov was in charge of the mathematics section. He wrote and edited school textbooks and manuals, prepared articles for teachers and pupils, involved himself purposefully in school mathematics programs, took part in committees on scholastics, and so on.

The difficult search for optimal systems of teaching mathematics in school, which continues even now, shows how complicated the problem is to which Andreĭ Nikolaevich gave himself entirely. He was happy with successes, but was greatly pained when his concepts and positions on questions of teaching school mathematics were misunderstood (though he never complained nor sought sympathy).

It may be freely said that school mathematics was a topic of his constant interest and concern for his whole life, beginning in 1922 when he started teaching at the Potylikha experimental-model school. On November 22 and 28, 1937 there were two meetings of the Moscow Mathematical Society devoted to a discussion of a plan put forth by Aleksandrov and Kolmogorov for a new elementary algebra text, which was published in 1939 (*Algebra*, part 1, [1939;1]), in which the authors formulated a number of principles used by them as a basis for writing the book:

> We have attempted everywhere to combine clearness of presentation with sufficient thoroughness and irreproachable logic. We began with the conviction that our book will be a correct and reliable guide for the student not only for a first acquaintance with the subject but also for further study of mathematics ... The authors have striven to give the students a complete and clear understanding of the meaning of all the operations performed. In particular, we have made great efforts to ensure that operations with symbolic expressions are not perceived as detached from arithmetic operations with numbers.

In 1941 Aleksandrov and Kolmogorov published the articles "Irrational numbers" [1941;9] and "Properties of inequalities and the concept of approximate computations" [1941;8] in the periodical *Matematika v shkole*. They were reprinted in 1961 in the book *Topics in the teaching of mathematics in secondary school* [1961;4].

To demonstrate the concreteness of Andreĭ Nikolaevich's work in the 1970's and the degree of his personal involvement in the affairs of mathematical education we present some personal reports written by him.

Report on the work during 1970 of A. N. Kolmogorov, member of the Academy of Pedagogical Sciences of the USSR.

1. At the physics and mathematics boarding school of Moscow State University I headed a group of mathematicians dealing with teaching methods, gave a lecture course, and was involved in general affairs of the school as chairman of the council of trustees. On the basis of school materials I prepared a mathematics textbook[10] for physics and mathematics schools (from the Academy of Pedagogical Sciences, V. A. Gusev and A. A. Shershevskiĭ are in the authors' collective), for which I wrote several chapters. During 24 days of the summer I was completely occupied with the summer school, through which final selections for the boarding school were made.

2. As deputy editor-in-chief of *Kvant* I supervised the mathematics section of the periodical and wrote for it a number of notes and a long article on modern understanding of the concept of function (*Kvant*, 1970, nos. 1 and 2).

3. A program for a new course for pedagogical institutes, "Scientific bases for a school course in mathematics" was developed, which was accepted by the Ministry of Education of the USSR. My next article of a series of articles on the theme of this course was published in *Matematika v shkole.*

4. *Educational materials on geometry for 5th grade* [1970;5] were written together with R. S. Cherkasov and A. F. Semenovich. With the same co-authors and with F. F. Nagibin I worked on a geometry textbook for grades 6–8. A trial edition of the textbook for the sixth grade has been printed, and one for the seventh grade is submitted to the publisher. I took part in a conference of teachers doing experimental work (in Vladimir).

5. As chairman of the mathematics section of the Committee on the Content of Secondary Education for the Academy of Sciences and the Academy of Pedagogical Sciences of the USSR I reviewed in detail textbooks for grades 4–6 prepared at the Academy of Pedagogical Sciences under the editorship of A. I. Markushevich.

6. On finishing my work for the above committee I headed the mathematics section of the Academic Methods Council of the Ministry of Education of the USSR and prepared a report, "A system of basic concepts and notation for school mathematics".

January 5, 1971

Report on the work during 1974 of A. N. Kolmogorov, member of the Academy of Pedagogical Sciences of the USSR.

1. Work on textbooks for schools of general education:

1.1. Together with O. S. Ivashev-Musatov and S. I. Shvartsburd I finished revising the ninth-grade textbook by B. E. Veĭts and I. G. Demidov (edited by me) on algebra and the foundations of analysis. The book is intended for use in public schools from the fall of 1975.

1.2. Together with A. F. Semenovich and R. S. Cherkasov I worked on a revision of a geometry textbook for grades 6–8. This will be finished in 1975.

2. Work involving the physics and mathematics boarding school of Moscow State University:

Lectures for the ninth grade in the spring semester; directing the summer school in Pushchino and the selection of students entering the physics and mathematics boarding school.

---

[10] *A mathematics curriculum for mathematics and physics schools*, by Kolmogorov, Gusev, A. B. Sosinskiĭ, and Shershevskiĭ, published by Moscow University Press in 1971 [1971;11].

At age 70 (April 25, 1973).

Material was gathered for a textbook for physics and mathematics schools and a manual for elective studies in general school (on the basis of experience in our summer schools).

3. Work as a head of the mathematics section of the Academic Methods Council of the Ministry of Education of the USSR.

4. Editing of the mathematics part of the student periodical *Kvant*, and taking part in the work of the editorial board of the periodical *Matematika v shkole*.

December 12, 1974

Kolmogorov's legacy in questions of scholastic education, which has hardly been studied at all, is largely unpublished or scattered about in various editions and is an inexhaustible source of ideas and thoughts, always clear and precise, expressed in unexpected form, catching the essence of the problem in an amazingly clear and incisive way. One important task of his students and successors will be to preserve this legacy, give it a systematic form, and make it generally accessible.

In this respect the broad cricle of readers may find useful the small book *Mathematics: science and profession* by Kolmogorov, published in 1988 as no. 64 of the *Kvant Library* series ([1988;1]; compiled by G. A. Gal′perin). The article [X-1] of A. M. Abramov describes in detail the diverse sides of Andreĭ Nikolaevich's activities in mathematics education.

Kolmogorov's scientific-organizational work in the 1970's and 1980's was connected with Moscow State University and the Steklov Institute of Mathematics.

From February 1, 1976 to January 1, 1980 he headed the Mathematical Statistics Branch and from January 1, 1980 the Mathematical Logic Branch in the Mechanics and Mathematics Department of Moscow State University. On October 3, 1983 he made the transition to work permanently at the Steklov Institute, where he headed the Mathematical Statistics and Information Theory Department, while continuing to head the Mathematics Section and the Mathematical Logic Branch at Moscow State University.

From 1973 to October 15, 1985 Kolmogorov was President of the Moscow Mathematical Society, and from 1982 he was editor-in-chief of the journal *Uspekhi Matematicheskikh Nauk.*

The fourth Soviet–Japan Symposium on Probability Theory and Mathematical Statistics was held in Tbilisi on August 22–29, 1982. Despite feeling unwell, Andreĭ Nikolaevich took part in the symposium and read a lecture, "On the logical, semantic, and algorithmic foundations of probability theory". (The text of this lecture was published under the title "On the logical foundations of probability theory" [PS-53].)

Taking part in this symposium were 45 Japanese and 270 Russians. The Japanese delegation was headed by K. Itô. The presence of Kolmogorov and Itô was a great event for the participants of the symposium.

At the beginning of the 1980's the Presidium of the Academy of Sciences resolved to publish selected works of Kolmogorov, and Andreĭ Nikolaevich devoted several years to this project. He himself compiled a list of his works that ought to be included, subdivided them into cycles according to the subject, wrote and dictated his own commentaries to them, and very carefully went through the commentaries written to various cycles mainly by his students. In 1985 the first volume appeared [MM] = [1985;2], in 1986 the second [PS] = [1986;3], and in 1987 the third [IA] = [1987;1].

In the preparations, Kolmogorov, the associate editors (Nikol′skiĭ and Yu. Prokhorov), the compilers (Tikhomirov and Shiryaev), and the editor (V. I. Bityutskov) were greatly helped by his students and successors, who responded at once when asked to translate a number of papers from foreign languages into Russian, to write commentaries, and to help in editing and proofreading.

On April 25, 1985, on his 82nd birthday, Andreĭ Nikolaevich dictated the following text for the **Afterword** ([IA], English p. 275):

> The three books of selected works presented to the reader include essentially all my work in mathematics, classical mechanics, the theory of turbulent motion, probability theory, mathematical logic, and information theory. Not included is my work on teaching methods and the history of mathematics, poetics, and articles of a general nature.
>
> What I have done in some directions seems to me sufficiently self-contained and complete, so that at the age of 82 I take satisfaction in leaving my accomplishments to my successors.
>
> In other directions matters are different, and what I have published seems to me only fragments of work I can only hope will be done by others in the future. What has already been done in

> these directions is described in many cases in the commentaries by a group of my students, to whom I express my heartfelt appreciation.

Andreĭ Nikolaevich Kolmogorov founded a whole series of scientific schools, many led by his students. The scholarly atmosphere of high standards and spirituality he engendered, the ability to stimulate the creative capacities of others, to find for each a problem or task suited to that person's strengths, the extraordinary generosity with which he shared his ideas—these are unforgettable for all his students.

Among Kolmogorov's students are the following.

Academicians of the Russian Academy of Sciences: V. I. Arnol′d, A. A. Borovkov, I. M. Gel′fand, A. I. Mal′tsev, M. D. Millionshchikov, V. S. Mikhalevich, S. M. Nikol′skiĭ, A. M. Obukhov, Yu. V. Prokhorov, Ya. G. Sinaĭ.

Academician of the Ukrainian Academy of Sciences B. V. Gnedenko.

Academician of the Uzbek Academy of Sciences S. Kh. Sirazhdinov.

Corresponding Members of the Academy of Sciences of the USSR: L. N. Bol′shev, A. S. Monin, B. A. Sevast′yanov.

Doctors and Candidates of Sciences, Research Collaborators: A. M. Abramov, V. M. Alekseev, M. Arato, D. A. Asarin, G. M. Bavli, G. I. Barenblatt, L. A. Bassalygo, Yu. K. Belyaev, V. I. Bityutskov, E. P. Bezhich, A. V. Bulinskiĭ, I. Ya. Verchenko, V. G. Vinokurov, V. G. Vovk, G. A. Gal′perin, A. N. Dvoĭchenkov, N. A. Dmitriev, R. L. Dobrushin, E. B. Dynkin, V. D. Erokhin, I. G. Zhurbenko, V. N. Zasukhin, V. M. Zolotarev, O. S. Ivashev-Musatov, V. V. Kozlov, M. V. Kozlov, A. T. Kondurar′, L. A. Levin, V. P. Leonov, R. F. Matveev, P. Martin-Löf, Yu. T. Medvedev, L. D. Meshalkin, R. A. Minlos, Yu. P. Ofman, Yu. S. Ochan, A. A. Petrov, B. Penkov, M. S. Pinsker, A. V. Prokhorov, Yu. A. Rozanov, M. Rozenblat-Rot, V. M. Tikhomirov, L. N. Tulaĭkov, V. A. Uspenskiĭ, M. K. Fage, S. V. Fomin, G. E. Shilov, A. N. Shiryaev, F. I. Shmidov, B. M. Yunovich, A. A. Yushkevich, A. M. Yaglom.

On Andreĭ Nikolaevich's 80th birthday, April 25, 1983, his students Aleksandr Mikhaĭlovich Obukhov and Akiva Moisevich Yaglom read a poem dedicated to him. It is impossible to better express the relation between students and their teacher:

Chords of congratulations ring
This day, anniversary of your birth,
Among dear friends and colleagues,
Representatives from far-off towns.

How much there is in this hall
Of love and warmth and esteem,
Feelings toward you earned by your work,
By the great effort of all your powers
In struggles for the triumph of science,
By your concern for all—your generation and the young.

Your scientific descendants now are many!
At times you criticized us,

For laziness, immodesty, more sins,
For ignorance of the secrets of "chi-square",
And perhaps for other things as well.

It is pleasant to remember all that is past:
The treacherous stump in the snowdrift,
When on a fine winter day
You invited us to Komarovka
And entertained us with a ski race!

To all corners of the world we have spread,
Kept warm forever by your consideration.
Your descendants, dancing about you,
Await your command, "Go forward!"

In April of 1986, marking his birthday, Andreĭ Nikolaevich invited some of his students to the country house in Komarovka. In speaking of their teacher all noted the invariable youthfulness of spirit that was characteristic of him. A serious illness made speech difficult for him, and he could not express what he wanted to. But the next day he dictated the following to Tikhomirov.

Reply to my students

You have already spoken here of my supposedly inexhaustible youthfulness. I am grateful for your appraisal, but I must make some reservations with regard to it. Old age is still an objective phenomenon from which there is no escape. A happy old age . . . , how can it be realized? Either by declining to accomplish anything new, or by accepting in essence a barren old age. If that were not the case, then an old man could look at this period as bright and joyous, though there would be an unavoidable sadness connected with it about not being able to do various things. And this applies not just to cold swims and athletic achievements.

In my doctor's opinion I am in fairly good condition. Nevertheless, the amount of work I do is already considerably less, which introduces regrettable limitations.

In my case I regard my scientific career as finished in the sense of obtaining new results. This grieves me, but I bow to inevitability.

In recent years my activities have evolved in a different direction, toward school reform, so important for our country. Here I think that if old age does not prevent me, I can still make myself very useful and even invaluable in working on textbooks for ordinary school and for young people fascinated by science. Both directions of activity entice me, and I would like to take part in them energetically and with youthful fervor. But time passes, months go by in which this or that was planned but is put off . . .

Therefore, it is especially important to choose activities in which I am most difficult to replace. If I concentrate on textbooks for talented children, then I cannot take part in creating textbooks

> for ordinary schools. And now you see me at such a crossroad. If I agree to work actively and with sufficient scope in one direction, then I cannot do this in the other direction. These kinds of emotional experiences are especially aggravating in old age ...
>
> I therefore value so much my young assistants, many of whom I have invited today.

The First World Congress of the Bernoulli Society (September 8–14, 1986) was a great event in probability theory and mathematical statistics. Unfortunately, for reasons of health Andreĭ Nikolaevich was unable to take part directly in the work of the congress. Immediately after the opening ceremonies there was a forum report by Kolmogorov and Uspenskiĭ entitled "Algorithms and randomness" (given by Uspenskiĭ) [1987;3] which not only took up general questions of applicability of the mathematical theory of probability to real world phenomena having a random character and showed how the theory of algorithms and recursive functions led to a precise mathematical meaning for comparison of the "complex" and the "random", but also contained a program for further investigations.

Before beginning this report Uspenskiĭ read the "Welcoming speech to the participants of the First World Congress of the Bernoulli Society", the full text of which was published in the journal *Teoriya Veroyatnosteĭ i ee Primeneniya* [1987;2].

Kolmogorov's merits were highly valued. He was awarded the title of Hero of Socialistic Labor (1963), seven Orders of Lenin (1944, 1945, 1953, 1961, 1963, 1973, 1975), the "Golden Star" medal (1963), the Order of the Red Banner for Labor (1940), the Order of the October Revolution (1983), and many medals.

In 1941 he was awarded the Stalin (State) Prize, and in 1965 the Lenin Prize.

In 1939 he was elected a member of the Academy of Sciences of the USSR, and in 1966 a member of the Academy of Pedagogical Sciences of the USSR.

In 1949 he was honored with the Chebyshev Prize of the Academy of Sciences, and in 1987 the Lobachevskiĭ Prize of the Academy of Sciences.

The high status of Andreĭ Nikolaevich in world science was expressed in his election to membership in numerous academies, universities, and societies:

1955—Honorary Doctor of Sciences from the University of Paris.
1956—Corresponding Member of the Rumanian Academy.
—Foreign Member of the Polish Academy of Sciences.
—Honorary Member of the Royal Statistical Society of Great Britain.
1957—Honorary Member of the International Institute of Statistics.
1959—Honorary Member of the American Academy of Arts and Sciences, Boston.
—Member of the Leopoldina German Academy of Naturalists, German Democratic Republic.
1960—Honorary Doctor of Sciences from Stockholm University.
1961—Foreign Member of the American Philosophical Society, Philadelphia.
1962—Honorary Doctor of Sciences of the Indian Institute of Statistics, Calcutta.
—Honorary Member of the American Meteorological Society.
—Honorary Member of the Mathematical Society of India.
—Honorary Member of the London Mathematical Society.
1963—Foreign Member of the Royal Academy of Sciences of the Netherlands.
1964—Member of the Royal Society of London.

1965—Honorary Member of the Rumanian Academy.
—Honorary Member of the Hungarian Academy of Sciences.
1967—Member of the National Academy of Sciences of the USA.
1968—Foreign Member of the French Academy of Sciences.
1973—Doctor of Sciences Honoris Causa Hungary.
1977—Honorary Member of the International Academy of the History of Science.
—Foreign Member of the Academy of Sciences of the German Democratic Republic.
—Member of the Society of the Order "Pour le mérite", Federal Republic of Germany.
1983—Foreign Member of the Finnish Academy of Sciences.

In 1963 Kolmogorov was awarded the Balzan Foundation International Prize in Mathematics (Fondation Internationale Balzan).[11]

In 1980 Kolmogorov was awarded the Wolf Foundation International Prize in Mathematics "for profound and original discoveries in Fourier analysis, probability theory, ergodic theory, and dynamical systems."[12]

In the summer of 1987 Andreĭ Nikolaevich, who had been suffering from a disorder of the motor system (Parkinson's disease) for a number of years, declined in health so much that he agreed (on Saturday, August 22) to be placed in a special clinic for preventative purposes, in order to carry out a complete regimen of medical analyses (including computer tomography) and to see whether his barely functional eyesight could be restored. The illness progressed rapidly, and at the beginning of October serious impairment was observed also in the lungs, so that he had to be moved to the pulmonary section. There Andreĭ Nikolaevich and I had the following brief conversation: "Where am I now?" he asked. "In the pulmonary section."—"Why?"—"They have found something wrong in your lungs."—"And what will come of this?"—"You'll have to stay here a while so that you can return home with healthy lungs."—"Well then, that's all right ... "

This was in essence the very last conversation with Andreĭ Nikolaevich. In the next few days his temperature began to oscillate sharply, his blood pressure changed sharply, and breathing became difficult. The destructive powers of the

[11]Eugene Balzan (1874–1953) was an Italian journalist and admistrative manager who worked for a long time for the Italian newspaper *Corriere della Sera*. In 1933 he moved to Switzerland. After his death in 1953 his daughter Angela Lina Balzan transferred the inheritance from her father to start a prize fund called the "Balzan Prize". The prize was awarded first in 1961 to the Nobel Foundation. In 1963 Kolmogorov shared the prize with Pope John XXIII (Peace Prize), Karl von Frisch (Biology), Paul Hindemith (Music), and Samuel Eliot Morison (History). The Balzan Foundation Prize in Mathematics was awarded to Enrico Bombieri in 1980 and to Jean-Pierre Serre in 1985.

[12]Among the founders and endowers of the foundation of the large Wolf family was Ricardo Wolf and his wife. Doctor R. Wolf (1887–1981) was an inventor, diplomat, and philanthropist. Born in Hanover, he emigrated to Cuba before the First World War and after completing his education (as a chemist) in Germany. Cuba became his second homeland. There he invented a method for extracting metal from the residues of metallurgical processes. He supported both morally and financially the revolutionary movement in Cuba led by Castro. From 1959 to 1961 he was Cuba's ambassador to Italy, and from 1961 to 1973 the ambassador to Israel. The Wolf Foundation Prize in Mathematics has been awarded to I. M. Gel'fand and C. L. Siegel (1978), J. Leray and A. Weil (1979), H. Cartan (1980), L. Ahlfors and O. Zariski (1981), H. Whitney and M. G. Kreĭn (1982), S. S. Chern and P. Erdős (1983–1984), K. Kodaira and H. Lewy (1984–1985), S. Eilenberg and A. Selberg (1986), and K. Itô and P. D. Lax (1988).

The last picture: Anna Dmitrievna and Andreĭ Nikolaevich in their Moscow apartment (May 1987).

illness began to tell. At 2:09 p.m. on October 20 a straight line took the place of the usual shape of the graph of his heart rhythm on the oscillator screen. Andreĭ Nikolaevich's heart had stopped.

The life of a great contemporary scholar—Andreĭ Nikolaevich Kolmogorov—came to an end.

His official obituary (*Izvestiya*, October 23, 1987) read as follows:

> **Andreĭ Nikolaevich Kolmogorov's whole life was an unparalleled feat in the name of science. He was a model of nobility, unselfishness, and moral purity in the service of his socialist homeland. A. N. Kolmogorov has entered the galaxy of the great Russian and world scholars.**

Epilogue

In a single article it hardly seems possible to encompass completely and exhaustively the multifaceted life and creative work of a brillant personality such as Andreĭ Nikolaevich Kolmogorov. Though this is a lengthy article, much was left outside its scope: his work with students, the walks and trips with him so wonderful in their intellectual content and emotional effect, the musical evenings at the famous Komarovka, the visits to old Russian towns, where Andreĭ Nikolaevich was an excellent excursion guide, and so on.

Like no one else, Andreĭ Nikolaevich was able to drive a wedge to the essence of a problem and grasp its main point, compelling one to look at it in a new way. I recall how, at a preparatory session (in 1986) for organizing a Congress of the Bernoulli Society in Tashkent, Andreĭ Nikolaevich, seemingly lost in his thoughts,

The memorial plaque on the wall of the Moscow University building where Kolmogorov lived from 1953 to 1987.

suddenly roused himself and began to speak. There was immediate quiet, and everyone heard his question: "And what is the distribution of ages of the invited speakers of the congress?" (A computer diagram showed that most were in their forties.)

Following a mainly chronological scheme, the author has tried not only to give a detailed description of the basic scientific accomplishments and discoveries of Kolmogorov (with much stress on the probabilistic and statistical aspects), but also to accompany the description with comments demonstrating the influence of his ideas and work on the formation and development of many directions in research.

In the preparation of this paper I used various sources, including much material dedicated to Kolmogorov that has appeared in the journals *Uspekhi Matematicheskikh Nauk* and *Teoriya Veroyatnosteĭ i ee Primeneniya*, and commentaries on his work in the three volumes [MM], [PS], and [IA].

Andreĭ Nikolaevich's wife Anna Dmitrievna was of invaluable assistance in my work on this paper, both by her advice and by her permission to use their personal archives. (Anna Dmitrievna did not outlive Andreĭ Nikolaevich even a year: she died on September 16, 1988.)

Discussions with Yu. V. Prokhorov, V. A. Uspenskiĭ, Ya. G. Sinaĭ, V. M. Tikhomirov, A. M. Abramovich, and A. V. Prokhorov made it possible to describe more completely and accurately Kolmogorov's life and creative paths.

I would like to hope that the articles in this collection give the reader a broad and diverse representation of a most unique personality of the twentieth century: Andreĭ Nikolaevich Kolmogorov.

On A. N. Kolmogorov

V. I. Arnol'd

I have always wanted to know how Andreĭ Nikolaevich passed from one topic to another: his investigations of various subjects changed whimsically, in an apparently unpredictable way. For example, his work on small denominators in classical mechanics was not foreshadowed by any previous work but appeared quite unexpectedly in 1953–54. His topological papers had appeared just as unexpectedly in 1935.

I constructed for myself a theory of the origin of Andreĭ Nikolaevich's work on invariant tori: it began with his studies of turbulence. In the well-known work of Landau (1943) it was invariant tori—attractors in the phase space of the Navier–Stokes equation—that were used to "explain" the onset of turbulence. A stable equilibrium state (a point attractor) corresponds to the laminar flow observed for small Reynolds number. The Landau scenario for the transition to turbulence is a succession of bifurcations as the Reynolds number increases. A limit cycle first arises, then the attractor becomes a two-dimensional torus, and the dimension of the invariant torus grows as the Reynolds number increases further. In a discussion at the Landau seminar Andreĭ Nikolaevich remarked that a transition to an infinite-dimensional torus and even to a continuous spectrum can already take place for a finite Reynolds number. On the other hand, even if the dimension of the invariant torus remains finite for a fixed Reynolds number, the spectrum of a conditionally periodic motion on a torus of sufficiently high dimension contains so many frequencies that it is practically indistinguishable from a continuous spectrum. The question as to which of these two cases actually holds was asked more than once by Andreĭ Nikolaevich. A program[1] for the seminar on the theory of dynamical systems and

[1] Here is the complete text of the program (seminar topics):

1. Boundary value problems for hyperbolic equations whose solutions depend everywhere discontinuously on a parameter (see, for example, S. L. Sobolev, *Dokl. Akad. Nauk SSSR* **109** (1956), p. 707).

2. Problems of classical mechanics in which the eigenfunctions depend everywhere discontinuously on a parameter (there is a survey of such problems in Kolmogorov's report at the 1954 ICM in Amsterdam).

3. The monogenic functions of Borel and the quasi-analytic functions of Gonchar (in hopes of applications to problems of the types 1 and 2).

4. The onset of high-frequency oscillations when the coefficients of the higher derivatives tend to zero (the work of Volosov and Lykova for ordinary differential equations).

5. In the mathematical theory of partial differential equations with a small parameter multiplying the higher derivatives, studies have up to the present time been made of phenomena of the type of boundary layers and internal layers converging to surfaces of discontinuity of limit solutions or their derivatives "as the viscosity vanishes". In real turbulence the solutions get

hydrodynamics was posted on a bulletin board in the Mechanics and Mathematics Department of Moscow State University at the end of the 1950's (among other things, the program included the problem of proving the practical impossibility of long-term dynamic weather prediction because of its strong dependence on higher harmonics of the initial conditions). Andreĭ Nikolaevich chuckled about the tori of Landau: "He (Landau) evidently did not know about other dynamical systems."

The transition from the tori of Landau to dynamical systems on a torus would be a completely natural train of thought. In the final analysis I almost believed in my theory and (in 1984) asked Andreĭ Nikolaevich whether it was really so. "No," he answered, "I was not at all thinking of that at the time. The main thing was that there appeared to be hope in 1953. From this I felt an extraordinary enthusiasm. I had thought for a long time about problems in celestial mechanics, from childhood, from Flammarion, and then—reading Charlier, Birkhoff, the mechanics of Whittaker, the work of Krylov and Bogolyubov, Chazy, Schmidt. I had tried several times, without results. But here was a beginning."

This is how it was. At that time Andreĭ Nikolaevich introduced a mathematical practicum in the Mechanics and Mathematics Department, and selected problems for it. Among the problems he chose the investigation of the motion of a point mass along a torus symmetric with respect to the vertical axis. This is a completely integrable Hamiltonian system with two degrees of freedom, and, as a rule, the motion in it takes place along two-dimensional tori in phase space. The trajectories wrap around these tori in a conditionally periodic manner: the angular coordinates on them can be chosen so that they vary uniformly as the phase point moves.

At the time the theory of integrable Hamiltonian systems was not, as it is now, a fashionable area of mathematics. It was regarded as a hopelessly obsolete, outmoded, and purely formal area of analytical mechanics. To occupy oneself with such an "uncurrent" topic was regarded as reprehensible for a mathematician, as a concession to the pressure of external circumstances (it was presumed that mathematicians must sum prime numbers, generalize Lebesgue integrals, investigate continuous but nondifferentiable groups).[2] Andreĭ Nikolaevich said with a laugh that the French write "Celestial mechanics" with a capital letter, but "applied" in lower case. And he always had a certain contempt for all forms of "mathematical elitism", whatever its source (be it Bourbaki or the Steklov Institute of Mathematics).

Thus, Andreĭ Nikolaevich observed that in the "integrable" problems of the mathematical practicum the suitably defined phases on the torus vary uniformly

---

worse in an everywhere dense fashion. It is proposed that a mathematical study be made of this phenomenon at least on model equations (the Burgers model?).

6. Questions of stability of laminar flows. Asymptotically vanishing stability (at least on model equations).

7. A discussion of the possibilities of applying ideas in the metric theory of dynamical systems to real mechanical and physical problems. Questions of the stability of the various types of spectrum. Structurally stable systems and structurally stable properties (in this last direction almost nothing is known for systems with several degrees of freedom!).

8. Consideration (at least on models) of the hypothesis that in the setting of the end of item 5 the dynamical system becomes a random process in the limit (the hypothesis of the practical impossibility of long-term weather prediction).

[2]Fréchet said to me in 1965, " ... Kolmogorov, isn't he the young fellow who constructed an integrable function with almost everywhere divergent Fourier series?" All the subsequent achievements of Andreĭ Nikolaevich—in probability theory, topology, functional analysis, the theory of turbulence, the theory of dynamical systems—were of less value in the eyes of Fréchet.

in time. He at once asked himself whether this is so if the system on the torus is not integrable, but only has an integral invariant (preserving a measure with positive analytic density). He answered this question in a 1953 note about systems on the torus—the first note in which small denominators appear. This note is not complicated in a technical aspect (although it already contains some lemmas necessary for the fundamental work of 1954). Andreĭ Nikolaevich's conclusion is as follows: it is almost always possible to introduce phases varying uniformly in time, but sometimes (when the ratio of the frequencies is abnormally well approximable by rational numbers) mixing is possible (the image of a small disk under the action of the phase flow is spread out over the whole torus).

The remark about mixing, which relates to a pathological case (occurring infinitely rarely), does not seem to be especially important. But that is what became the source of Andreĭ Nikolaevich's famous work on small denominators published in 1954, in which he proved that invariant tori are preserved under a small change of the Hamiltonian.

Kolmogorov's arguments (which he gave in a report at the 1954 ICM in Amsterdam) were as follows.

In integrable systems the motion along an invariant torus is always conditionally periodic (it is possible to introduce phases that rotate uniformly in time). Consequently, mixing does not occur in integrable systems. To see whether the phenomenon he discovered has applications in mechanics, Andreĭ Nikolaevich decided to look for motions along tori in nonintegrable systems, where mixing could in principle be observed.

But how can one find an invariant torus in the phase space of a nonintegrable system? It is natural to begin with perturbation theory, treating a system close to being integrable. Diverse variants of perturbation theory have been discussed repeatedly in celestial mechanics, and later in early quantum mechanics.[3]

However, all these perturbation theories lead to divergent series. Andreĭ Nikolaevich understood that the divergence can be overcome if instead of expansions in powers of a small parameter one uses Newton's method in a function space (about which he had read not long before in Kantorovich's paper, "Functional analysis and applied mathematics" in *Uspekhi Mat. Nauk*).

Thus, Kolmogorov's "method of accelerated convergence" was devised not at all for those remarkable applications in classical problems of mechanics to which it led, but for investigating the possibility of realizing a special set-theoretic pathology in systems on the two-dimensional torus (mixing).

Andreĭ Nikolaevich did not solve the problem he himself posed on realization of mixing on weakly perturbed invariant tori, because, on the tori he found, his method automatically constructs angular coordinates that vary uniformly as the phase point moves. As far as I know, the question of mixing from which all this work of Kolmogorov grew remains unanswered today.

This technical question is insignificant in comparison with the results obtained; nobody remembers it any longer. Physicists say (I heard this from M. A. Leontovich) that new physics most often begins with a refinement of the last decimal place. As we have just seen, new mathematics can also be born from a refinement

[3]Especially thoroughly in Born's book, "Vorlesungen über Atommechanik", of which an amusing translation into Russian was published in the 1930's in Khar′kov: for example, one finds трехизмерительные разновидности (Dreidimensionale Mannigfaltigkeiten).

of small technical details of previous work. Already from this it is clear that the planning of fundamental research is bureaucratic nonsense (and most often a fraud).

* * *

Although Andreĭ Nikolaevich himself regarded the hopes that appeared in 1953 as the main stimulus for his work, he always spoke with gratitude about Stalin (following the old principle of saying only nice things about the dead): "First, he gave each academician a quilt in the hard year of the war, and second, he pardoned my fight in the Academy of Sciences, saying, 'such things happen also here'." Andreĭ Nikolaevich also tried to speak kindly about Lysenko, who had fallen into disfavor, claiming that the latter had sincerely erred out of ignorance (while Lysenko was in power, the relation of Andreĭ Nikolaevich to this "champion in the struggle against chance in science" was quite different).

Repeating what Khodasevich said of Gor′kiĭ, one can say of Kolmogorov that he was at the same time one of the most obstinate and one of the most unstable of people.

"Some day I will explain everything to you," Andreĭ Nikolaevich used to tell me after having done something contrary to his principles. Seemingly, pressure was exerted on him by some evil genius whose influence was enormous (the role of the group transmitting the pressure was played by well-known mathematicians). He hardly lived to the times when it became possible to speak of these things, and, like almost all people of his generation who lived through the 1930's and 40's, he was afraid of "them" to his last day. One should not forget that for a professor of that time not to tell the proper authorities about seditious remarks made by an undergraduate or graduate student not infrequently meant being accused the next day of having sympathy with the seditious ideas (in a denouncement by the very same student–provocateur).

* * *

Andreĭ Nikolaevich said that he could never think intensely about a mathematical problem with full effort for more than two weeks. And he thought that any single discovery could be presented in a four-page *Doklady* note, "because the human brain is not capable of creating anything more complicated at one time." According to him, he maintained a lively interest in a topic of his investigations only as long as it was unclear in which direction the answer to the question lay ("like going along a razor's edge"). As soon as the situation became clear, Andreĭ Nikolaevich tried to get done with writing out the proofs as quickly as possible, and began looking for some apprentice, as it were, to whom he could hand over the whole area of research. At such times it was not a bad idea to keep at a distance from him.

One can distinguish three stages in the development of any area of science. The first is the pioneering stage: the breakthrough into a new area, a striking and usually unexpected discovery, often running counter to established notions. Then follows the technical stage, which is long and laborious. The theory is overgrown with details, it becomes cumbersome and difficult to understand, but in return it encompasses an ever greater number of applications. Finally, in the third stage there emerges a new and more general view of the problem and of its connections with other questions apparently remote from it: a breakthrough into a new area of research has been made possible.

Andreĭ Nikolaevich on vacation.

For the mathematical work of Andreĭ Nikolaevich it was characteristic that he was a pioneer and discoverer in many areas, at times solving problems that had stood for two hundred years. He tried to avoid the technical work in generalizing the theory constructed (incidentally, he said that Jews do especially well at this stage, and he said this rather with admiration, since he perceived his instinctive aversion to this kind of work as a deficiency[4]). On the other hand, Andreĭ Nikolaevich's achievements were remarkable at the third stage when it is necessary to comprehend the results obtained and envisage new paths, at the stage when fundamental generalizing theories are created.

* * *

An example of an unexpected breakthrough of Kolmogorov into a new area is the work in topology he published in four notes in *Comptes Rendus* and reported at the Moscow Topology Conference in 1935. In these papers he constructed a theory of cohomology (simultaneously with but independently of J. Alexander). He did not work in topology after this, but when Milnor's papers on differentiable structures on spheres appeared, they made a very strong impression on him. After Milnor's report at the Leningrad Congress in 1961 Andreĭ Nikolaevich asked me to analyze the proofs and tell him what was going on (I was then a graduate student). I tried to comply. I began learning from V. A. Rokhlin, S. P. Novikov, and D. B. Fuks (and even was an opponent of Novikov at the defense of his Candidate's dissertation on differentiable structures on products of spheres). But my attempts to explain something to Andreĭ Nikolaevich were not crowned with success. He said to me:

[4] "Due to age and laziness, after doing something worthwhile I at best write it down at once but then usually abandon the search for improvements and extensions" (from a letter of March 8, 1958).

> My papers in topology have never been understood in a proper way. You see, I started out from physical concepts—from hydrodynamics and electromagnetic theory, and not at all from combinatorics. The cohomology *groups* I then introduced were mastered and are now used by everyone. But more was done in those notes: I constructed not only groups, but a *ring*! This ring is much more important, and I think that if topologists also master it, they can obtain much that is interesting.

Apparently, all his information about the development of topology after 1935 Kolmogorov got from P. S. Aleksandrov and his students. In any case, Kolmogorov's appraisal of the cohomology ring is remarkable: it contains a penetrating analysis of his work and a prognosis of the significance of the cohomology operations that has proven to be correct. (This comment on Andreĭ Nikolaevich's statement above is due to Rokhlin, who in this case showed a tolerance unusual for him; as for me, during the 1960's I tried militantly with naive intransigence to tell my teacher what had really been going on in topology for the preceding thirty years.)

* * *

But Andreĭ Nikolaevich had his own unalterable points of view on everything. For example, he said to me that spectral sequences were contained in the Kazan work of Pavel Sergeevich Aleksandrov. And that after the age of sixty one should not do mathematics. (This conclusion was apparently based on his experience with mathematicians of preceding generations.) Thus, my attempts to explain homotopic topology to him ended just as unsuccessfully as the attempts to teach him to bicycle or to water ski. Andreĭ Nikolaevich dreamed of becoming a buoy keeper after turning sixty, and for a long time tried to select for himself a suitable part of the Volga. But when the time approached, buoy keepers had already gone from rowboats to motorboats, which he hated, hence the project had to be abandoned. So Andreĭ Nikolaevich decided to return to the profession of schoolteacher, which is where he had started at one time.

* * *

The last mathematical work about which Kolmogorov told me (probably in 1964) had a "biological" origin. It had to do with the minimal cube with room for a "brain" or "computer" of N elements ("neurons") of fixed size, each of them joined with at most k others by means of "wires" of fixed thickness. The number k is fixed, and N tends to infinity. It is clear that a very simple "brain" (like a "worm" of N elements connected in sequence) can be put in a cube with radius of order $\sqrt[3]{N}$. The grey matter of the brain (the bodies of neurons) is distributed over the surface, and the white matter (the connections) is inside. This fact led Andreĭ Nikolaevich to the hypothesis that the minimal radius has order $\sqrt{N}$, and a cube of smaller radius does not have enough room for any sufficiently complex brain (the words "sufficiently complex" can be given a precise mathematical meaning).

In the final analysis this turned out to be the case (Andreĭ Nikolaevich's original estimates contained superfluous logarithms, and the definitive result without logarithms was joint work with Barzdin).

Of course, Kolmogorov understood perfectly well that his theorems had little relation to the structure of a biological brain, and therefore they were not mentioned in the paper about the brain. But the source of the whole theory was actually all the reflections on grey and white matter. It is interesting to note that this paper has remained little known even to specialists, perhaps because the mathematical exposition is too serious. When I mentioned it in a paper in *Physics Today* dedicated to Kolmogorov (October 1989), I received a sudden deluge of letters from American engineers who were apparently working in miniaturization of computers, with requests for a precise reference to his work.

* * *

I was recently in the mountains near Marseilles, and I again made a tour of the calanques, a remarkable system of fjords in a half-kilometer sheer precipice in the maritime Alps. Andreĭ Nikolaevich had showed me this place in 1965: uninhabited mountains five kilometers from Marseilles, a marked trail where arrows led down beyond the precipice. There happened to be a ledge there. If you stepped out on it, you could see a next ledge, and in this way descend little by little to the sea. So recounted Andreĭ Nikolaevich—and now the Université de Luminy has been built at this place.

It was always very interesting to talk with Andreĭ Nikolaevich, and I regret not writing down his stories. Fortunately, some of his letters remain. I believe that the several passages below give a fairly clear idea about their author and his world views.

Moscow, March 28, 1965.

I was very glad to receive your letter of February 14 upon my return from the Caucasus, where I had gone on March 5, returning on the 23rd. There were five of us traveling together (Dima Gordeev, Lenya Bassalygo, Misha Kozlov, and Per Martin-Löf, my 22-year-old Swedish research assistant). In Bakuriani it snowed at first for six days, which did not prevent us from traveling around. In particular, Per and I managed the great slope in Tsagveri along the ravine of the Chernaya River. Dima Gordeev, on the other hand, persistently trained for eight hours a day on a slalom hill. Then S. V. Fomin arrived and brought clear weather. On the first sunny day we walked on the slopes of the local Tskhara-Tskhara ridge, and in three hours there at an altitude of about 2400 meters all my young companions had become so sunburned (walking some in swimsuits and some without anything on) that as a consequence they could hardly sleep the next two nights. On the fourth sunny day we hiked along the top of the same ridge (at 2800 meters). This was timed well, since on the next day the mountains were covered with clouds, out of which an unpleasant wind blew. Misha Kozlov, Martin-Löf, and I gave reports in Tbilisi, and all of us went sightseeing and drinking with the local mathematicians. After that, we still had time for two day-long excursions:

a) In Betaniya, not far from Tbilisi, where there were bunches of spring flowers in the forest near patches of old snow (our common blue snowdrops, tiny cyclamens, crocuses, early irises). Our goal, though, was a church from the 12th century with some frescoes.

b) In Kintsvissi, not far from Gori on the slopes of the Trialetskiĭ range, where there is a truly magnificent painting from the beginning of the 13th century, and its impression on one, as the work of a great and completely individual (though anonymous) artist, is comparable to the impression of the Dionysian frescoes in the Ferapontovo monastery. The last part of the way to the memorial itself turned out to be rather difficult, so besides us only G. S. Chogoshvili made it there. As I understand, during the four-hour wait for us the rest of them enjoyed carousing in the nearest village accessible by car. Then in addition there was a festive supper in the Intourist restaurant near the little house in Gori where Stalin was born. From Gori we at once left for Moscow (in Tbilisi our skis were loaded into the train accompanied by a young Tbilisi mathematician delegated for the task. After our departure our companions probably returned to the same restaurant more than once with him).

But you are due for some skiing around Easter (that is, during the two weeks from April 18 to May 2). According to the guidebook for Savoie or Dauphiné, there you can choose for yourself the ski station you want; at this time the very highest one is desirable (1700–2000 m). You have to reserve beforehand a place at a hotel of whatever rating (beginning with a hostel with double bunks). It will probably be quite satisfactory for you to rent skis, unless you want to buy some to bring here ...

I don't want to say bad things about **, but I also don't want to defend him from your speculation on the matter of his capacity for regarding as interesting only areas of mathematics that he himself is interested in or has at least mastered. But I do want to defend myself. At present I have to concern myself with finding time to do all that remains for me to do, and my plans are quite extensive and diverse. Therefore, I am rather stingy in my efforts to learn things I do not intend to use in my own work, and sometimes stingy even in the less demanding efforts required, for example, to understand survey reports (or, say, your explanations). In my younger friends here there is often a lack of understanding of inevitable age differences; for instance, the desire absolutely to teach me to ride a bicycle or to water-ski.

But I do not see in myself any tendency, arising from such self-restraint, to deny the *objective* interest and significance of new directions. Sometimes I suspend judgement, sometimes I even actively support and recommend the young to study things that in their general impression seem to me to be significant and promising, though they may go outside the bounds of my own repertoire. But if I do more actively and spiritedly defend the significance of directions that I value because of my knowledge of their structure (which is sometimes hidden from those passively reading the finished papers) and perspective, that seems to me both natural and legitimate. For example, our "small denominators", and many others.

Please pass on my special greetings to Leray and his wife and children. With them I have also formed more unconstrained and personal relations than with other French mathematicians. It was so with Schwartz, too, and in a different way with Favard, and of comparatively less prominent people it was so with A. Rényi (the theory of probability and statistics with engineering and physical applications). I would be glad to receive material that would help me write an appropriate obituary for Favard. I know his mathematical work well enough, but not his pedagogical and public activities nor his personal biography. Both are rather interesting (including his active help to Spanish emigrants, and in general very unexpected activities for a mathematician) ...

I am very pleased by the concluding phrase with corrections ... I (regret) blame myself that I (offended) distressed you ... The second correction is undoubtedly right, since it is not so easy to "offend" me. Replacing "regret" by "blame myself" apparently means that it is not your way to regret something ... Uryson's grave is in a little place called Batz near le Croisic; it seems that mademoiselle Cornu—a spinster who takes care of the grave—is still alive.

A. N. Kolmogorov.

Moscow, October 11, 1965.

... It is only now that I get around to answering your letter of August 29 from Chamonix, because I was very busy at the beginning of September, and then I traveled to Yugoslavia (Belgrad, an excursion to "old Serbia" in search of some frescoes from the 13th century, Zagreb and an excursion to the shores of the Adriatic Sea).

I have indeed spent quite a lot of time observing the views and customs of the most diverse circles in France and other countries, but some of what you wrote was interesting to me. I was in France as a young man on equal footing with students only in 1930–31, and although in 1958 I lived for several days in a skiers' hostel with double bunks, I was nevertheless seen as a professor by those around me (which is what you are, of course, but this is not yet written on you).

I will write to Fréchet without fail. But for the present I am extremely pressed for time. I am ever more deeply entangled in school affairs: at an ordinary school in Bolshevo a collaborator and I are trying to teach the fundamentals of differential calculus in the 9th grade and also to introduce the elements of set theory (under the theme "the geometric meaning of equations and inequalities"). They made me the chairman of the mathematics section of the committee that in effect will work out curricula and order appropriate textbooks. This is a rather important matter, and there is hope to actually get something done.

Of the 101 graduates of the boarding school only 44 wanted to enter Mechanics-Mathematics Department, and 32 were accepted (about 70%; from the school number 7 (Kronrod) it was about 60%, and much less from the other schools). On the other hand, *all* our candidates were accepted in the Physics and Technology Institute; apparently, there our preparation is more suitable (there was not less competition there). In the Physics Department the boarding school students were even a bit less successful (about 60%) because of the extreme formalism of the requirements in mathematics following Novoselov's standard text, and perhaps also because of hostility toward our institution.

I received the biography of Favard. I have not yet made any use of it, but I ask you to pass on my thanks, and I hope that something about Favard appears here.

There are many places more attractive than le Croisic in Bretagne. It is deserted there everywhere in the late fall, so you can set out traveling without reserving hotel rooms in advance. I hope you do visit Batz and le Croisic. I stayed there in the "Hôtel de l'Océan" on the very seashore. But every "Maison de Jeunesse" is available to you if they are open out of season.[5]

I know the paper on complexity of algorithms about which you wrote. This is a whole area of research, though not large, but it needs essential refinement: Turing machines do not provide a suitable apparatus here. One can give a reasonable definition of the "minimal possible complexity" that is unique up to a bounded factor under broad natural assumptions. When the true complexity is of order T, Turing machines sometimes give T^2. Now it has been made sufficiently simple ...

... I would welcome any participation by you in the writing of public school textbooks, but I think that the authors' collective must definitely be connected with experimental teaching in the public schools. For algebra in grades 9 and 10 my collective includes besides me Shershevskiĭ, who is now teaching in the boarding school but has a great deal of experience in ordinary schools, and a certain Suvorova, who in Bolshevo is already trying the test sections written.

In my presence there was a call to I. G. Petrovskiĭ from Paris about your taking part in a Journal devoted to the new *Investigative mathematics*, which must be something that stands in contrast to the old journals. We explained that we had not yet been informed of the existence of "investigative mathematics" as a new science, but we knew Malgrange and Tits as excellent mathematicians ...

* * *

This was the journal *Inventiones Mathematicae*. As for writing textbooks, I categorically declined to take part in this both because of my desire to do mathematics and because of serious disagreements with Andreĭ Nikolaevich (who was inclined to regard all schoolchildren as brilliant mathematicians like himself).

I recall how once (in the mid-1950's), having gathered his students (undergraduate and graduate) at his home at Christmas, he delivered his opinion about mathematical abilities. According to his theory, the earlier the stage of general human development at which a person stops, the higher his mathematical abilities. Our most brilliant mathematician, said Kolmogorov, stopped at the age of four to five, when children love to tear off the legs and wings of insects. Andreĭ Nikolaevich

---

[5]I went to Batz, unexpectedly for myself, after boarding a train departing from the Montparnasse railway station one October evening. In le Croisic the train arrived at one o'clock in the night. The little town was dead. The porter did not let me in the Hôtel de l'Océan, since he had not believed that I was alone and not a representative of a bunch of gangsters. Sheltered from the bright moon and from the now fairly cold wind that smelled of iodine, I spent the night under a thuja tree in a garden near pill-boxes of the Atlantic Wall, which were partly converted into villas (in any case the surrounding empty villas seemed to resemble preserved pill-boxes in their architecture). In the morning I came to Batz, and found mademoiselle Cornu in her tobacco shop, surrounded by many cats. Uryson's grave by the wall of the cemetery had been thoughtfully kept tidy (P. S. Uryson had drowned before the eyes of mademoiselle Cornu in 1924).

A. N. Kolmogorov in Komarovka.

regarded himself as having stopped at the level of thirteen years, when boys are very curious and interested in everything in the world but not yet distracted by adult interests (I remember that he estimated the level of Aleksandrov as sixteen or even eighteen years).

In any case, Andreĭ Nikolaevich always assumed that the person with whom he happened to be talking was equal to him in intellect, probably not because he incorrectly appraised the reality ("to most students it makes no difference what is said in lectures: they simply memorize the formulations of certain theorems for the examination", he said about the mechanics-mathematics students at Moscow University), but because he was well brought up (and perhaps considered such confidence in the listener to be helpful and uplifting). This is probably the reason that his wonderful lectures were so incomprehensible to most students (besides, formally his lectures were very far from the standard stupidifying dictation that predominated even then in the teaching of mathematics and is so effectively ridiculed in Feynman's "Surely you're joking, Mr. Feynman").

"Mathematics can be taught really well only by a person who himself is fascinated by it and perceives it as a living, evolving science," said Kolmogorov. In this sense his lectures, for all their technical deficiencies, were remarkably interesting for those who wanted to understand the ideas and not just chase after the signs and indices (among his lectures that I had occasion to attend were ones on Galois fields, dynamical systems, the Euler summation formula, Markov chains, information theory, and so on).

Perhaps his approach to teaching was affected by the unrestrained graduate-school existence which he later remembered as his happiest days. A graduate student then had to pass 14 examinations in 14 different mathematical disciplines. But an examination could be replaced by an independent result in the corresponding area. Andreĭ Nikolaevich said that he thus did not take a single examination, but instead wrote 14 different papers on different topics with new results. "One of the results," he added, "turned out to be false, but I understood this only after I passed."

Andreĭ Nikolaevich was himself a splendid dean. He said that one must forgive talented people for their talent, and he saved more than one mathematician who is now well-known from being expelled from the university. Taking a scholarship away from an obstreperous student, this dean himself secretly helped the young man get through the difficult time. The level then reached by the department was never again reached and hardly will ever be reached again.

"Es ist eigentlich wie ein Wunder", wrote Einstein, remembering his own student years, *"dass der moderne Lehrbetrieb die heilige Neugier des Forschens noch nicht ganz erdrosselt hat; denn dies delikate Pflänzchen bedarf neben Anregung hauptsächlich der Freiheit; ohne diese geht es unweigerlich zugrunde. Es ist ein grosser Irrtum zu glauben dass Freude am Schauen und Suchen durch Zwang und Pflichtgefühl gefördert werden könne. Ich denke, dass man selbst einem gesunden Raubtier seine Fressgier wegnehmen könnte, wenn es gelänge, es mit Hilfe der Peitsche fortgesetzt zum Fressen zu zwingen, wenn es keinen Hunger hat, besonders wenn man die unter solchem Zwang verabreichten Speisen entsprechend auswählte."**

Andreĭ Nikolaevich distinguished himself from other professors I knew by his total respect for the personality of the student, of whom he always expected to hear something new and unexpected.

He certainly loved to teach and give lectures, whether or not there were any obvious results. In particular, he regarded as very unfortunate the cancellation of the lecture series for schoolchildren (which had existed under the Moscow Mathematical Society until Dean O. B. Lupanov removed the Society from control of the Moscow Mathematical Olympiad and of school circles).

* * *

The following letter gives an idea of the pedagogical workload of Andreĭ Nikolaevich (who had just been removed from the office of dean because of related agitation in the department due to the Hungarian Uprising).

Kislovodsk, May 31, 1957.

... You have not yet responded to me on the matter of a circle or seminar for first-year students. Nevertheless, I will not open up anything for the first-year level without you, since my program already has the following:

1) more active leadership of our permanent seminar in the Probability Theory Branch than in past years;

2) regular meetings of coworkers and graduate students of the Steklov Institute and of the Branch on diverse applied work (besides graduate students, we have taken from the current fifth-year people Aĭvazyan, Gladkov, Kolchin, and Leonov as junior researchers under definite applied topics);

**Translator's note.* The following is P. A. Schlipp's translation of Einstein: "It is, in fact, nothing short of a miracle", wrote Einstein, remembering his own student years, "that modern methods of instruction have not yet entirely strangled the sacred curiousity of inquiry; for the delicate little plant, aside from stimulation, stands mainly in need of freedom; without this it goes to wrack and ruin without fail. It is a very grave mistake to think that the enjoyment of seeing and searching can be promoted by means of coercion and a sense of duty. To the contrary, I believe that it would be possible to rob even a healthy beast of prey of its voraciousness, if it were possible, with the aid of a whip, to force the beast to devour continuously, even when not hungry, especially if the food handed out under such coercion were to be selected accordingly."

3) the "random processes" course—this is a required course for fourth-year students of our specialization, and I think that you would not be harmed by attending;

4) a seminar on dynamical systems and random processes for the graduate students Alekseev, Meshalkin, Erokhin, Rozanov, and for you (there will also be other participants, but these are ready now to work fairly intensely and systematically);

5) Tikhomirov's seminar for 3rd- and 4th-year students on selected topics in probability theory and combinatorics (?!), which you are not prohibited from attending, but which I want to keep in a format completely accessible to a general audience.

Nonetheless, if the whole enterprise takes place, I will try to keep my promise to meet weekly in a circle or seminar for first-year students and to bring enough problems, and also to protect you from a possible tendency toward training young boys in irresponsible and empty chatter, as happens (for all their interest) in the circles of T. V. ... They write me that apparently all Moscow students will be mobilized in July to help the Moscow militia (?!) in connection with the festival, but could this be some kind of pasquinade? ...

* * *

Toussuire (Savoie), April 2, 1958.

... For the past two days my life in France has seemed very nonacademic. Yesterday there was a 24-hour strike of the railroads, together with the Paris subway and buses. However, we—residents of the suburbs—were served by a huge number of military trucks that transported us to Paris. Moreover, the public gave the equivalent of the cost of a subway ticket to the soldiers carrying us, who were therefore terribly cheerful and polite. This morning the travel bureau nevertheless procured a ticket here for me. In view of Easter vacations and yesterday's strike, the train no. 609 I was to leave on became so long that (calling itself a single train!) it stood on three tracks, and the numbers of the cars were distributed according to the laws of chance from 1 to 55. I found my no. 17 just as its section started off, but fortunately I occupied my numbered seat even though the aisles were packed with unnumbered passengers. By some special courtesy (I don't know how to explain it) of a waiter I got a very delicious *grand déjeuner* in the dining car even though it was the second shift (there were four in all). The eight-hour journey was interesting: first the flat plains of rural France, with wooden barges pulled along canals by automatic tractors, then the tunnels and rocky mountains, the Rhône, the extraordinarily beautiful Lake Bourget, and, finally, our valley with a little mountain river and snow-covered peaks. The train was going to Italy, and at the end my compartment consisted solely of Italians, extremely dirty and even smelling of garlic.

In St. Jean en Maurienne I had to raise a great commotion to get the Italians to unbarricade the door, which was blocked by trunks. I was able to jump out, and at once found a small company of people looking for a bus to Toussuire. A bus was found, but since there were only six of us, an automobile was substituted in its place (for the same price), and we were quickly taken to an altitude of 1800 m.

Toussuire consists of ten or so houses, including 5 or 6 small inns with 10–15 rooms each. In some of them there are also "dormitories". The rooms turned out to be all locked, and I stayed in a dormitory until Monday. I rented skis for 10 days, and have already walked around. The weather is wonderful, an almost full moon is now shining. Tomorrow I shall take some bread, cheese, butter, and bananas (in quantity) and go out the whole day to wander along some small ridges at an altitude of 2200–2400 m. But today I was stopped with the most friendly words by a communist (there are several of them among the employees, guides, and ski instructors), who compelled me to drink two wineglasses of some very strong potion. For my room and board, including the dormitory and very nourishing food, I pay 1400 francs, that is, slightly more than for my room with private bath in Paris. I have not yet met the other travelers. They consist of: 1) young people of student type; 2) unassuming intelligent families with children from 4 to 17, all ardent skiers, of course ...

Now something about your attacks on me ...

... I regard formal rigor to be *obligatory*, and I think that in the final analysis, after a lot of work (which is usually *helpful* for a final understanding), rigor can always be combined (in the exposition of *important* results, that is, in essence, *simple* results) with complete simplicity and naturalness. The only way to realize these ideals is strictly to require logical clarity even where it is burdensome for the present.

... I never have the time (or energy) to write properly. The diversity of my mathematical and nonmathematical activities (if the latter, like a deanship, is regarded as something useful) provides somewhat of an excuse for this state of affairs, but I am always aware of how I state everything poorly and in a fragmentary manner. However, from here I will bring a few model pedagogical writings for publication both in French and in Russian.

You can also look at my Amsterdam report (in regard to a summary account without proofs) ...

Paris, April 15, 1958 (in my "office" at the Poincaré Institute).

... I stayed exactly eight days in Toussuire (since I left on the ninth day with the same bus I arrived with). The weather was extremely capricious, with snow several times each day, sunshine, and at times a dazzling golden fog. It was impossible to go anywhere very far away because of the fogs and the possible avalanches after 30–40 cm of new snow. It was only the last day that I discovered from morning ski tracks that a party of five had nevertheless taken about a six-hour route to the highest nearby point (Pointe d'Ouillon, 2436 m), from which there is, according to the guide books, a known descent along a ridge to another mountain (Monte Cartier, 2250 m), and then considerably farther below Toussuire (but just to Toussuire it would be necessary to climb along some rocks). Following the tracks, I did this whole very easy but fairly intricate route (in the sense that if you strayed from the trail you might come upon rocks or an avalanche-prone slope), mostly in the sun, but three times in a fog and in huge snowflakes (which did not, however, evoke any desire to put clothes on over my shorts). Thus, I did not catch up with my predecessors, who at the end took the usual route and descended to a bridge across a mountain river; however, in my Russian wildness I got across considerably higher along some rocks, after throwing my skis over. I ended up in Toussuire at 14:50, ordered a magnificent farewell lunch with a more expensive brand of wine (I usually take the standard $\frac{1}{4}$-liters of red wine on tap), drank some coffee, paid my bill at the inn, gave the rest of my Téléski tickets to some youngsters, returned my skis, and got on the bus at exactly 16:30.

As a result of the sunshine, I lost all the skin on my face, but in a fairly painless way, and I myself got suntanned as much as possible in eight days (without peeling).

Then I spent two more days as a guest of Favard in Grenoble, where we also went to the mountains (in a car), sat in a cafe, and watched his 10-year-old daughter ski on a little hill by the cafe. Grenoble itself, the dense spruce forests thick with snow in the nearby mountains, the castle we visited, and the art museum in Grenoble were also rather interesting, and my reception in Favard's family was truly cordial.

Here I have already given two lectures after the Easter break, tomorrow I give a report in the probability seminar, and now I am going to the "mathematical tea" that takes place on Wednesday of every week at 16:45 ...

* * *

Calcutta, April 16, 1962.

The numerous servants working in the garden of Professor Mahalanobis observe that the room for honored guests is occupied by a grey and suntanned man who does not speak English, and who gets up at sunrise and strolls silently in the garden. In addition, before my arrival they were ordered to clean the pond so that I could bathe in the mornings (nothing came of it). Thus, a 10-year-old girl in a bright shawl draped with beads quite persistently wanted me to answer that I was a Hindu, and my "I am Russian" did not convince her—perhaps she thought that this was some kind of special Indian nationality.

Now it is 3:30 a.m. Still dark. Since it is rather cool toward morning, I shut off the airconditioning and open a window. The birds are singing. I will go to sleep again.

At 6:30 a handsome youth, barefoot and in a blue nightshirt, will come and place a little table with tea and fruit in my room (beside the bed, if I am lying down). At 7:00 an American student with whom I swim in the pool at the student dormitory will come by.

At 8:00 we will have a real breakfast, with Mahalanobis and all the house guests. Since Madam Mahalanobis is sick, an English geologist named Pamela Robinson is the hostess at the table.

Then I will go to our consulate and clear up the details of my return trip in the travel bureau.

In Bombay I saw terrible contrasts between the magnificent hotels in the center and the poverty in the settlements on the outskirts. Calcutta is more traditional and poorer on the average, but the times here are apparently good, and I have not yet found any starving people. As for beggars, inexperienced foreigners might exaggerate their number. Barefoot men come to the museum of sculptures with their wives, infants in arms. The attendants only see to it that the families do not sit on the floor to eat. It is natural for a European to take all these people as beggars from their appearance, but they are just looking at the gods assembled here.

* * *

August 9, 1969. The research vessel "Dmitrii Mendeleev".

... Yesterday I spent the first half of the day with Maurício Peixoto ("x" is pronounced like "sh" in Portugese). We discussed the question of a symposium in Brazilia in August of 1971 with Smale, you, and Sinaĭ taking part. It should be not even a symposium, but something like a summer school for a whole month in a place even more attractive than Rio, in the words of Peixoto. In Rio I was taken everywhere by Erlen Viktorovich Lenskiĭ, who is there for a year and speaks fluent Portugese with his colleagues and with ordinary people (which is more difficult, as

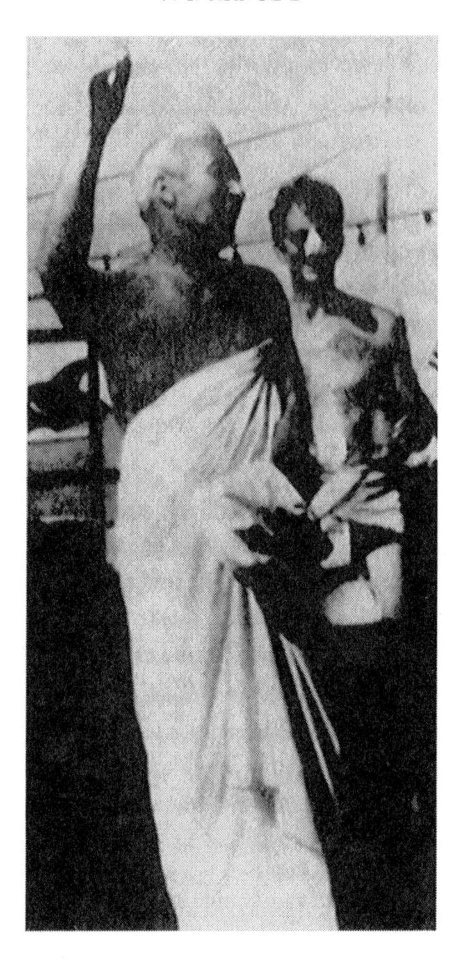

The Neptune Ceremony (research vessel "Dmitrii Mendeleev", 1969).

you know). I taught him to swim without being bothered by the rain in the "winter" here (the temperature of the water was never less than 18°). The four days in Rio were very interesting, but filled with turmoil, so the return to our measured life aboard ship was also pleasant. For the most part I am occupied with revealing the cases where our instruments register the spectrum of vibrations of the rope used to tow the thermoanemometer rather than the turbulence, and other such things.

My participation in the activities of the expedition was even reflected in poems composed on the occasion of the Neptune ceremony upon crossing the equator:

> First to be christened is of course
> Academician Kolmogorov.
> He gives the scientists a scolding,
> They pull the handle of the instrument
> And let loose all the devils
> For the two-thirds law ...

The devils treated me with relative softness, and since I was robed in a navy bedsheet as a Roman toga, only it suffered from the soot on the machine oil with which they were smeared.

However, there is also more cultural entertainment. We even have programs printed on a rotary press for musical evenings of Vivaldi, Bach, or Schumann.

Besides Rio we were in Reykjavik and made a grand excursion in buses to see the geisers and waterfalls (that is, grand with regard to the impressions it made; the trip took one full day). Then we made an unforeseen stop at Conakry (to send a sick person to Moscow by airplane). We now have a month's supply of food; the fuel was taken on in Kaliningrad for the whole trip, and we get fresh water from a water purifier. But before returning we shall certainly make another stop in Gibraltar, from where this letter will probably be sent ...

* * *

The last decade of Andreĭ Nikolaevich's life was darkened by serious illness. First he began to complain about his eyesight, and the usual 40-kilometer ski tours along the Vorya had to be shortened to 20-kilometer tours along the Skalba.

But even then, during our last ski tour, the almost completely blind Kolmogorov leapt over the bank with his skis onto the ice of the Klyaz′ma River. Later, in the summer, it became difficult for him to contend with the waves in the ocean, though in the fall he still managed to escape from the strict supervision of Anna Dmitrievna and the doctors over the fence of "Uzkoe"* in order to swim in the pond (and he taught me where it was most convenient to climb over the fence in order to get into Uzkoe from Yasenevo; Andreĭ Nikolaevich never was someone to worry a lot about behaving himself properly, though, and it was not without pride that he would tell of his fight with some militiamen at the Yaroslavl train station).

In the last years his life was very difficult. Sometimes he had to be carried in one's arms. Anna Dmitrievna, a nurse named Asya Aleksandrovna Bukanova, his students, and graduates of the boarding school created by him watched over him around the clock for several years.

At times Andreĭ Nikolaevich could say only a few words in an hour, but nonetheless it was always interesting to be with him. I remember how, several months before his death, he talked about how surprisingly slowly the tracer shells flew near Komarovka during the war, and how he lived on a sofa in the building of the Presidium of the Academy of Sciences in the Neskuchnyĭ Gardens, after returning to Moscow from Kazan in 1942 at the request of the Artillery Command.

I remember his account of a winter climb of the Brocken in the 1930's: proudly skiing down the mountain in shorts, Andreĭ Nikolaevich encountered two young people with a camera. They asked him to stop and come over. Instead of photographing him as he expected, the young people asked *him* to photograph *them*.

* * *

Up until recent times there was a picture in the Mechanics and Mathematics Department of Moscow University that showed M. I. Kalinin** talking with professors, lecturers, and graduate students in the old building of the university on Mokhovaya Street. There it was easy to see Kolmogorov, S. A. Yanovskaya, V. V. Golubev, V. F. Kagan, P. S. Aleksandrov, and others. Here is what Andreĭ Nikolaevich told about the event.

At the time Kalinin's daughter was friends with a graduate student in mechanics, and Mikhail Ivanovich had come to get to know the department. He gave a

---

* *Translator's note*: A sanatorium for members of the Academy of Sciences.

** *Translator's note*: The Chairman of the Presidium of the Supreme Soviet of the USSR, 1923–1946.

little speech, and then asked all present to talk about their concerns. Each of them started to speak about things that bothered them: graduate students about the shortage of domitory rooms, especially for families, someone about the necessity of sending graduates to provincial universities, and Pavel Sergeevich Aleksandrov about holes in the bathroom roof. In his concluding words the All-Union Elder said: well, I see that you are all at sixes and sevens. As for the graduate students, you should first get a job and a place to live, and only then get married. And as for the bathroom, for that you have the People's Commissariat of Education ...

* * *

There is another story of Andreĭ Nikolaevich that has stuck in my mind—about Hermann Weyl. According to Kolmogorov, Weyl loved the songs of the Russian Cossacks. In the music room of his Göttingen apartment, which occupied a whole floor, Weyl would sit close to the radio, his back to his guests, listening while leaning his elbows on the radio ... There was also a room for ping-pong. In general one sensed an excess of means in comparison with the level of the usual professor; this came from his wife, who belonged to a higher musical-artistic circle going back somehow to Wagner ...

From Andreĭ Nikolaevich's stories about Hadamard:

Hadamard was a passionate collector of ferns. When he came to Moscow, Kolmogorov and Aleksandrov took him boating (on Obraztsovskiĭ Pond on the Klyaz′ma, I think [V.A.]). Suddenly Hadamard saw something on the bank and asked them to put in to shore there quickly. He crossed over to the prow of the boat, and when it drew near the shore, he became so excited in reaching for the bank that he fell into the water. It turned out that an unusual species of fern was growing there, one he had sought everywhere for many years. Hadamard was completely happy. But shortly afterward he was supposed to be taken to a reception with the president in the Presidium of the Academy of Sciences (I believe the president then was Komarov [V.A.]).

Hadamard had to change, and put on Pavel Sergeevich's clothes. But this was rather noticeable (Hadamard was much taller). At the reception people asked Hadamard: "Herr Professor, what happened to you? You aren't in your own clothes—you didn't fall into the water, did you?" To this Hadamard haughtily replied: "Why do you think that a mathematics professor cannot have any other kind of adventure than that?"

Andreĭ Nikolaevich visited Hadamard for the last time when the latter was 90 years old, I believe. Among other things they talked about school olympiads—for a long time France has had something analogous to an olympiad, the *Concours Général*, in which the best (in each subject separately) graduates of secondary schools in the whole of France take part at the same time. The problems are chosen from problems compiled by the best teachers in all of France. The teachers send problems to Paris, and the ministry can judge the quality of its teachers from the quality of these problems (a practice we might do well to adopt). The results of the competition determine the first-place mathematician among the graduates of a particular year, then the second-place, third-place, ... , thousandth-place, and so on.

Hadamard vividly recalled the *Concours Général* in which he took part. "I was second," he said, "and the one who was first also became a mathematician. But much weaker—he was always weaker." It was clear that Hadamard's "defeat" in the *Concours Général* was a painful memory for him even now!

For Kolmogorov mathematics always remained in part a sport. But when on one of his anniversaries (in a report to the Moscow Mathematical Society) I compared him with a mountain climber who made first ascents, contrasting him to I. M. Gel′fand, whose work I compared with the building of highways, both men were offended. " ... Why, you don't think I am capable of creating general theories?" said Andreĭ Nikolaevich. "Why, you think I can't solve difficult problems?" added I. M.

Andreĭ Nikolaevich himself passionately loved music and was ready to listen endlessly to his favorite records, of which he had many, both in Komarovka and in Moscow. For me he always put on the Schumann Quintet, and this made a happy occasion of even the difficult periods on watch when he could hardly speak.

There were also comic incidents.

Anna Dmitrievna had a helper, the elderly and intelligent Galina Ivanovna, who took care of things in the kitchen of the Kolmogorovs' apartment in Professor Zone "L" of Moscow University (the term "zone" had been used since the times when the building was constructed by prisoners).

She returned home late in the evening and did not have time to buy food for herself. Therefore, she asked Anna Dmitrievna to arrange for her to get a permit into the main university building (which was guarded by militiamen), where she could buy something for supper. After consulting with the chairman (B.V.G.), Anna Dmitrievna answered decisively no: for a person with *such* a family name there was nothing to do. Galina Ivanova asked me to help her.

"My God," I asked, "what on earth is your family name?"

"Marx," she answered.

Apparently, with such a name it was supposed to be as hard just to get a permit into the university building as to be admitted into it as a student. (Fortunately, before long a chairman was found with a broader viewpoint of things.)

* * *

Sometimes the illness receded, as it were, and Andreĭ Nikolaevich could speak for longer periods of time. True, it was not especially easy to understand his distinctive diction even before the illness. The story goes that during one of Kolmogorov's anniversaries I. M. Gel′fand mentioned his visit to Komarovka. Pavel Sergeevich Aleksandrov immediately confirmed that Israil′ Moiseevich *actually was* at Komarovka and had even rescued a *cat* who had gotten closed up in the oven, which was beginning to heat. Legend has it (and this is quite likely) that Israil′ Moiseevich added the following comment: "Yes, I did discover a cat in the oven, but by that time I had heard the mewing already for half an hour; I just didn't interpret it correctly."

* * *

Andreĭ Nikolaevich was even more proud of his athletic accomplishments than of his mathematical achievements.

> In 1939, already Academician-Secretary, I decided to see how far I could swim in the icy waters of the Klyaz′ma, and I returned to Komarovka, on skis, with a temperature so high that they feared for my life at the hospital on Granovskiĭ Street (where the

> Academician-Secretary was supposed to be treated). In this way I came to understand that my capabilities were not unlimited. Yet at the age of seventy and at the beginning of winter I ran from the university to swim in the Moscow River, toward the Neskuchnyĭ Gardens. The embankment was so slippery with ice that it was impossible to climb out, and there was no one around. I looked for a place to climb out for a longer time than then on the Klyaz′ma, just barely got out—and did not come down with any illness.

* * *

Kolmogorov recalled with pleasure the trips of his youth in the north. The longest was the Vologda River–Sukhona–Vychegda–Pechora–Shugor–Sos′va–Ob′–Biĭsk (and then barefoot along the Altaĭ). In a trip along the Kuloĭ and the Pinega he succeeded in rigging a sail that had defied the efforts of the local fishermen, and after that Andreĭ Nikolaevich was admitted as an equal (which meant that they began to swear at him just as they did at each other).

One of the last lengthy conversations with Andreĭ Nikolaevich was about the future of humanity. He always looked doubtfully at the list of former editors on the cover of *Mathematische Annalen.* "How will the cover look in 500 years?" he had asked Hilbert. Moreover, he doubted that our culture could even exist for such a long time, principally because of the demographic catastrophe predicted by Malthus. Andreĭ Nikolaevich dreamed of a new structure of society in which the richness of the life of the mind would triumph over the instincts. However strange and naive these ideas were, it is difficult to argue seriously with them: humanity has been rather late in listening to the warnings of thinkers, and Andreĭ Nikolaevich took it as his duty to remind people of this at the end of his long and happy (in spite of everything) life.

In Memory of A. N. Kolmogorov

S. M. Nikol'skiĭ

Andreĭ Nikolaevich Kolmogorov was one of the most outstanding scientists of our epoch. His research work exerted a considerable influence on the development of many areas of mathematics and its applications in this twentieth century of ours. In the theory of functions, probability theory and statistics, functional analysis and topology, geometry, logic, the philosophy of mathematics, hydrodynamics, genetics, classical mechanics, mathematical physics—everywhere he left behind him fundamental work on which the investigations of many researchers over the whole world are now based.

Kolmogorov was a magnificent and gifted mathematician who brilliantly solved a number of problems that had stood for a long time in mathematics before him, and he himself posed many new important problems.

Below I confine myself to a very brief and incomplete survey of Kolmogorov's contributions to the theory of functions of a real variable. Then I shall pass to purely personal remembrances about Kolmogorov and to some characteristics of his organizational activities in academics.

Kolmogorov began his research during his student years in the area of the theory of functions while attending the well-known seminar of Professor N. N. Luzin at Moscow State University.

Topics in this seminar were fundamental questions in the theory of functions of a real variable: set theory, measure theory and integration, and the theory of convergence (usually almost everywhere) of trigonometric series and Fourier series. Andreĭ Nikolaevich carried out important investigations of all these questions during the 1920's.

Luzin posed the following problem: does the Fourier series of a Lebesgue-integrable function converge almost everywhere or not?

In 1923 Kolmogorov constructed an example of a periodic function $f \in L$ whose Fourier series diverges everywhere. This result of his has always been regarded as fundamentally important. Furthermore, the masterly abilities he used in constructing the example were universally admired.

Kolmogorov also obtained sufficient conditions for the Fourier series of a function f in L_2 to converge almost everywhere.

These results were surpassed only in the 1960's by Carleson, who proved that the Fourier series of any function $f \in L_2$ converges to it almost everywhere. Hunt later extended this positive result of Carleson to L_p $(1 < p < \infty)$.

This article is the text of a report delivered at a session dedicated to the memory of A. N. Kolmogorov at the Conference on Random Processes in Rome (July 2, 1988).

At present the results of Kolmogorov, Carleson, and Hunt constitute a foundation in the theory of Fourier series. Investigations of this kind are now being carried out for orthogonal series of functions of several variables.

I mention that the negative result of Kolmogorov has now been generalized (by Bochkarev) to arbitrary orthogonal systems of functions that are uniformly bounded.

It was in the 1920's also that Kolmogorov investigated the theory of operations on sets and the theory of the Denjoy integral. This integral, despite its great generality, turned out to be suitable in a number of concrete studies, for example, in the theory of integral equations. Investigations in measure theory led Kolmogorov, in particular, to the creation of an axiomatic theory of probability.

Andreĭ Nikolaevich possessed the power of abstract thinking to perfection and set great value on it, yet his thoughts were directed toward concrete goals—toward the practical. It is not surprising that he regarded numerical estimates to be significant in cases important for mathematical analysis. He valued absolutely sharp estimates, but in the cases when it was difficult to obtain a sharp estimate he tried to get an asymptotic estimate which at worst was sharp in the sense of order.

Below I want to dwell on some investigations of this kind in the 1930's.

An estimate of approximations by Fourier sums. Kolmogorov proved the following result: The least upper bound of the deviations of functions f from their Fourier sums $s_N(f,x)$ of order N, over all periodic functions f with derivative $f^{(r)}$ not exceeding 1 in modulus, can be expressed by the asymptotic equality

$$\sup_{|f^{(r)}(x)|\le 1} |f(x) - s_N(f,x)| = \frac{4}{\pi^2}\frac{\log N}{N^r} + O(N^{-r}).$$

Due to Kolmogorov's influence there are now in the mathematical literature many results important for analysis that can be characterized as follows. Suppose that the functions f in a given class $\mathfrak{M}$ are approximated in a normed space H ($\mathfrak{M} \subset H$) by certain polynomials $u_N(f) \in H$. Find an exact expression or an asymptotic expression for the least upper bound of the deviations of $u_N(f)$ from f:

$$\sup_{f\in\mathfrak{M}} \|f - u_N(f)\| = ?$$

The Kolmogorov problem on widths. Suppose that in some normed space H of functions we are given a set $\mathfrak{M}$ of functions and a natural number N. It is required to find in H functions $\psi_1, \dots, \psi_N$ such that the least upper bound of the distances (best approximations) from the functions $f \in \mathfrak{M}$ to the linear subspace spanned by $\psi_1, \dots, \psi_N$ is as small as possible:

$$d = \inf_{\psi_k} \sup_{f\in\mathfrak{M}} \inf_{c_k} \left\| f - \sum_{k=1}^{N} c_k\psi_k \right\|.$$

Kolmogorov called the number d the width of the set $\mathfrak{M}$. He solved this problem in the space L_2 when $\mathfrak{M}$ is the class of functions defined on a closed interval and having rth-order derivatives bounded in norm ($\|f^{(r)}(x)\|_{L_2} \le 1$). He found the exact values of d in the periodic and nonperiodic cases.

In the periodic case the desired functions are the classical trigonometric functions

$$1, \cos x, \sin x, \cos 2x, \sin 2x, \dots,$$

and in the nonperiodic case they are the eigenfunctions of a certain Sturm–Liouville problem.

Many results of this kind have now been obtained by other authors in diverse cases that are important for mathematical analysis. Sometimes the statement of the problem has been modified. For the most part, systems of various classical functions have been obtained as functions $\psi_1, \dots, \psi_N$ solving the problem.

However, fairly interesting cases have recently been discovered (by Kashin) when functions that are definitely not classical appear as solutions. For example, nontrigonometric functions in the periodic case.

The Kolmogorov inequality. This is the inequality

$$M_k \leq c_{nk} M_0^{\frac{n-k}{n}} M_n^{\frac{k}{n}},$$

where

$$c_{nk} = \frac{A^{(n-k)}}{\left(A^{(n)}\right)^{\frac{n-k}{n}}},$$
$$M_k = \sup_{-\infty<x<\infty} |f^{(k)}(x)|,$$

$A^{(k)}$ being the corresponding Bernoulli number ($0 \leq k \leq n$; $k, n = 0, 1, 2, \dots$).

The novelty and the difficulty of this result of Kolmogorov is that the constant c_{nk} in it was found with absolute precision.

Other authors were able to use and refine the method invented here to get similar results in different cases needed for mathematical analysis.

I have mentioned only some of Kolmogorov's results obtained in the 1920's and 1930's. They have fostered a large number of papers, which continue to appear even now.

In the 1950's, when he was not more than fifty, he obtained a remarkable result that required of him great virtuosity and a deep penetration into the problem. It is said that Kolmogorov regarded the proof of this result as the most technically difficult of all the mathematical results he obtained.

The formulation is as follows: an arbitrary continuous function on the n-dimensional cube can be represented as a finite sum of superpositions of continuous functions of one variable.

The first steps toward this unexpected result were investigations first by Kolmogorov and then by him and his outstanding student V. I. Arnol′d jointly and separately. They finished with the result given above, which was obtained by Kolmogorov.

I first met Andreĭ Nikolaevich Kolmogorov in the autumn of 1931 in Dnepropetrovsk (Ukraine), where he had gone together with his inseparable friend Pavel Sergeevich Aleksandrov to give lectures at the university.

I was then twenty-six and Andreĭ Nikolaevich twenty-eight. I was a lecturer at Dnepropetrovsk University, and he was a professor at Moscow University.

At the time a great transformation was taking place in our country, also in the area of higher education.

In Dnepropetrovsk with its population of 100,000 there were then three institutions of higher education before 1930, but there became ten of them in 1930. A large number of mathematics teachers were urgently needed. To prepare them

Nikol′skiĭ, Mal′tsev, and Aleksandrov on the 1938 trip.

it was decided to invite prominent mathematicians from scientific centers of the country. Kolmogorov and Aleksandrov responded to these invitations.

To be sure, they were also interested in the nature of the area—first and foremost the Dnieper River, with Dnepropetrovsk spread out on its banks. Here the Dnieper abounds in green islands and magnificent sandy beaches. Kolmogorov and Aleksandrov loved these places and began to visit them regularly, mainly in the spring and autumn. When the lectures were over, they spent time on the Dnieper. They swam much, and became tan in the sun. They loved the sun: to spend a long time under the burning summer sun was a great pleasure for them. They loved water and swam no matter what the temperature.

Of course, they did use the beach for thinking. However, it happened that on the beaches they were accessible to local student youths. Relaxed conversations on scientific and nonscientific topics took place there.

Later, I also was not infrequently a participant in such conversations and gradually became closer to these prominent people. I always attended their lectures, and that was something for me to talk about.

I greatly enjoyed water sports, rowing in particular, and this also drew us together. Together we plied the rapid waters of the Dnieper with our oars.

Later, in 1938, Aleksandrov, Kolmogorov, and two students of his, Mal′tsev and I, traveled a distance of about 1500 km down along tributaries to the Volga and along the Volga itself. At the time Pavel Sergeevich was a corresponding member of the Academy of Sciences of the USSR, but Andreĭ Nikolaevich was not yet an Academician. In the summer of 1939 we continued our journey along the Volga and traveled 600 km.

The lectures of Kolmogorov and Aleksandrov were very interesting, of course. Aleksandrov gave his with great oratorical skill, while Andreĭ Nikolaevich struggled to be understood. The Muscovites often said about Kolmogorov that his reports were difficult to understand. My observations on this account show that Andreĭ Nikolaevich usually divided each of his reports into two parts. The first and shorter part was the overture, which he usually played clearly, so that it was understandable to all. But the second part, involving details of the proofs, could be difficult. Apparently, Andreĭ Nikolaevich overestimated his Muscovite listeners. The same

Kolmogorov and Nikol′skiĭ on the 1938 trip.

thing happened in conversations—in manner he was very simple and democratic, and if the person with whom he was talking did not understand him, it was not because of a haughty unwillingness to explain on Kolmogorov's part, but because of an exaggerated notion about the capabilities of that person. As I have said, in Dnepropetrovsk he struggled mightily to be understood, and he attained his goal in general.

The first cycle of his lectures (in 1931) was devoted to number theory: Dedekind cuts. At that time the theorem on the completeness of the axioms of the real numbers was a new thing for us. Not even fairly complete analysis courses, Vallée Poussin for example, included this.

During the 1930's we were fortunate enough to hear from Andreĭ Nikolaevich himself about his remarkable results, in fresh form just as they were obtained and even in a sufficiently elementary presentation, mainly in the theory of functions. The results of which I have spoken were also topics of his lectures.

It should be mentioned that in the last years of his trips to Dnepropetrovsk (1937–1940) Kolmogorov organized a seminar on the theory of approximation of functions by polynomials at the university, distributing to the participants topics for independent investigation. I helped him, and led the seminar in his absence. By this time I had become interested in approximation theory under his influence and had obtained a number of results in this theory.

The war interrupted the operation of the seminar, but after the war it resumed. For some time I traveled there even from Moscow. A strong school in the theory of approximation of functions evolved from this seminar. A few of its representatives (A. F. Timan, V. P. Motornyĭ) now work in Dnepropetrovsk, while others moved to other cities and there established centers of approximation theory that are no less strong (V. K. Dzyadyk and N. P. Korneĭchuk in Kiev, Yu. A. Brudnyĭ in Yaroslavl).

What I have related about Kolmogorov's activities in Dnepropetrovsk is certainly a very small part of his total activities, and is meant simply as an example. I have given this particular example because it is connected with my own personal experiences.

Kolmogorov exerted a very great influence on the young. He had many students in diverse areas of mathematics, especially in the theory of functions, functional analysis, probability theory, and logic. More than a few of them now occupy high positions in the world of science. I believe that they all thank their good fortune to have been brought together with this great mathematician who also had exalted human qualities such as unpretentiousness, sympathy, and fairness.

In 1940 I traveled to Moscow to work on my doctoral dissertation at the Steklov Institute of Mathematics. My advisor was Andreĭ Nikolaevich. I subsequently remained at the Steklov Institute in the branch concerned with the theory of functions. At the time Andreĭ Nikolaevich headed the probability theory branch as well as the probability theory branch of Moscow State University. In general he was directly responsible for organizing the state of probability theory and statistics, and for some time also logic. But as we know, this did not keep him from being interested in many other areas of science and obtaining profound results there.

Andreĭ Nikolaevich and I together with his friend Pavel Sergeevich maintained the best possible professional and just friendly relations for as long as they lived. Later I was formally equal to them from the point of view of our positions on the academic ladder, and we often had to work together in fulfilling our academic duties, and sometimes also support each other in the complex vicissitudes of academic democracy.

With the exception of his last six years, Andreĭ Nikolaevich was healthy, robust, physically fit, and capable of working incredibly hard. He and Aleksandrov shared a country house near Moscow where they usually went for three or four days of the week for work and relaxation. Two days of the week they were in Moscow, where they lectured, held seminars, and took part in meetings.

From time to time I visited them as a guest at their house. I usually arrived on the train at 1:00 p.m. for lunch. Up to this time my hosts had already been working as a rule, even though it was Sunday. After lunch we walked in the forest, and then had dinner. After dinner Pavel Sergeevich and I continued with our conversation, while Kolmogorov spent time with his students. Then there was tea, to which the students were also invited. No later than 8:00 p.m. I (and the students) hurried through the woods to the train for Moscow. And our hosts went to sleep no later than 11:00 p.m., so they could get up early, do their gymnastic exercises, take a swim in the river when it was not frozen over, and begin working.

There were of course exceptions: they might start on a trip instead of working. They usually spent their vacation time on long boat trips (rowing), but Andreĭ Nikolaevich also found time for mountain climbing.

Andreĭ Nikolaevich did a great amount of work for society, and he was very serious about this work. I shall mention only isolated aspects of this side of his activities.

He was elected to the Academy of Sciences in 1939 when he was thirty-six. At that time he first had to go through selections of the sections of the Academy, which included academicians who were physicists and mathematicians, and then he had to pass through final selections at a general meeting of all academicians.

It is interesting that the physics and mathematics section of the Academy, after electing Kolmogorov as a member, immediately chose him as chief secretary, whose responsibilities included heading the general leadership of all the physics and mathematics institutes of the Academy. Andreĭ Nikolaevich stayed in this post for three years. The war was going on during the second half of this period.

In 1953 Kolmogorov became dean of the Mechanics and Mathematics Department of Moscow State University. Here he showed great initiative. With his participation the academic curricula in the department were changed significantly. The changes meant that special consideration was given to attracting students to research beginning from their early years.

At Andreĭ Nikolaevich's initiative, improvements were made in the teaching of general required mathematical subjects; for example, the roles of functional analysis and logic were strengthened.

The physics and mathematics boarding school founded by Andreĭ Nikolaevich for mathematically gifted children from the provinces has been a great achievement for our country. The school was made up, in particular, of the most distinguished participants of the mathematical olympiads. Kolmogorov was constantly visiting his school, giving lectures, and going on trips with the students.

Those who finish the Kolmogorov school enter the best of our institutes of higher education. There are by now 400 Candidates of Sciences (PhD's) from alumni of the school. Among them are more than a few prominent scholars.

But Andreĭ Nikolaevich also gave much of his attention and energy to the teaching of mathematics in our public schools.

In recent times there has been a dominant tendency to create unified curricula and textbooks for all schools in the Soviet Union. At the request of the Ministry of Education, Andreĭ Nikolaevich directed the preparation of such textbooks in mathematics, and fulfilling these complex duties caused him much agitation.

In any case, a complete cycle of the projected textbooks was prepared in the 1960's. Several of the books were written with the direct participation of Kolmogorov, and they were used on a large scale for about ten years. We now find ourselves in a new wave of school reform that may change somewhat the established transformations.

This process began to take place when Andreĭ Nikolaevich was already seriously ill. Parkinson's disease had begun inexorably to destroy him, and glaucoma had made him blind. During his last two years he could neither see nor speak. At this time some of his grateful students were with him day and night to his final days.

In conclusion I want to say that the Academy of Sciences of the USSR has commissioned me to express gratitude to the Accademia Nazionale dei Lincei and to the organization committee of this conference, which arranged this talk devoted especially to the memory of our great compatriot Andreĭ Nikolaevich Kolmogorov. The Academy also thanks all of you that are present.

Remembrances of A. N. Kolmogorov

YA. G. SINAĬ

In 1956 A. N. Kolmogorov had several students at once from our class: Yu. Rozanov, A. Shiryaev, V. Leonov, V. Erokhin, and me. As far as I recall, V. Tikhomirov had become his student a little earlier. In this (academic) year Andreĭ Nikolaevich was dean of the Mechanics and Mathematics Department for a short time and therefore knew many students altogether, but he had a special relationship with our class. It was not without reason that to his very last years he invariably took part in the reunions of the graduating class of 1957 held every five years.

At the beginning of the academic year Kolmogorov organized a seminar on probability theory, and we all started working in it. For him this was a period of intensive investigations in the area of information theory in its most diverse aspects. At the same time he began a major program of investigations on estimating ε-nets of compacta in function spaces and on estimating ε-widths, with Tikhomirov as his main collaborator. Later, everything relating to this work was summed up by their long joint survey paper in *Uspekhi Matematicheskikh Nauk*.

By the end of the fifth year there is often a feeling among students that they have already mastered the whole of mathematics. However, in Andreĭ Nikolaevich's seminars we (I, at least) always felt ignorant. From some mysterious depths he would draw completely unexpected facts, and then give analogies and explanations that became clear only much later. A little example: In discussing properties of typical realizations of stationary random processes Andreĭ Nikolaevich noted that they are entire functions of a complex variable and proposed studying fairly subtle characteristics of such functions. He introduced a distinctive sporting spirit into the work of the seminar, in which those solutions were especially valued that involved some kind of crafty guess. I do not remember that there were discussions of difficult theorems whose proofs could take up one or several meetings of the seminar, and almost nothing was said about abstract concepts nor ways of giving the greatest possible generality to this or that result, although Kolmogorov valued both these qualities of mathematical creativeness very highly (see below).

In the fall of 1957 I became a graduate student under Andreĭ Nikolaevich. At the same time he began a famous course of lectures on the theory of dynamical systems, which later was continued as a seminar. Much has already been written about this seminar. Among those present besides us were V. M. Alekseev, V. I. Arnol′d, L. D. Meshalkin, M. S. Pinsker, M. M. Postnikov, K. A. Sitnikov, and many others. The first part of the course definitely had a probabilistic bias, although in presenting the von Neumann theory of dynamical systems with pure point spectrum Kolmogorov made use of Pontryagin's theory of characters. For probabilists these were completely unfamiliar objects, of course, but he used them as freely as everything else. Before beginning the lecture he asked the listeners who was familiar with the theory of characters, and only Postnikov and Sitnikov raised their hands. Later in the course he presented the theorem that was to become the basis for the famous KAM theory, together with a complete proof. In early 1958 Andreĭ Nikolaevich departed to spend half a year in France and left Meshalkin and me a program for preparation for the examination in classical mechanics, which included this proof. In December 1957, in one of his last lectures, he talked about his work on the entropy of dynamical systems, which made up an epoch in ergodic theory and the theory of dynamical systems. Much has already been written about the history of this work.

During our post-graduate work, in 1959, Shiryaev, our late friend Leonov, I, and one of my relatives started on a trip to the Crimea. Andreĭ Nikolaevich very much loved the Crimea, had been there many times, and knew the region well. At the time he and Anna Dmitrievna were vacationing in Rabochii Ugolok near Alushta, and there was an agreement that we would go down to see him. As soon as we had appeared, he was very glad to see us and on the spot agreed to go away with us for a whole day in the direction of Gurzuf. This turned out to be a very long and difficult day. We came to the Ayu-Dag cliff, went down to its promontory, swam out to a rock in the sea, and there Andreĭ Nikolaevich gave us a lecture on the geological structure of the Crimea, on the laccoliths, lava domes, and so on, which I remember to this day. By evening we found ourselves in Gurzuf, having gone about 30 kilometers on foot, and we got back by a passing car late at night, at 10 or 11. I asked Andreĭ Nikolaevich whether he was tired, and he answered that if he had something to eat, then he could do the whole thing over again. I think that he was not exaggerating his capabilities.

In a few days we started in the same company for the Crimean preserve and the Great Canyon. It was the first time I had taken a trip with Kolmogorov lasting several days. The difference in our ages was not at all felt. It was as if someone of our own age were traveling with us, except that he was just more experienced and very inquisitive. As a rule, Andreĭ Nikolaevich found the trail before we did, and he knew the names of the plants and trees. When we were going along the Great Canyon, he taught us to dive into the pools of mountain water we came to on the trail. Shortly before this, Shiryaev had become a colleague of Kolmogorov at the Steklov Institute, and in front of the next pool Andreĭ Nikolaevich said to Alik (in jest, of course): "Dive in, Alik, or I'll fire you!"

In the fall of the next year there was a large conference on probability theory and mathematical statistics in Vilnius. To this day I do not know whether it was a chance occurrence favoring people who worked in probability theory, or a trick of my friend Shiryaev, but it happened that, quite unexpectedly for me, Kolmogorov, Gnedenko, Shiryaev, and I ended up in the same sleeping compartment. Before this Kolmogorov had returned from a meeting of the Rectorate, where the sharp increase in the number of abortions in the dormitories of Moscow University had been discussed, and he was quite taken up with his idea that the most rational measure would be to allow contraceptives to be sold at the university. He asked Alik and me from which time problems of sex had started to distract us from mathematics. We both shyly kept quiet, apprehensive for our family life. After some time Andreĭ Nikolaevich and I found ourselves in the corridor of the train car, and he began to discuss a problem in the theory of dynamical systems. As always,

Cramér and Kolmogorov (Sukhumi).

this was a long monologue about things he had thought about repeatedly and for a long time, and I was under great tension trying to follow the course of his thoughts. Since I had also reflected about closely related topics, I was able to interject a few words at some moment. The reaction was unexpected. He said, "I see that you understand all this better than I do," and he never again returned to these questions. It seemed to me that at the time of his long tirade he understood something he had not thought about before, and then the topic lost interest for him. This episode is also interesting in that it shows a certain aspect of Kolmogorov that I had occasion to observe more than once. He very highly (often unjustifiably so) valued in other people qualities that he thought he himself lacked. This can explain the fact that he supported work in which the main achievement was to prove a well-known or intuitively clear result in the most general possible setting. With extremely rare exceptions, his creative work in mathematics was very concrete, and striving toward the abstract was not typical of him. Nevertheless, he often valued it in others.

After finishing my post-graduate studies, I encountered Andreĭ Nikolaevich less often except for his last years, when I was among those who helped him in his everyday life. At that time I once went to "Uzkoe",* where he and Anna Dmitrievna were. During a stroll I began to ask him about his papers on turbulence before the war. These papers have enjoyed an extraordinary reputation. The concept of the "Kolmogorov spectrum" and other such concepts are now familiar to every physicist. It is all the more surprising that the work was done by a mathematician

** Translator's note*: A sanatorium for members of the Academy of Sciences.

who had devoted a considerable portion of his time to abstract areas of mathematics. I found his answer to me striking. He said that he deduced his similarity laws by analyzing experimental results over half a year. At the time his apartment was piled high with stacks of papers, and he was literally crawling over the floor examining them. I had occasion one more time to discuss turbulence with Andreĭ Nikolaevich. This was on the Caucasian seacoast, when Alik Shiryaev and I were accompanying him and H. Cramér after a conference in Tbilisi on probability theory. The conversation was of rare concreteness. Kolmogorov was arguing in terms of the concrete equations of state and discussing the role of the (thermodynamic) entropy. It seemed to me at the time that I was talking with a professional physicist.

Connected with the personality of Andreĭ Nikolaevich was a paradox like one in set theory. In those years many wanted to be like him, to imitate his intonation, gestures, and so on. But an amazing characteristic of Kolmogorov was his uniqueness. And it is not possible to imitate the inimitable.

The Influence of Andreĭ Nikolaevich Kolmogorov on My Life

P. L. UL′YANOV

On October 12–22, 1987 there was a school on the theory of functions in the village of Byurakan, not far from Erevan. Here we heard that Andreĭ Nikolaevich Kolmogorov had died on October 20. Many of us wanted to go to the funeral, but there was a week of dense impenetrable fog in Moscow, and planes did not fly. We took a train. A few days after the funeral I was walking in the Novodevich′e Cemetery where Andreĭ Nikolaevich lies.

I did not study officially under Kolmogorov; I was not one of his students, post-graduate students, nor research assistants. But at the same time my research and pedagogical work at Moscow University was directly connected in large part with him. He exerted influence on my circle of interests, and on the immediate turning points of my life and career. All this had to do with my life in Moscow after 1950, when I first became acquainted with Andreĭ Nikolaevich. I would like to tell about some events of that time.

In 1950 I graduated from the Mechanics and Mathematics Department of Saratov State University and was recommended for post-graduate work in that department. My diploma work advisor was Nikolaĭ Petrovich Kuptsov. I wrote my diploma work on a new proof of Carathéodory's theorem on the boundary correspondence under a conformal mapping of domains, using results about boundary properties of analytic functions. Shortly before I finished at the university, Petr Vasil′evich Sokolov, a graduate student in the same department who lived in the same dormitory as I, suggested that I go to Moscow University and try to pass the graduate-school entrance examinations there and study under D. E. Men′shov. What I knew of Men′shov as a mathematician I had learned from G. N. Polozhiĭ's lectures to us on the theory of functions of a complex variable and from Privalov's book *Introduction to the theory of functions of a complex variable.* Moreover, my advisor N. P. Kuptsov had been a graduate student under Men′shov. I thought over this idea for some time and finally decided in July of 1950 to go to Moscow and submit my documents to the Scientific Research Institute of Mechanics and Mathematics of Moscow University. The director of this institute was Corresponding Member of the Academy of Sciences Vyacheslav Vasil′evich Stepanov, and his deputy was Associate Professor Sergeĭ Fedorovich Lidyaev. My application made no mention of a possible advisor, since I had not previously talked with anyone; in fact, I could not have found anyone there in July.

In September of that year I again came to Moscow University to take the entrance examinations and was sent to the university dormitory at 32 Stromynka

Street. (On my first arrival in Moscow in July I had slept several nights at the Paveletskiĭ Railway Station because I could not buy a ticket from Moscow to Saratov, and I did not guess, nor was it suggested to me, that I could ask to stay in a dormitory during those days.) I passed three entrance examinations for graduate school (mathematics, history of the Communist Party of the USSR, and a foreign language) with perfect marks. Men′shov, L. S. Pontryagin, and S. F. Lidyaev took care of mathematics. Professor Nina Karlovna Bari, whom I did not know, was named as my advisor. She worked in the theory of functions of a real variable, especially trigonometric series. I did have some knowledge of this area. At the university I took a course in the theory of functions of a real variable given by A. E. Liber, a specialist in differential geometry, and then I read the textbooks *Introduction to the theory of functions of a real variable* [1] by Aleksandrov and Kolmogorov and *The theory of functions of a real variable* by Natanson. I was greatly impressed by Luzin's dissertation *The integral and trigonometric series* (printed by Lissner and Sobko, 1915), which I studied in the summer of 1950. The end of 1949 saw the appearance of Natanson's monograph *The constructive theory of functions*, part of which I at once read. This book contains Jackson's inequality

$$E_n(f) \leq 12\omega\Big(\frac{1}{n}, f\Big), \tag{1}$$

where $E_n(f)$ is the best approximation of a function $f \in C(0, 2\pi)$ in the metric of C by trigonometric polynomials of order n, and the modulus of continuity is

$$\omega(\delta, f) = \sup_{|t_1 - t_2| \leq \delta} |f(t_1) - f(t_2)|.$$

At the beginning of 1950 I wanted to lower the number 12 in (1). Using the approximation of f by the integrals

$$U_n(x) = C_m \int_{-\pi}^{\pi} f(t) \left(\frac{\sin n\frac{t-x}{2}}{\sin \frac{t-x}{2}} \right)^{2m} dt,$$

I could lower the constant 12 in (1) to a number less than 5. This result, together with the new proof of Carathéodory's theorem constituted the paper "On the principle of continuity of a conformal mapping and on Jackson's theorem" that I submitted to the graduate school of Moscow University in July of 1950. The mathematics examiners for the graduate school did not look at my paper, and I therefore did not begin working on the indicated questions for the time being. Later I learned that improving the constant in (1) was connected with the Faber problems from the 1930's, and then in the 1960's N. P. Korneĭchuk proved the best possible inequality

$$E_n(f) \leq \omega\Big(\frac{\pi}{n}, f\Big).$$

I should perhaps mention that I still do not know the exact constant in the inequality (1). In 15 or 20 years I again returned to the theory of approximations in imbedding theorems, in Haar series, and in other questions.

By all the above I want to emphasize that I was also interested in studying the theory of functions of a real variable and in taking part in the well-known seminar of Bari and Men′shov on these topics. The plan I had to follow in my graduate studies was drawn up with Bari. With respect to social work, they elected me to the office of the Leninist Young Communist League of the Soviet Union for graduate students of the Mechanics and Mathematics Department of Moscow University, where I led the sports section, since in Saratov I had graduated from a three-year sports school in skiing and track and field.

On July 22, 1950 V. V. Stepanov died, and Academician A. N. Kolmogorov was named director of the Scientific Research Institute of Mechanics and Mathematics of Moscow University beginning in 1951. He was in this post until 1953, and from 1954 to 1956 and from 1978 to 1987 he headed the mathematics section of the Mechanics and Mathematics Department of Moscow University.

Andreĭ Nikolaevich had an unusual way of working with the graduate students of the institute. He often met with students from different branches of the department, was interested in what they were doing, asked about research problems, gave a variety of advice, and spoke about diverse new problems. For many this was extremely important. As for me, at first I only took graduate school examinations and submitted reports, went to various seminars, and sometimes took part in ski races (including the competitions for the championship of Moscow). Andreĭ Nikolaevich heard about this and invited me to his country house in Komarovka (near Bolshevo) at the beginning of 1951. This was a large two-story house where Pavel Sergeevich Aleksandrov often was with Kolmogorov and his wife Anna Dmitrievna. I was there several times. Sometimes I stayed overnight if there was a lengthy discussion in the evening or if a long ski tour was planned for the next day. I remember that a variety of conversations took place both at the house (especially in the library, and then Andreĭ Nikolaevich often sat on the library ladder) and during ski tours, when we went slowly. At my first encounter with him at the country house Kolmogorov let me look over some new journals, in particular, *Fundamenta Mathematicae*, where there was a paper by Sokolowski about trigonometric series that were conjugates of Fourier series of functions of two variables. We discussed this paper. Then he put on records, and we listened to music. After a late supper, I spent the night in a room on the ground floor. On the table were bowls with nuts and fruit. At first I did not take any. But the next day Anna Dmitrievna asked me why I did not seem to notice them. In the morning I usually do exercises, and I began doing that also here. Hearing about this, Pavel Sergeevich had me pump water into the storage tank for 15–20 minutes, "to develop your arms." It seemed to me that Pavel Sergeevich regarded himself as responsible for managing the household, more so than Andreĭ Nikolaevich. For example, once we had to wax our skis, and Kolmogorov and I started heating them on the electric stove. But some wax fell on the stove, and smoke and fumes rose in the room. Kolmogorov said that we had better get out of there quickly so that Pavel Sergeevich did not find us with this, because he would not exactly praise us. After that we took another electric heater and waxed the skis in the corridor.

On ski outings Andreĭ Nikolaevich liked to choose new routes; someone had to make new tracks, and this was always hard. Both Andreĭ Nikolaevich and Pavel Sergeevich were good skiers. But I still took part in ski competitions at the time, and I was a young man and fairly strong, so they made me break trail. On our first tour I quickly became tired and then fell behind. During a rest stop Andreĭ Nikolaevich asked me why I did not go ahead. I answered that with such a load I needed a good breakfast and some food with me, which I did not have. During the rest I borrowed some sandwiches from the others on the tour (including Yuriĭ Mikhaĭlovich Smirnov) and said that I could now go much better and even lead, which I did. Later, every time I stayed overnight and there was skiing the next morning Anna Dmitrievna fed me a nutritious breakfast. And what wonderful dinners (or suppers) she treated us to after the tours!

I would also like to mention a conversation that was not pleasant for me. Sometimes many people were at the country house. Often there was Vadim Ivanovich Bityutskov (on *Soviet Encyclopedia* matters), Sergeĭ Vasil'evich Fomin (on matters connected with a joint book on the theory of functions and functional analysis), and others. During one dinner Andreĭ Nikolaevich said that Sergeĭ Vasil'evich was a good skier and had previously been in competitions, and he asked about the results. Sergeĭ Vasil'evich said that he did 10 kilometers in about 46 minutes (at the time 50 minutes was class III, 46 minutes was class II, and 43 minutes was class I). Out of youthful boastfulness I said that this was not a high result and that I had once shown such a time in my first competitions. Andreĭ Nikolaevich asked about my results. At the time I did 10 kilometers in 37 minutes and 30 kilometers in a few seconds over 2 hours. Even then I sensed the irrelevance of any kind of comparison; to this day the conversation evokes an unpleasant feeling in me.

At the time of the ski tours Kolmogorov questioned me about my athletic background. I said that I had begun to ski at about the age of five, and my first "skis" were made for me by my brother from curved staves (boards) of an old cracked barrel. Skiing on the slopes of a ravine, and later skating on frozen ponds and ice-covered roads were our customary winter pastimes. In the higher classes I studied in the district center Novye Burasy, 23 kilometers from my town. The Second World War was going on, and it was a hungry time. Therefore, I often went home for provisions, on foot in the summer and on skis in the winter. It could happen that one got caught in a snowstorm or even a real blizzard of the steppes, and there was a very real danger of getting lost and then simply sitting down and freezing to death after exhausting one's strength; there had been such cases in our town. All this made me learn to be a good skier. True, even after the war when at the university I was enrolled at the same time for three years in a sports school in Saratov. My favorite distance in competitions was 30 kilometers, that is, I was a long-distance skier.

When Andreĭ Nikolaevich settled in his apartment at Moscow University, he began to ski also from the university. I then lived on Lomonosov Prospect. We skied to the university and from there started on skis toward Vnukovo. When it was close to spring many on these tours wore only shorts. Andreĭ Nikolaevich and Pavel Sergeevich regarded this as good for the health, and thereby set an example for the young. Once in March Alekseĭ Fedorovich Filippov and I also stripped down and skied in only shorts. Everything would have been all right, except that Filippov and I were fairly lean, and such tours in freezing weather were not for us; we quickly dressed.

Sometimes Andreĭ Nikolaevich said that "Ul'yanov has the character of an athlete." To this day I do not know whether that was supposed to be good or not. Most likely the answer depends on the specific event one is talking about. Then I said that in sports (track and field, skiing, and so on) it is easier to recognize progress than, say, in mathematics. Everything described belongs to the 1950's, at the time I was a graduate student.

As for my research work, I first took various special courses, including Bari's "Trigonometric series" and Men'shov's "Supplementary chapters in the theory of functions" (it was about the integral, and then about summation of series), and I attended seminars on real and complex analysis.

In 1951 Bari gave me a problem. Namely, the following definition is familiar: for a given function $f(t)$ on $[0, 2\pi]$, a continuous function $F(t)$ on $[0, 2\pi]$ is called a primitive for f if $F'(t) = f(t)$ almost everywhere on $[0, 2\pi]$. In his dissertation Luzin proved that $f(t)$ has a primitive if and only if it is measurable and finite almost everywhere on $[0, 2\pi]$. He used this result in the dissertation for the representability of functions by a trigonometric series that is summable to $f(t)$ almost everywhere by the Riemann and Poisson methods (see [**2**], pp. 78, 236).

In 1940 Men'shov established that for any measurable function $f(t)$ that is finite almost everywhere on $[0, 2\pi]$ there exists a trigonometric seies converging to it almost everywhere on this interval.

The problem posed by Bari was to combine the Luzin and Men'shov theorems. While Luzin's theorem was not very hard to prove, and I was already familiar with the proof from his dissertation, Men'shov's theorem was very difficult to prove, and I had not yet had time to analyze the complete proof.

In one of Kolmogorov's talks with the graduate students in our department I mentioned the problem of combining the Luzin and Men'shov theorems. He said that this was an interesting problem. I told Bari about the conversation at a meeting when it came up that I had talked with Kolmogorov. She had at the time a simpler proof of Men'shov's theorem, and all this made it possible for her to quickly prove the following theorem:

For any function $f(t)$ that is measurable and finite almost everywhere on $[0, 2\pi]$ there exists a continuous function $F(t)$ on the same interval such that $F'(t) = f(t)$ almost everywhere on $[0, 2\pi]$, and the result of differentiating the Fourier series of $F(t)$ termwise is a trigonometric series converging to $f(t)$ almost everywhere (see [**4**]). She gave a report about this result at a meeting of the Moscow Mathematical Society on April 29, 1952, with the title "On primitives of functions and almost everywhere convergent trigonometric series". The result was later presented in her monograph [**5**] (Chapter XV, § 2).

In connection with the foregoing Bari advised me to study the paper [**6**] of Marcinkiewicz on convergence almost everywhere of Fourier series of functions $f(t) \in L^p(0, 2\pi)$ satisfying the condition

$$(2) \qquad \int_0^{2\pi} \int_0^{2\pi} \frac{|f(x+t) - f(x-t)|^p}{t} \, dt \, dx < \infty \quad (1 < p \leq 2),$$

and to try to make it local. This I did. I reported it in the seminar of Bari and Men'shov on the theory of functions, and at the end of 1952 Kolmogorov presented it for publication in the journal *Izvestiya Akademii Nauk* (see [**8**]).

In the case $p = 2$ the condition (2) is equivalent, as shown by Plessner, to the requirement

$$(3) \qquad \sum_{n=2}^{\infty} (a_n^2 + b_n^2) \log n < \infty,$$

where a_n and b_n are the Fourier coefficients of f.

Thus, for $p = 2$ our theorem is an extension to the local case of the well-known theorem of Kolmogorov–Seliverstov and Plessner. I mention that S. B. Stechkin, G. Alexits, N. K. Bari, and others were working on such theorems at the time (see [**5**], Chapter V, § 10).

Conditions of the form (2) or (3) on functions are often used in diverse problems. In this connection we established that under certain requirements the following conditions are equivalent:

$$\int_0^{2\pi}\int_0^{2\pi} \alpha(t)[f(x+t)-f(x-t)]^2 dt\, dx < \infty$$

and

$$\sum_{n=2}^{\infty}(a_n^2+b_n^2)\omega(n) < \infty, \quad \text{where} \quad \omega(n) = \int_{2\pi/n}^{2\pi} \alpha(t)\, dt.$$

At the insistence of the referee Stechkin, I rewrote these results several times, and they were published in the journal *Uspekhi Matematicheskikh Nauk*, where the editor-in-chief was Kolmogorov. Recently I needed such results, and I reread my paper and saw that [**7**] was nevertheless written in far from the best possible way.

In the academic year 1951/1952 Kolmogorov made a report in the seminar of Bari and Men′shov. I do not recall the title of the report, but it was about various integrals and their possible uses in diverse problems. It was here that I first heard the definition of the A-integral introduced by Kolmogorov. It is defined as follows:

A function $f(t)$ is said to be A-integrable on an interval $[a,b]$ ($f \in A(a,b)$) if

$$nE\{t : |f(t)| \geq n\} = o\left(\frac{1}{n}\right) \quad \text{as} \quad n \to \infty \tag{4}$$

and the following limit exists and is finite:

$$\lim_{n\to\infty}\int_a^b [f(t)]_n dt = J, \tag{5}$$

where $[f(t)]_n = f(t)$ for $|f(t)| \leq n$, $[f(t)]_n = n$ for $f(t) > n$, and $[f(t)]_n = -n$ for $f(t) < -n$. The number J is called the A-integral of f on $[a,b]$, $(A)\int_a^b f(t)\, dt = J$.

The report mentioned the use of this integral in probability theory, in the theory of functions, and especially in the theory of conjugate trigonometric series and conjugate functions. I focused my attention on the applications to conjugate functions and worked on this in both my Candidate's dissertation ("An application of the A-integral to trigonometric series, and some local theorems on convergence of Fourier series", 1953) and my Doctoral dissertation ("An integral of Cauchy type. Convergence and summability", 1959), where A-integration was used also in complex analysis.

As for the definition of the A-integral, it should be mentioned that an integral was defined by Titchmarsh in the form of the limit (5) in 1929 (see [**5**], Chapter VIII, § 18) and called the Q-integral by him. But it turned out that the Q-integral does not have the property of additivity. In general form the idea of defining the A-integral as the limit (5) under the condition (4) was expressed by Kolmogorov (see [**9**], pp. 56–58, or [**10**], pp. 73–75) in a probabilistic form as a generalized mathematical expectation. The A-integral then has the property of additivity.

We recall that if $f \in L(0,2\pi)$, then the conjugate function is defined to be

$$\overline{f}(x) = -\frac{1}{\pi}\lim_{\varepsilon\to+0}\int_{\varepsilon}^{\pi}\frac{f(x+t)-f(x-t)}{2\tan(t/2)}dt, \tag{6}$$

which exists for almost all x (the theorems of Luzin and Privalov), and the conjugate series is defined to be the series

$$\sum_{n=1}^{\infty} a_n \sin nx - b_n \cos nx, \tag{7}$$

where a_n and b_n are the Fourier coefficients of f.

The function $\overline{f}(x)$ is not necessarily Lebesgue-integrable on $[0, 2\pi]$. What is more, it need not be Denjoy integrable, not even in the wide sense. In 1925 Kolmogorov [**11**] showed that $\overline{f}(x)$ satisfies the condition (4) on $[0, 2\pi]$, and in 1929 Titchmarsh proved that the conjugate function $\overline{f}(x)$ is in $Q(0, 2\pi)$.

In the summer and fall of 1952 I was occupied with studying the properties of conjugate functions, and I obtained a number of results. Some of them follow.

1) *If $f \in L(0, 2\pi)$ and the functions $g(t)$ and $\overline{g}(t)$ are bounded on $[0, 2\pi]$, then $\overline{f}g \in A(0, 2\pi)$, and*

$$(A)\int_0^{2\pi} \overline{f}g\,dt = -(L)\int_0^{2\pi} f\overline{g}\,dt. \tag{8}$$

This equality is an extension of a formula of M. Riesz (for $f \in L^p$ and $g \in L^q$ with $1/p + 1/q = 1$ and $1 < p < \infty$) to the case $p = 1$. It follows from (8), in particular, that if $f \in L(0, 2\pi)$, then the conjugate series (7) is the Fourier series of $\overline{f}$ in the sense of A-integration. Hence if $\overline{f} \in L(0, 2\pi)$, then the conjugate series (7) is the Fourier–Lebesgue series of $\overline{f}$. But this is a theorem of Kolmogorov [**12**] from 1927; it was later proved also by Smirnov and Titchmarsh (see [**5**], Chapter VIII).

2) *If $f \in L(0, 2\pi)$, then for almost all $x \in [0, 2\pi]$*

$$f(x) = \frac{1}{2}a_0 + \frac{1}{\pi}\lim_{\varepsilon\to+0} as\left\{(A)\int_{-\pi+\varepsilon}^{x-\varepsilon} + (A)\int_{x+\varepsilon}^{\pi-\varepsilon}\right\}\overline{f}(t)\frac{1}{2\tan((t-x)/2)}dt, \tag{9}$$

that is, in some sense the inversion of the formula (6). In essence, the asymptotic limit and A-integration have arisen in (9).

3) *If $\{a_n\} \in V_0$ (for example, $a_n \downarrow 0$), then the functions*

$$f(t) = \sum_{n=1}^{\infty} a_n \cos nt \quad \textit{and} \quad \overline{f}(t) = \sum_{n=1}^{\infty} a_n \sin nt \tag{10}$$

make sense for all $t \not\equiv 0 \pmod{2\pi}$. Series of the form (10) have been studied by many authors (Hardy–Littlewood, Kolmogorov, Young, Sidon, Boas, Denjoy, and others). The functions $f(t)$ and $\overline{f}(t)$ can fail to be Lebesgue-integrable. We show that they are always A-integrable, the series (10) are the A-Fourier series of f and $\overline{f}$, the formula (8) is valid for the A-integrals, and for all $x \not\equiv 0 \pmod{\pi}$ we have the equalities

$$\overline{f}(x) = -\frac{1}{\pi}(A)\int_0^{\pi}\frac{f(x+t) - f(x-t)}{2\tan(t/2)}dt,$$
$$f(x) = \frac{1}{\pi}(A)\int_0^{\pi}\frac{\overline{f}(x+t) - \overline{f}(x-t)}{2\tan(t/2)}dt.$$

From the results 3) I then wrote a paper, which Kolmogorov for the first time presented on my behalf to the journal *Doklady Akademii Nauk SSSR* [**13**]. Detailed proofs of 1)–3) were given in *Matematicheskiĭ Sbornik* and *Uchenye Zapiski Moskov.*

Gos. Univ. (see also [**5**], Chapter VIII, § 18, Chapter X, § 4). I am pleased to say that most of my papers in *Doklady* and *Izvestiya AN SSSR* were presented by Kolmogorov. My paper "Some results about series in the Haar system", which was the last one presented to *Doklady* by Kolmogorov (in 1981), dealt with the uniqueness of Haar series and with properties of the Fourier–Haar coefficients of a superposition of functions.

At the beginning of 1953 S. F. Lidyaev, Deputy Director of the Institute of Mechanics and Mathematics, told me he had been present during a lengthy conversation at the institute between Kolmogorov and Bari about me. Andreĭ Nikolaevich insisted that the graduate student Ul'yanov should prepare his dissertation for the defense. Nina Karlovna objected that Ul'yanov had almost a whole year of graduate work in front of him, and it did not make sense to hurry. Then they discussed my results, and Andreĭ Nikolaevich suggested himself as an opponent. After hearing this from Lidyaev, I was unexpectedly glad at such an evaluation by Kolmogorov and troubled and vexed by Bari's reaction, and I said that I was not ready to finish work on my dissertation. To this Lidyaev responded with an interesting phrase: "Young man, Andreĭ Nikolaevich knows what you can do better than you do." Soon Bari told me that it was time to put my Candidate's dissertation into form. It contained results about A-integration and local convergence theorems. Thus, with regard to a defense of my Candidate's dissertation Bari did not take the initiative. On the other hand, I would like to remark that in 1958 she was the first to speak to me about the necessity of putting my Doctoral dissertation into form, and she invited S. M. Nikol'skiĭ, V. Ya. Kozlov, I. F. Lokhin, and S. B. Stechkin as opponents. The latter dissertation also contained topics related to A-integration, and one of the results was that every L-integral of Cauchy type is a Cauchy A-integral. More precisely, if a finite domain $\mathcal{G}$ is bounded by a sufficiently smooth contour Γ and $f \in L(\Gamma)$, then the integral of Cauchy type

$$F(z) = \frac{1}{2\pi i}(L)\int_\Gamma f(u)\frac{du}{u-z} \qquad (z \in \mathcal{G})$$

admits the representation

$$F(z) = \frac{1}{2\pi i}(A)\int_\Gamma \frac{F_i(u)}{u-z}du,$$

where $F_i(u)$ are the angular interior limit values of the function $F(z)$.

Kolmogorov's problems on integrals of type A have been used by many mathematicians both for studying integrals and for applications in real analysis, complex analysis, and functional analysis and other topics. The following are among those who have worked in this area: I. A. Vinogradova, O. D. Tsereteli, I. L. Bondi, G. A. Khuskivadze, A. G. Dzhvarsheĭshvili, S. Nakaniskhi, A. I. Rubinshteĭn, F. S. Vakher, T. P. Lukashenko, A. B. Aleksandrov, A. A. Talalyan, T. Salimov, G. G. Gevorkyan, E. Yu. Terekhina, A. V. Rybkin, and others.

In the spring of 1953 I wrote my (Candidate's) dissertation, and the defense was set for June 10, 1953. The opponents were A. N. Kolmogorov and D. E. Men'shov. Anatoliĭ Illarionovich Shirshov's defense of his Candidate's dissertation in algebra took place on the same day (his advisor was A. G. Kurosh). The defense was in the old building of Moscow University on the third floor, that is, no. 9 on Mokhovaya Street (now this is no. 20 Marx Prospect). At Kolmogorov's suggestion, supported by Men′shov, the Academic Council qualified the dissertation as outstanding. Shirshov's dissertation was also deemed outstanding at the suggestion of the opponents. Shirshov, like me, lived in the dormitory at 32 Stromynka. In honor of the defense we decided to have a festive dinner at the dormitory. Attending were close friends and comrades of us, including our fellow graduate students. Our teachers and older comrades were also there (Kolmogorov, Stechkin, Postnikov, and others). A close relative of Bari had just died, and thus she and Men′shov went to the funeral. Gennadiĭ Ivanovich Barykov, a graduate student in mechanics and a Hero of the Soviet Union in the Second World War, was in charge of the dinner. Although graduate students of that time (as yet now) were people of modest means, products such as butter, a variety of sausages, black and red caviar, and also wine, were much cheaper, and thus on our table there was wine and there were snacks. Barykov saw to it that delicious meat pies were prepared in the kitchen. Of all wines Andreĭ Nikolaevich preferred Don champagne. He proposed a toast to the scientific mom of Petr Lavrent′evich (that is, to Bari). There was much talk about mathematics, and, of course, there were toasts to Andreĭ Nikolaevich. By and by, as all became less inhibited, Sergeĭ Borisovich Stechkin stood up and asked Kolmogorov: "We heard from Pavel Sergeevich Aleksandrov that whoever sits with Andreĭ Nikolaevich for 10 minutes gets problems for a Candidate's dissertation, and whoever sits with him for 30 minutes gets problems for a Doctoral dissertation. But Ul′yanov has sat with you for more than 3 hours. What happens then?" Kolmogorov answered that he had never heard anything of the kind from Pavel Sergeevich. Late in the night Kolmogorov departed for Komarovka, and, to see that everything would be all right, we asked his student Anatoliĭ Vasil′evich Martynov, who had been in the war and was a strong fellow, to accompany him to the country house. Martynov was late for the return train and arrived on foot in Moscow early on the morning of June 11.

Let me go on to how I ended up working at Moscow University, not something I expected. I must admit that I did not even think of the possibility. At the end of 1952 some of our graduate students started looking for a job; I did, too. At the time Nina Karlovna Bari did not think of this, and the chairman Dmitriĭ Evgen′evich Men′shov hardly knew me at all. The first possibility seemed for me to go and work at Saratov University, where I had studied, where they remembered me, and where I could be invited. There was much in favor of this possibility, but there was also one large drawback. I will explain this in more detail. Saratov University was opened at the beginning of the twentieth century and was situated in four new buildings on the main and oldest street, Moscow Street (now V. I. Lenin Prospect). In the same area, on the neighboring Tsyganskaya Street (now I. S. Kutyakov Street), were the university's dormitory no. 1, Saratov Prison, and a vegetable oil factory. I lived for four years in dormitory no. 1, and I knew it very well. Many students, including me, worked as loaders at the vegetable oil factory. Oilcakes were made there from soy beans and sunflower seeds. At the time there was food rationing, and many were hungry. For part-time work they most often hired students, because students did not steal oilcakes from the factory. Earning extra money at this factory was a big help for students to cover living expenses. On January 26, 1943 one of the founders of genetics, Academician Nikolaĭ Ivanovich Vavilov, died at Saratov Prison. Only in 1970, shortly after his rehabilitation, was there a monument erected at Voskresenskoe Cemetery in Saratov, at the expense of relatives. On the monument is written, "Academician Vavilov Nikolaĭ Ivanovich

1887–1943" (sculptor K. A. Suminov). Nearby stands a monument from 1939 dedicated to Nikolaĭ Gavrilovich Chernyshevskiĭ (1828–1889) (sculptor P. F. Dunduk, architect M. V. Krestin). On the arch around the monument is written, "Chernyshevskiĭ was ... a revolutionary democrat, he was able to influence all the political events of his epoch in a revolutionary spirit, carrying the idea—over the obstacles and barriers of censorship—of peasant revolution, the idea of the struggle of the masses for the overthrow of all the old regimes" (V. I. Lenin). On the other side of the arch are the words: "I have served my Motherland well, and I have the right to its gratitude", N. G. Chernyshevskiĭ 1888.

In the third building of Saratov University, in Gorky Auditorium, I attended lectures on mathematical analysis (Prof. G. P. Boev), algebra (Prof. N. G. Chudakov), physics (Profs. P. V. Golubkov and V. I. Kalinin), and other subjects. Only later did I learn that Nikolaĭ Ivanovich Vavilov had been a professor in the Agronomy Department of Saratov University and had on June 4, 1920, at the Third All-Russian Congress of Selectionists, presented a report, "The law of homologous series in hereditary variability", which laid the foundation for the development of genetics in the USSR. There is now a memorial plaque about this near the auditorium. After Vavilov's report there was a great ovation, and then the plant physiologist V. R. Zelenskiĭ spoke from the tribune: "This congress has become an historical event. Here biologists greet their Mendeleev." It has been decided to erect a monument to Vavilov in Saratov at the intersection of Mikhaĭlovskaya Street (now N. I. Vavilov Street) and Peace Lane.

A request was sent to the Ministry of Education from Saratov University that I be given a job in the Mechanics and Mathematics Department there. At that time very few apartments had been constructed, the university did not have any available, and they could only put me in the dormitory on Tsyganskaya, which, though it was located on the next street from the university, was scarcely fit for normal dwelling. Earlier there had been a tobacco warehouse in the building, the first story was so low that the windows were on the level of the pavement, and the blotches of dampness on the walls reached almost to the ceiling. It was said that the higher humidity was needed to preserve the tobacco. But these conditions were poorly suited for people: some of our students who had lived there became sick and withdrew from the university. In general, to endure such conditions one had to be in sound health. Moreover, a common kitchen and toilets were there. And at that time I was already married, and we expected a child. So the question of a dwelling was something I could not give a low priority. Some of us graduate students had begun to arrange a job at the Moscow Institute of Physics and Technology (in Dolgoprudnyĭ). The Department of Higher Mathematics there was headed by S. M. Nikol'skiĭ. I was not acquainted with him, but I had heard his reports on imbedding theory at the Steklov Institute. In January 1953 I went to the Institute of Physics and Technology, met Nikol'skiĭ in a large auditorium for student examinations, introduced myself, and talked about my work. After this I went to their personnel director, who talked with me for a long time and said finally that he would come to our job assignment meeting at Moscow University. As for a dwelling, he promised that they would give me a room not far from the institute in about a year. This suited me, since the rooms were in new houses. In May 1953 there was a job assignment meeting for graduate students in the Scientific Research Institute of Mechanics and Mathematics. At the meeting were the director of the institute A. N. Kolmogorov and the dean of the Mechanics and Mathematics Department Yu. N. Rabotnov.

When they took up my case, Kolmogorov proposed that I keep working in the department, which was quite unexpected for me. Here the question of a place to live came up at once. Andreĭ Nikolaevich thought that I was alone and would live in a room for graduate students in the university building in the Lenin Hills. But I was no longer alone. I had a family, and thus the problem about living quarters had to be solved somehow. They let me walk around and think about things while the meeting went on. I had to choose one of three variants: Moscow University; the Moscow Institute of Physics and Technology whose personnel director was sitting at the meeting with a claim on me and with a promise of a future room; and Saratov University. I decided to go to work where there would be a place to live, and my order of preference was the university, then the institute, and in the extreme case Saratov with its dormitory almost unfit for living.

At the end of the job assignment meeting they sent for me, and Dean Rabotnov said that they could give me a room in a communal apartment of the residential building of Moscow University on Lomonosov Prospect, the construction of which was almost finished. Of course, I agreed with pleasure, and from September 1, 1953 I began working in the Theory of Functions and Functional Analysis Branch. It should be mentioned that Dean Rabotnov did not know me at all, and it was undoubtedly Kolmogorov who interceded for me about the room.

On April 23, 1954 I made my first report at a meeting of the Moscow Mathematical Society, on the topic of "The A-integral and its application to the theory of trigonometric series". At Kolmogorov's suggestion I was elected on June 1, 1954 as a member of the Moscow Mathematical Society, whereupon I received a salutary congratulation from Pavel Sergeevich Aleksandrov, President of the Society.

In 1954 Men′shov, Bari, and Stechkin wrote a review of the papers presented at the competition for the Moscow Mathematical Society's Prize for Young Mathematicians.

The review begins with: "The papers of P. L. Ul′yanov, the content of which was reported to the Mathematical Society at the 'Lomonosov Lectures' in April of 1954, are devoted to the so-called A-integral and its applications to the theory of trigonometric series. These papers constitute part of his Candidate's dissertation (evaluated by the Mathematical Section of the Academic Council as outstanding) and earned thanks at the Lomonosov Lectures. Some of the results were published in *Doklady Akademii Nauk SSSR* **90** (1953), no. 1, and detailed expositions of the rest have been sent to the printer in the form of two papers: one is submitted to *Matematicheskiĭ Sbornik* and will be published at the end of 1954 or beginning of 1955, and the other will appear in *Uchenye Zapiski Universiteta* in 1954. Here in brief is the content of these papers ... " The papers were then presented, and at the end there appears: "We think that Ul′yanov certainly deserves to be awarded the Moscow Mathematical Society's Prize for Young Mathematicians."

In this connection Nina Karlovna Bari sent me a letter on January 3, 1955.

Dear Petr Lavrent′evich,

Perhaps the review we wrote about you will prove to be useful some day. Please keep my remaining extra copy, else I will put it somewhere and then have to painstakingly search for it.

Do not forget to present Kolmogorov with an offprint of your last article: he is interested in this, and he spoke in your favor at the meeting where the prizes were awarded.

Best wishes,
N. Bari

On March 1, 1955, at a meeting of the Moscow Mathematical Society, President Aleksandrov gave out the Prizes for Young Mathematicians for the year 1954 to M. M. Postnikov ("Investigations in the homological theory of continuous mappings"), L. A. Skornyakov ("Alternative division rings" and "Topological projective planes"), and P. L. Ul′yanov (for papers on the A-integral and its applications to the theory of trigonometric series). I reported my results to the society on April 23, 1954. The prize certificate was signed by President P. Aleksandrov, Vice-President A. Kolmogorov, S. Sobolev, and members of the board of directors.

After the awarding of the prizes I made the report "On the extension of functions", which also concerned conjugate functions and Cauchy A-integrals.

During the 1950's I had many conversations with Andreĭ Nikolaevich on research topics. He said in this connection that it was hardly worthwhile to work for years on the same questions. He suggested that I become interested in new problems. For example, in 1953 he had a small seminar. I do not recall its name (perhaps the theory of dynamical systems or ergodic theory), but the first meetings were in the old building of Moscow University (9 Mokhovaya Street). He invited me to these meetings, which began at 8 o'clock in the morning. I lived in the dormitory on Stromynka Street, where one went to bed very late in the graduate students' rooms. To get to the meeting on time I had to get up at about 6 o'clock in the morning. I went to the meetings several times, and I found that I understood very little because of my ignorance of the topic and my difficulty in understanding Kolmogorov's statements; moreover, often I simply nodded off to sleep. Due to all this, I did not start going to the seminar, which I very much regret. V. M. Alekseev, S. F. Fomin, and others attended this seminar. I remember also my conversation with Andreĭ Nikolaevich at his apartment at Moscow University, where he told me about nomogram problems. Of course, I did not study solely the theory of the A-integral. I was then doing research on orthogonal series, imbedding theorems, series in Haar systems, the theory of approximation of functions, and so on. But as I understood, this did not correspond to what Kolmogorov expected of me. Perhaps I lacked sufficient boldness and power to do research in fairly different areas in mathematics. There was an address of Kolmogorov at a meeting of the Moscow Mathematical Society where he deprecated some of our research (true, he did not give the names of the researchers). Once Andreĭ Nikolaevich had high praise for the deep investigations of Men′shov on real and complex analysis. But at the end he said that if Men′shov were to apply such investigative powers in new areas of mathematics, then his influence on mathematics would greatly increase.

In my post-graduate years Andreĭ Nikolaevich enlisted my help with mathematics examinations for those entering Moscow University. This was my first time doing this sort of thing. Such work was then regarded as an honor. Once the Mechanics and Mathematics Department proposed two written papers (one in algebra with trigonometry, and the other in geometry) and an oral examination in mathematics. I was assigned one group (of about 20 people), for which I alone checked all the written papers, and then until late I alone conducted the oral examinations. All the written papers with grades "2" and "5" were looked at by Andreĭ Nikolaevich, and the grades had to be approved by him. I took a very long time for the oral examinations and coordinated the grades with the written papers. Often I myself did not know what grade should be given, since there was competition. Properly speaking, my examinations took too long and might have been complained about, but then there were almost no appeals, though the number of appeals has grown considerably in recent years. Part of the applicants were then school graduates who had studied from textbooks written under the guidance of Kolmogorov. Andreĭ Nikolaevich certainly expended much of his power and energy for school mathematical education. He was probably the first to begin introducing broad changes in the old secondary school mathematics curricula. This was a necessary and important issue. For a number of years I was chairman of the committee on entrance examinations in mathematics at Moscow University, and we studied new textbooks for secondary school, since questions and problems for applicants were made up from them. It seemed to me that the first textbooks that came out for school contained fairly many errors, and I spoke to him about that. To all appearances, Kolmogorov's first assistants in this project were not chosen in the best way.

For many years I was connected with the publication of papers in various mathematical journals. Kolmogorov sent me papers from *Doklady Akad. Nauk SSSR*, *Izvestiya Akad. Nauk SSSR*, and other journals to get my opinion of them. As an example I shall tell about the publication of a paper by Vladislav Fedorovich Emel′yanov, a young mathematician from Saratov University. Two of his papers in *Doklady* were presented by Andreĭ Nikolaevich. In order that the reader understand what follows I recall some definitions. Let E be a measurable subset of $[0, \infty)$, and take a number $p \in [1, \infty)$. A system $\{\varphi_n\}$ of elements of a Banach space B is called a *basis* if every element $f \in B$ has a unique expansion as a series in $\{\varphi_n\}$ convergent to f. Let $\{\varphi_n\}$ be a given system of functions on E. A set $J \subset [1, \infty)$ of numbers is called the set of *basis indices* of the system $\{\varphi_n\}$ if $\{\varphi_n\}$ forms a basis for the closure in $L^p(E)$ of its linear span precisely when $p \in J$.

The system $\{\varphi_n\}$ is complete in $L^p(E)$ if the closure in $L^p(E)$ of the linear span of $\{\varphi_n\}$ coincides with $L^p(E)$. A set $J \subset [1, \infty)$ is called the set of *completeness indices* of the system $\{\varphi_n\}$ if $\{\varphi_n\}$ is complete in $L^p(E)$ precisely when $p \in J$. A definition of this type is also given for minimal systems in Banach spaces.

For $E = [0, 1]$ a number of results were known.

For example, if $\{\varphi_n\}$ is a basis in $L^p(E)$ and in $L^q(E)$ with $1 \leq p < q < \infty$, then $\{\varphi_n\}$ is a basis in $L^r(E)$ for all $r \in (p, q)$ (see [**14**], p. 12). These questions had also been considered in papers by S. F. Gerasimov, B. V. Ryazanov, and A. N. Slepchenko.

Suppose now that $E = [0, \infty)$. Emel′yanov obtained several results, wrote a paper in *Doklady*, and asked Kolmogorov to present it. In this connection Andreĭ Nikolaevich wrote me on December 23, 1972:

V. F. Emel′yanov has sent me the enclosed note for presentation in *Doklady*. He writes that he reported some of the results in the seminar led by you and Dmitriĭ Evgen′evich. Emel′yanov's results are elegant and apparently require very subtle constructions for their proofs. I would be willing to present the paper to *Doklady*. But on the whole the note is written in a rather childish language, and the formulations themselves are ambiguous in some respects ...

You yourself could probably think of some advice to the author about the style of exposition. Would you help him edit the note better?

Respectfully yours,
A. Kolmogorov

In the text there is a small hint about improving the style of the exposition. All this was done. I wrote my suggestions to Emel′yanov, a rewritten paper was sent, this version was then presented by Andreĭ Nikolaevich to *Doklady*, and it was published [**15**]. Let me formulate some results from this paper for the spaces $L^p(E)$ with $E = [0, \infty)$.

THEOREM 3. *A set $J \subset [1, \infty)$ is the set of completeness indices of some countable (orthogonal) system of functions on $[0, \infty)$ if and only if J is a G_δ-set in $[1, \infty)$.*

THEOREM 7. *A set $J \subset [1, \infty)$ is the set of minimality indices of some countable system of functions if and only if J is an $F_{\sigma\delta}$-set in $[1, \infty)$.*

THEOREM 9. *A set $J \subset [1, \infty)$ is a set of basis indices if and only if J is an F_σ-set in $[1, \infty)$.*

THEOREM 11. *For a countable system of functions on $[0, \infty)$ the set of all $1 \leq p < \infty$ for which the system forms a basis of the space $L^p(0, \infty)$ is representable as the intersection of some G_δ-set with some F_σ-set.*

I liked these results of Emel'yanov, which differ sharply from the properties of the spaces $L^p(0, 1)$.

I wrote a nice review, and in this connection I received the following reply from Andreĭ Nikolaevich:

Dear Petr Lavrent'evich,

Emel'yanov's results sound beautiful. Theorems 2, 3, 4, 6, 8, 9, and 10 seem at first glance quite surprising.

In the development of Theorem 11 it would be interesting to find a definitive characterization of the sets of those p for which a certain system of functions forms a basis in L^p.

In any case it is a pleasure for me to present the note to *Doklady*, although I regard this whole circle of problems as having by now degenerated into a kind of sport.

Respectfully yours,
A. Kolmogorov

Thus, the results obtained were beautiful, from the results a new problem was posed in connection with Theorem 11, and Kolmogorov gladly presented the paper, yet at the end he criticized it and compared it with a form of sport. It seems to me that Andreĭ Nikolaevich's statements are clearly contradictory, but he always supported interesting and beautiful results. (I remark that I have not yet seen a definitive solution with regard to Theorem 11. It has always seemed interesting to me to know which spaces have this or that system as a basis.)

Kolmogorov very carefully read through the papers he presented in the theory of functions, making annotations concerning both the theme and the exposition. All this played an important role, especially for younger mathematicians.

I recall how Andreĭ Nikolaevich celebrated important anniversaries of his life. For example, on May 5, 1953 there was a conference of the Section of Physical and Mathematical Sciences, the Academic Council of the Mechanics and Mathematics Department of Moscow University, the Moscow Mathematical Society, and the Steklov Institute of Mathematics, dedicated to the fiftieth birthday of Andreĭ Nikolaevich Kolmogorov. The following reports were read: "The role of A. N. Kolmogorov in the mathematical life of our country" by I. G. Petrovskiĭ and "A. N. Kolmogorov as a mathematician" by P. S. Aleksandrov, I. M. Gel'fand, and A. Ya. Khinchin.

After this, various organizations greeted Kolmogorov, and some telegrams were read. I appeared with congratulations to him from the graduate students of the Scientific Research Institute of Mechanics and Mathematics, of which he was director. About Andreĭ Nikolaevich's talk I remember only his thoughts that the anniversary speeches were inclined to exaggerate the merits of the hero of the day, and that certain facts actually looked rather different from the way they seemed now to some of the speakers. This conference was held in the old building of Moscow University.

Kolmogorov's sixtieth birthday was marked at Moscow University in the Lenin Hills. For his outstanding merits in the area of mathematics he was awarded the title of Hero of Socialist Labor. He appeared at the conference with a talk in which he spoke about some of his results. Then there was a festive dinner in the dining room on the second floor of Moscow University. The tables were arranged on the perimeter of a rectangle with a small inside aisle; the guests sat on the the outside, while Andreĭ Nikolaevich, passing from one group to another, went along the inside, took a chair, sat down, accepted congratulations, and conversed. In this way he approached us. I sat together with Men′shov and Lyusternik. We of course congratulated him. Men′shov and Lyusternik recalled some of the events of past years. But I asked Andreĭ Nikolaevich how it happened that, on the one hand, he disparaged the theory of functions of a real variable, while on the other hand, he chose for his anniversary talk only three of his many outstanding results from various areas of mathematics, and two of them were from the theory of functions of a real variable: "On the divergence almost everywhere of Fourier–Lebesgue series" and "On the representability of continuous functions of n variables as sums of superpositions of continuous functions of one variable". Wasn't there a contradiction here? You see, Andreĭ Nikolaevich undoubtedly loved real variables, though he sometimes spoke abusively of the theory. At first he was silent, and then he stood up with displeasure, said to me, "Up to this time no one has obtained anything better," turned, and went on to other mathematicians. My neighbors stared at me with condemnation. I think that Andreĭ Nikolaevich quickly forgot my tactlessness.

On 25–26 April 1983 Moscow University hosted a meeting of the Moscow Mathematical Society and the Mechanics and Mathematics Department dedicated to the eightieth birthday of Andreĭ Nikolaevich Kolmogorov. Reports were given on approximation theory, dynamical systems, stochastic mechanics, and limit theorems in probability theory. They were published in *Uspekhi Matematicheskikh Nauk* **38** (1983), no. 4. I gave the report "A. N. Kolmogorov and divergent Fourier series."

For the last few years Kolmogorov had been seriously ill, but on his eightieth birthday he followed the reports very well. My report included the statement of a new theorem of K. S. Kazaryan: *There is a uniformly bounded orthonormal system with respect to which every Fourier series converges on some set of positive measure.* Andreĭ Nikolaevich at once remarked that I had left out something from the theorem. Indeed, in the given formulation the theorem was almost obvious. I should have added the completeness of the system, and then it became considerably more interesting.

After the meeting there was a reception for Andreĭ Nikolaevich; all mentioned his merits in science and in pedagogy, especially his combining of the theoretical and the practical. In talking with Anna Dmitrievna I said that I always felt bad about disturbing Andreĭ Nikolaevich with various questions. She replied, "he loved and loves you, and you could and can always turn to him."

The eightieth birthday celebration at the conference hall of Moscow University on April 25, 1983. Left to right: V. A. Kotel'nikov, A. N. Kolmogorov, A. D. Kolmogorova, and S. L. Sobolev.

From this one can conclude how he loved his doctoral students, how he was delighted by their achievements, and how he was proud of their successes. For I was not even regarded as one of his doctoral students.

At the beginning of May 1978 I turned fifty, and the occasion was marked on May 12, 1978 in the Academic Council of the Mechanics and Mathematics Department of Moscow University, and before this in the Theory of Functions and Functional Analysis Branch. Andreĭ Nikolaevich came and gave a talk, though it was mostly on my athletic achievements. Later I invited him for supper, but he declined, since he was going to Obukhov, someone else from Saratov, who had turned sixty. In this way I learned that Academician Aleksandr Mikhaĭlovich Obukhov from Saratov, the prominent specialist in the physics of the atmosphere, had been a student of Kolmogorov. It is very unfortunate that he died on December 3, 1989. Andreĭ Nikolaevich also had a student from Saratov named V. N. Zasukhin, who defended his dissertation in 1941 and was killed in battle during the Second World War.

At the end of the 1970's Men′shov stopped going to the Academic Councils of the department and stopped attending the meetings of his seminar on the theory of functions of a real variable. At the time he was more than 85 years old. In this connection the question arose as to the head of the Theory of Functions and Functional Analysis Branch in the Mechanics and Mathematics Department of Moscow University. Kolmogorov was head of the mathematics section of the department, and he proposed in 1979 that I be considered for the head of the branch. This was supported by the dean of the department, Corresponding Member of the Academy of Sciences A. I. Kostrikin. I have been at the post from that time.

In 1981 I was elected a corresponding member of the Academy of Sciences of the USSR. As always, the nominated candidates were discussed at a meeting with A. P. Aleksandrov, President of the academy. Nikol′skiĭ was there, and he later told me that when the subject came to the specialists nominated in the theory of functions, Kolmogorov said that the time had come to elect Ul′yanov. From all appearances Andreĭ Nikolaevich's statement played a large role in the fact that they then elected me as a corresponding member.

December 9, 1983 was the centennial of the birth of Academician Nikolaĭ Nikolaevich Luzin (December 9, 1883 to February 28, 1950). In this connection we gathered material about Luzin and his school. Of his doctoral students Men′shov, Aleksandrov, and Kolmogorov were still living.

Dmitriĭ Evgen′evich Men′shov actively took part in the writing of a paper, "On the 70th anniversary of the Luzin seminar on the theory of functions". Among the participants of the Luzin seminar were the students P. S. Aleksandrov, D. E. Men′shov, and A. Ya. Khinchin, who made up the first generation of Luzin's students together with M. Ya. Suslin, V. N. Veniaminov, and V. S. Fedorov, who joined within a year. From 1920 N. K. Bari, V. I. Glivenko, L. G. Shnirel′man, and N. A. Selivanov worked in the seminar, from 1921 A. N. Kolmogorov, M. A. Lavrent′ev, and L. A. Lyusternik, and from 1922 L. V. Keldysh, P. S. Novikov, E. A. Leontovich, I. N. Khlodovskiĭ, and G. A. Seliverstov. The paper detailed the investigations in the seminar, along with the new seminars that branched off from it (on this see the paper [**16**]).

Because of illness P. S. Aleksandrov did not take part in the work on reminiscences of N. N. Luzin.

As for Kolmogorov, he took part most ardently. First, on June 8, 1983 he gave his student V. A. Uspenskiĭ an interview in connection with the centennial of Luzin's birth. This interview, "A student on his teacher", was published in *Uspekhi Matematicheskikh Nauk* **40** (1985), no. 3, pp. 7–8 (it was printed earlier on September 7, 1983 in the newsletter *Puti v nauku* (Paths into Science) of Kemerovo University). Andreĭ Nikolaevich became a student of Luzin in his second year at the university. In some publications it has been asserted that Kolmogorov himself came upon the topics of his investigations and, properly speaking, did not have a teacher. The published interview allows one to say that Andreĭ Nikolaevich did not himself think so. Here are some excerpts from the interview.

> Each student came to Nikolaĭ Nikolaevich Luzin in his Arbat apartment once a week in the evening—on a day of the week permanently set aside for him. My day was common with Petr Sergeevich Novikov, Lyudmila Vsevolodovna Keldysh, and Igor′ Nikolaevich Khlodovskiĭ. The sessions consisted of conversations between Luzin and us four on research topics. The intensive work with students was one of the novelties cultivated by Nikolaĭ Nikolaevich ... The possibility of consulting with Luzin, of discussing with him results that were not completely finished was very important ... My advisor in graduate school was Luzin as before ... He contributed to mathematics as the author of first-class work in the metric and descriptive theory of functions and the descriptive theory of sets ...

In the same year 1983 Andreĭ Nikolaevich told me that in the first place a student should honor the memory of his teacher. He apparently had in mind his own teacher. It is interesting to compare this with the rumors of disagreements in the Luzin school. I think it is hardly worthwhile now for us to try to judge who was right and who was wrong. But it must be said that it is never a pretty affair when there are bitter and uncompromising arguments between a teacher and students.

Second, Andreĭ took an active part in the organization of a joint meeting of the Moscow Mathematical Society and the Mechanics and Mathematics Department of Moscow University dedicated to the centennial of Luzin's birth. It was held on December 13, 1983.

Agenda for the meeting:

1. Introductory speech by the president of the Moscow Mathematical Society, A. N. Kolmogorov.
2. S. M. Nikol′skiĭ, "The All-Union school on the theory of functions, dedicated to the centennial of the birth of Academician N. N. Luzin" (Kemerovo, September 9–19, 1983).
3. P. L. Ul′yanov, "On Luzin's work in the metric theory of functions".
4. E. P. Dolzhenko and G. Ts. Tumarkin, "N. N. Luzin and the theory of boundary properties of analytic functions".
5. V. A. Uspenskiĭ, "N. N. Luzin and descriptive set theory".

Andreĭ Nikolaevich was present at this meeting, and he tried to give his talk, but due to his illness and agitation the address did not take place. The rest of the reports were made, and some of them were published in *Uspekhi Matematicheskikh Nauk* (1985), where the interview "A student his teacher" with Kolmogorov also appears, together with Lebesgue's Foreword to the book *Lectures on analytic sets and their applications* by Luzin.

Third, Andreĭ Nikolaevich supported the idea of giving Luzin's name to the large auditorium 16-10 in the Mechanics and Mathematics Department of Moscow University. It is now called the "Luzin auditorium"; next to it hangs a memorial plaque in honor of Luzin.

The All-Union school was timed to coincide with the centennial. Metal badges were made for the participants with a picture of mathematical symbols and the inscription:

N. N. Luzin

100

years

1983

One of the badges was given also to Andreĭ Nikolaevich, and he wore it on the lapel of his coat for several months.

I visited Andreĭ Nikolaevich in the academic hospital on Lenin Prospect. The conversation began to touch upon university matters, and he asked me about the specialists in the theory of functions who were members of a specialized council on dissertation defense, of which he had a list. He was chairman of the council, and we talked about new members. I answered him, but I thought that he knew as much about them as I did. Then the conversation turned to everyday matters. For a long time we discussed how to fix up and heat a country house, and which regions near Moscow were preferable for such houses. To my surprise it turned out that he knew quite well many places near Moscow, in particular, the regions of the Zelenogradskaya and Pravda stations, not far from which I had acquired a country house at one time in the town of Lesnoe. It happened that in past years Andreĭ Nikolaevich had been in places where I now am very often, on foot in summer and on skis in winter. He remembered the name of the Skalba stream that flows through there, and he described the picturesque steep banks overgrown with a beautiful pine forest. There was a short conversation about the theory of functions, which I did not try to make longer, since I felt that Andreĭ Nikolaevich was sick and that it was extremely difficult for him to talk. Once a barber came to the hospital ward to give Andreĭ Nikolaevich a haircut and shave while I was there. The barber asked him, "Is this your son?" Andreĭ Nikolaevich replied: "My scientific son."

Various events can be recalled and described about Kolmogorov. Many of his students as well as his successors have used his problems and often his methods for the development of mathematics. In time there will probably be more detailed books concerned with the life and creative work of this great Russian mathematician A. N. Kolmogorov. The things he did, like what Lobachevskiĭ and Chebyshev did, have adorned the development of mathematics both in our country and in the whole world.

It is a pleasure for me to acknowledge that such an outstanding man of science as Andreĭ Nikolaevich Kolmogorov exerted a direct influence on my life, on my research and my work at Moscow University.

References

1. P. S. Aleksandrov and A. N. Kolmogorov, *Introduction to the theory of functions of a real variable*, 2nd ed., GTTI, Moscow–Leningrad, 1933. (Russian)
2. N. N. Luzin, *The integral and trigonometric series*, GITTL, Moscow–Leningrad, 1951. (Russian)
3. D. E. Men′shov, *On the representation of measurable functions by trigonometric series*, Mat. Sb. **9** (1940), 667–690. (Russian)
4. N. K. Bari, *On primitive functions and trigonometric series converging almost everywhere*, Mat. Sb. **31** (1952), 687–702. (Russian)
5. N. K. Bari, *Trigonometric series*, Fizmatgiz, Moscow, 1961; English transl., Vols. I, II, Macmillan, New York, 1964.
6. J. Marcinkiewicz, *Sur une nouvelle condition pour la convergence presque partout de séries de Fourier*, Ann. Scuola Norm. Sup. Pisa **8** (1939), 239–240.
7. P. L. Ul′yanov, *On some equivalent conditions for convergence of series and integrals*, Uspekhi Mat. Nauk **8** (1953), no. 6, 133–141. (Russian)
8. P. L. Ul′yanov, *A generalization of a theorem of Marcinkiewicz*, Izv. Akad. Nauk SSSR **17** (1953), 513–524. (Russian)
9. A. Kolmogoroff, *Grundbegriffe der Wahrscheinlichkeitsrechnung*, Springer, Berlin, 1933.
10. A. N. Kolmogorov, *Foundations of the theory of probability*, Russian transl. of [9], ONTI, Moscow–Leningrad, 1936; English transl., Chelsea, New York, 1950.
11. A. Kolmogoroff, *Sur les fonctions harmoniques conjuguées et les séries de Fourier*, Fund. Math. **7** (1925), 24–29.
12. A. Kolmogoroff, *Sur un procédé d'integration de M. Denjoy*, Fund. Math. **11** (1928), 27–28.
13. P. L. Ul′yanov, *On trigonometric series with monotonically decreasing coefficients*, Dokl. Akad. Nauk SSSR **90** (1953), 33–36. (Russian)
14. P. L. Ul′yanov, *Some questions in the theory of orthogonal and biorthogonal series*, Izv. Akad. Nauk AzSSR Ser. Fiz.-Mat. i Tekhn. Nauk **1965**, no. 6, 11–13. (Russian)
15. V. F. Emel′yanov, *On certain features of Fourier series of functions defined on an infinite interval*, Dokl. Akad. Nauk SSSR **213** (1973), 522–524; English transl. in Russian Math. Dokl. **14** (1974).
16. D. E. Men′shov, V. A. Skvortsov, and P. L. Ul′yanov, *On the 70th anniversary of the Luzin seminar on the theory of functions*, Uspekhi Mat. Nauk **40** (1985), no. 3, 219–225. (Russian)

A Few Words on A. N. Kolmogorov

P. S. Aleksandrov

The unusual breadth of Kolmogorov's creative interests and the vast range and variety of the mathematical domains in which he has worked during various periods of his life make him outstanding not only among Soviet mathematicians, but all over the world; one can even say that in this respect he is unrivalled among mathematicians of our time. In the many mathematical disciplines in which he has worked he obtained really basic and important results the proof of which often required overcoming great difficulties and therefore great creative efforts. This could be said even about the results he obtained as quite a young man in set theory and function theory both descriptive and metric, for example, his theory of operations on sets and his famous example of a divergent Fourier series.

Later came his work on the general theory of measure, both abstract, that is, "properly general", and geometric; and then began his fundamental work in various branches of probability theory, which undoubtedly made him the foremost representative of this discipline the whole world over.

His first papers on mathematical logic and the foundations of mathematics also appeared early; much later they were joined by his studies on information theory.

Kolmogorov's contribution to topology is very large. It suffices to recall that simultaneously with the outstanding American topologist Alexander and entirely independent of him, Kolmogorov arrived at the notion of cohomology and founded the theory of cohomological operations, that is, he obtained results that essentially transformed the whole of topology. The deep connections between topology and the theory of ordinary differential equations, celestial mechanics, and also the general theory of dynamical systems are well known. These connections already appeared in the early works of Poincaré. Kolmogorov's ideas in the whole of this vast mathematical discipline, which were further developed by his numerous pupils, has determined to a considerable extent its present state. Finally, we must mention his research in mechanics proper, in particular, his famous work on the theory of turbulence, which had a direct impact on the experimental natural sciences. All we have said, and that is by no means all that could be said about Kolmogorov as a scientist, undoubtedly makes it obvious that in him we have one of the most outstanding representatives of contemporary mathematics in the widest sense of this word, including applied mathematics. His position in science is acknowledged

Reprinted from *Russian Mathematical Surveys* **38** (1983), no. 4, 5–7, by permission.

without question in the international world of science and manifests itself, in particular, in the fact that he occupies the first place among all Soviet mathematicians in the number of foreign academies and scientific societies that have elected him as a member and also of universities that have conferred upon him honorary doctorates. Among these there are: the Paris Academy of Sciences, the London Royal Society, the German Academy of Sciences "Leopoldina", the Netherlands Academy of Sciences, the Polish Academy of Sciences, the London Mathematical Society, the US National Academy, the American Philosophical Society founded by Benjamin Franklin, and the universities of Paris, Berlin, Warsaw, etc.

Kolmogorov was born on 25 April 1903 in the town of Tambov, where his mother Mariya Yakovlevna Kolmogorova had been delayed on her way from the Crimea. She died in childbed, and all the responsibility for his upbringing was taken on by her sister Vera Yakovlevna Kolmogorova, who in fact replaced his mother. Kolmogorov treated Vera Yakovlevna as his mother until her death in 1950 in Komarovka at the age of 87. From his mother's side Kolmogorov was of aristocratic descent: his grandfather Yakov Stepanovich Kolmogorov was a district head of nobles in Uglich. His father was the son of a clergyman. He was an agronomist with a highly specialized training or, what was called at the time "a learned agronomist".

My friendship with Kolmogorov occupies in my life quite an exceptional and unique place: this friendship had lasted for fifty years in 1979, and throughout this half-century it showed no sign of strain and was never accompanied by any quarrel. During this period we had no misunderstanding on questions in any way important to our outlook on life. Even when our views on any subject differed, we treated them with complete understanding and sympathy.

As I mentioned before, Kolmogorov had many pupils in various branches of mathematics. Some of them have become themselves outstanding specialists. His older pupils who have later on become members of the USSR Academy of Sciences are Sergeĭ Mikhaĭlovich Nikol′skiĭ (born in 1905) and the late Anatoliĭ Ivanovich Mal′tsev (born in 1910). Next come Boris Vladimirovich Gnedenko (born in 1912), Member of the Ukrainian Academy of Sciences, an outstanding specialist in probability theory Mikhail Dmitrievich Millionshchikov (1913–1973), Member of the USSR Academy of Sciences, and Izrail′ Moiseevich Gel′fand (born in 1913), Corresponding Member of the USSR Academy of Sciences, elected a Foreign Member of the US National Academy of Sciences and the Paris Academy of Sciences. Considerably younger but still belonging to the older generation of Kolmogorov's pupils are Aleksandr Mikhaĭlovich Obukhov (born in 1918), Member of the USSR Academy of Sciences, and Andreĭ Sergeevich Monin (born in 1921), Corresponding Member of the USSR Academy of Sciences.

They are followed by Vladimir Andreevich Uspenskiĭ, Vladimir Mikhaĭlovich Tikhomirov, Vladimir Mikhaĭlovich Alekseev, Yakov Grigor′evich Sinaĭ, and Vladimir Igorevich Arnol′d.

Especially large is the group of Kolmogorov's pupils working in probability theory and mathematical statistics. Among them are Yuriĭ Vasil′evich Prokhorov, Member of the USSR Academy of Sciences, Login Nikolaevich Bol′shev, Corresponding Member of the USSR Academy of Sciences, Sagdy Khasanovich Sirazhdinov, Member of the Uzbek Academy of Sciences, Vladimir Sergeevich Mikhalevich, Member of the Ukrainian Academy of Sciences, Boris Aleksandrovich Sevast′yanov, Yuriĭ Anatol′evich Rozanov, Al′bert Nikolaevich Shiryaev, and Igor′ Georgievich

Zhurbenko. Of course, this list is in no way complete. But as the title of my note makes clear, it is not an anniversary survey of the life and activities of A. N. Kolmogorov and altogether is in no way a traditional "jubilee article" and does not aspire to completeness.

March 1981

Memories of P. S. Aleksandrov

A. N. Kolmogorov

Pavel Sergeevich Aleksandrov died six months before my eightieth birthday. As editor of *Uspekhi Matematicheskikh Nauk* he thought it necessary in March 1981, to prepare for later publication in the periodical a short article about me. In it he wrote as follows:

"My friendship with Kolmogorov occupies in my life quite an exceptional and unique place: this friendship had lasted for fifty years in 1979, and throughout this half-century it showed no sign of strain and was never accompanied by any quarrel. During this period we had no misunderstanding on questions in any way important to our outlook on life. Even when our views on any subject differed, we treated them with complete understanding and sympathy."

Our close friendship began in 1929. As I now publish a description of my life with Aleksandrov in the early years of this friendship (1929–1931) I should like to preface my testimony to it by echoing in part Aleksandrov's words quoted above: for me these 53 years of close and indissoluble friendship were the reason why all my life was on the whole full of happiness, and the basis of that happiness was the unceasing thoughtfulness on the part of Aleksandrov.

We got to know one another shortly after my arrival at Moscow University in 1920. Our first encounters in 1920–1929 are described in Chapter I. The whole of Chapter II is devoted to our travels in the summer of 1929 along the Volga and in the Caucasus. Chapter III tells of our travels abroad (to Göttingen and Paris) in 1930–1931. In conclusion there are extracts from Aleksandrov's letters to me from the USA, where he stayed for about four months after leaving Europe in the Spring of 1931.

Chapter I
FIRST MEETINGS

I arrived at Moscow University with a fair knowledge of mathematics. Thanks to the book "*Novye idei v matematike*" (New ideas in mathematics) I knew in particular the beginnings of set theory. I studied many questions in articles in the Encyclopedia of Brockhaus and Efron, filling out for myself what was presented too concisely in these articles.

Reprinted from *Russian Mathematical Surveys* **41** (1986), 225–246, by permission.

At that time a student grant was of very little material value, but students of the second course received, in addition to a grant, a ration which consisted of a pood (16 kilos) of baked bread and a kilo of fat (whether it was vegetable fat or cow fat depended on the final word of science—it seems that there was at first vegetable fat and then fresh butter). Therefore the first thing I did was to check the minimum number of examinations required for moving into the second course (at that time attendance at lectures was not compulsory).

In the first year I chose to attend courses in set theory by Ivan Ivanovich Zhegalkin and on projective geometry (a subject in which I had been specially interested even before going up to the University) by Alekseĭ Konstantinovich Vlasov. Among the first courses I chose was one on the theory of analytic functions by Nikolaĭ Nikolaevich Luzin (this course was given that year by two lecturers—Luzin and Boleslav Kornelievich Mlodzeevskiĭ; a little later I found out that it was more fashionable to attend both these courses so as to be able to compare them, criticizing of course Mlodzeevskiĭ for lack of rigour). Luzin's lectures were attended by many older students and even by lecturers.

My first achievement was connected with Luzin's course and after it I received some attention. Luzin loved to improvise in his lectures and in the lecture on the proof of Cauchy's theorem it occurred to him to use the following lemma: let a square be divided into a finite number of squares; then for any constant C there is a C' such that for every curve of length not greater than C the sum of the perimeters of the squares touching the curve is not greater than C'. Two weeks later I went to the president of the students' mathematical group, Semen Samsonovich Kovner, with a short manuscript (this manuscript still exists, it is dated 4 January 1921), in which this assertion was refuted (there was a small gap in my proof at first but I quickly filled it). All this was reported to Luzin, who agreed with my observation and in his lectures he gave the correct proof of Cauchy's theorem.

Soon Pavel Samuilovich Uryson approached me, suggesting that I visit him regularly. He was trying to interest me in problems in topology, having in mind particularly the problem of the number of geodesics on closed surfaces. In 1921–1922 I was in constant touch with V. V. Stepanov and whilst attending his seminar I managed to solve one problem in which Luzin showed a lasting interest—I demonstrated the existence of Fourier–Lebesgue series with arbitrarily slowly decreasing Fourier coefficients. This led Luzin to suggest that I, together with a small group of students, call on him regularly, intending apparently that I should concentrate on the metrical theory of functions (the theory of the integral, the theory of Fourier series, and so on). As we shall see later I continued with this plan for some time, but before this I found some results in the descriptive theory of functions, which seemed to me extremely important: I began to work out a general theory of operations on sets. As my work on these lines did not fit in with Luzin's plans, I took the first draft of the theory of operations to Uryson and he took me along to Pavel Sergeevich Aleksandrov. This was very reasonable, since my general theorems about arbitrary operations on sets were natural generalizations of Aleksandrov's theorem on A-sets.

I remember the huge sheets of paper Aleksandrov pulled out from somewhere with plans for the formation of sets of ever higher classes, the contemplation of which finally led Aleksandrov to the result that all B-sets of any class are A-sets. These sheets of paper were scattered all over the floor and Aleksandrov, with me behind him, crawled along wanting to make intuitive the derivation of B-sets of high (even transfinite) order as the result of a single application of an A-operation.

At that time (the summer of 1921) I had obtained a result that attracted general attention in international circles of specialists in the metrical theory of functions—I constructed a Fourier series divergent almost everywhere. The metrical theory of functions hung on this, and in the years 1922–1925 I chiefly worked on the theory of trigonometric and orthogonal series. The ground for scientific collaboration with Aleksandrov was forfeited, but my descriptive papers of 1921–1922 lay there on Luzin's desk; he had found them to be methodologically incorrect, and they lay there untouched until 1926.

My personal contacts with Aleksandrov were also very limited at this time, although we met fairly often, for example at concerts in the Small Hall of the Conservatoire. We greeted each other but did not get into conversation. Apparently Aleksandrov's starched collars and general stiffness embarrassed me somewhat. Nevertheless through all this period I was conscious of his kind attitude to me and of his care. Aleksandrov arranged that all my work on descriptive theory was published. And it was he who set in motion the chance for me to stay on at the Institute of Mathematics and Mechanics at Moscow University after my post-graduate study.

At that time there was no strict limit on the number of years a research student could spend doing research (there was for example the case of a young mathematician who spent seven years as a research student!). As this lack of restriction gave a research student complete freedom to devote himself to his research, with no other obligations, I too preferred not to put an end to my studies. In 1928–1929 there was a much stricter control on the number of years a student had for research and in 1929 the number of people graduating was about 70—something that had never happened before. I was one of these graduates (I had spent four years as a research student).

Then came the question of where to continue my work. That year there was one vacancy at the Institute of Mathematics and Mechanics—the post of a senior researcher. Along with me, one of the older generation of mathematicians had a claim to the post; the director of the Institute, Dmitriĭ Fedorovich Egorov, was well aware of my scientific achievements, but he still thought it his duty to follow the guide-lines of age when appointing a scientific colleague. For me there was only one other attractive possibility. The Ukrainian Institute of Mathematics was opened in Kharkov in 1928 and the Director was Sergeĭ Natanovich Bernstein, who was then at the height of his international fame and of his standing in the USSR. A special building had already been built for the Institute but the staff still had to be found. Bernstein sent me a proposal suggesting that I cooperate with him at the Institute where, according to his plan, I should only begin work there after a year's experience abroad, and for this he took steps to get me a Rockefeller scholarship. However, Aleksandrov was very much against Bernstein's plan and finally managed to persuade Egorov to give preference to my application.

Chapter II
TRAVELS IN 1929. THE KOMAROVKA HOUSE

In the summer of 1929 the Society for Proletarian Tourism and Excursions advertised the availability in one of the cities on the Volga (Yaroslavl, Nizhnii Novgorod, Kazan, and so on) of boats, sails, and tents for hire at very low cost;

with them one could sail to any city lying further down the river. For example one could hire the equipment in Yaroslavl and hand it in in Samara. As I already had some boating experience, I decided to organize one of these trips. I estimated that a reasonable number to go on one of these was three. First I invited my close friend Gleb Aleksandrovich Seliverstov. It turned out however that he could not set out with me at the time which suited me best. (The idea of making the expodition early, in the season of white nights, attracted me very much.) We agreed that Seliverstov should join us later. Another name was suggested: an acquaintance of mine from high school, Nikolaĭ Dmitrievich Nyuberg. To this day I cannot be quite sure how I came to suggest that Aleksandrov should be the third member. But he accepted straight away.

On 16 June 1929 we set sail down the Volga from Yaroslavl. For Aleksandrov a sailing holiday of this kind was something new, but he quickly undertook to be our quarter-master and even before leaving Moscow he was buying delicacies of all sorts. And it is from the day we set out, 16 June, that Aleksandrov and I date our friendship which, as I have already said, lasted fifty-three years.

We got a boat of the "Ostashkov" type—the kind used by all the Volga buoy-keepers then. After some rearrangement of the boat, two pairs of oars were placed on the stern and on the bow, and this make it possible for one or even two members of the crew to sleep in the middle of the boat even while we were rowing. The sloping sail was fairly primitive, but it meant that we could still move with a side wind. For al three members of the crew we bought in Moscow "Jungsturm" suits, which were then very popular among the young. What we had in mind here was that our everyday garments would be shorts and sports shirts. Oddly enough we did not take any guide books with us, only a steamboat timetable, from which we could work out the distance covered and the distance to the nearest landing state. The only book we took was the Odyssey.

We did not fix our destination exactly. It was suggested that after handing in the boat we should go on to the Caucasus, since Aleksandrov was due to spend August on the Black Sea with some of his students (one of whom was to be Lev Semenovich Pontryagin). We both intended to work on mathematics in the Caucasus, so Aleksandrov also had in his luggage a portable typewriter and a folding table he had bought in Göttingen.

The typical scenery on the banks of a big Central Russian river, be it the Oka, the Dnieper, the Don, or the Volga, is pretty much the same. The river is usually divided into several arms, which flow between green water meadows and round sandy islands, overgrown with osier beds. The sands on the Volga are washed away, leaving an almost pure whiteness (my description relates to the 20's and 30's, before the construction of large reservoirs on the Volga).

This countryside is not without its own particular grandeur. Aleksandrov soon grew fond of it and in later years we have often sailed along the Belaya, the Kama, the Volga, and the Dnieper.

When we sailed along the Volga we usually chose our route along some Volozhka (that is the name in the Volga dialect for the minor channels). As there was a strong enough current in the Volozhka, we could determine whether it was carrying us into the main stream.

We usually pitched camp on a sandy island, on the upper or lower extremity, where the flow of the water is especially felt. During the first days of our journey we often swam at night; in the white summer nights, gliding along past the overgrown osier beds on the bank, the air filled with bird song, make a lasting impression on us. We wished that this could go on for ever.

Aleksandrov's appreciation of nature was not confined to any one phenomenon. He mainly appreciated very simple impressions, such as the sounds and colours of the white nights which I have just described. Time and again I happened to catch Aleksandrov at dawn by the river singing some melody without words.

Of course, we did not forget to go for a walk in Kostroma, with its business quarters, cloisters, churches, or to wander past the Nizhegorodskii citadel, or to climb onto the high precipices along the banks of the Volga, or to visit the home of the Ul′yanov family in Ulyanovsk. But dominating everything was the continuous movement downstream with short stops to buy food and prepare it (over camp fires and on the primus we had with us). It was very easy in 1929 to replenish our store of provisions at the small markets near the landing stages. Much of our time was, however, devoted to reading the Odyssey, to idleness, and of course to conversations "on every subject".

In Kazan, Nikolaĭ Dmitrievich Nyuberg left us according to plan. It was arranged that on a firmly agreed day in Sviyazhsk we would meet Seliverstov, who was to come with us in place of Nyuberg. Despite the fact that there had been a great storm the day before, we were standing on the platform of the railway station in Sviyazhsk at the agreed time, but there was no sign of Seliverstov.

After visiting Kazan citadel with its Sumbekin tower and the buildings of Kazan University, Aleksandrov and I set off to continue our journey to the place where the Volga and the Kama meet, past high banks rich in apple orchards—the home land of the Antonovka apple—and on the twenty-first day after we had set out from Yaroslavl we arrived safe and sound at Samara. There we handed in the boat and the equipment that went with it. We had covered 1300 kilometers.

Our journey continued on to the Caucasus. We went as far as Astrakhan in a two-berth cabin, first class, on the river steamer. The upper class cabins on the sea-going type of streamer, on which we sailed from Astrakhan to Baku, were extremely unattractive. We preferred to buy fourth class tickets, spread out on the deck the canvas of our tents, and declare that area our inviolable territory. We could sit and lie comfortably enough on the canvas.

After staying one day in Baku we went by train to the station at Akstafa, then by bus to Dilizhan, and from there on foot (with the typewriter and the folding table) to Lake Sevan. Naturally we were immediately attracted by the rocky islet, which with the drop in the level of Lake Sevan has now become a peninsula, and we wanted to settle there. This proved to be a simple matter. The cells of the monastery there stood empty and we occupied one of them. The permanent population of the island consisted then of the archimandrite of the monastery (who had a fairly big house), his maid (who looked after some cows), the head of the meteorological station with his small family, and finally of the "Captain", who did indeed command "the Sevan fleet", consisting of one motor boat and a few of the usual row boats. His picturesque figure has occasionally been described in literature (for example, by Marietta Shaginyan).

Every day the archimandrite opened the lower church (there were still two abandoned temples on the top of the hill), lit candles and in complete solitude recited the service. Obviously the head of the meteorological station carried out his duties. The captain at times brought honoured guests, for example, Sar′yan or the then President of the All-Russian Central Executive Committee of Armenia, but he was also ready to patronize humble tourists.

On the island we both set to work. With our manuscripts, typewriter, and folding table we sought out the secluded bays. In the intervals between our studies, we bathed a lot. To study I took refuge in the shade, while Aleksandrov lay for hours in full sunlight wearing only dark glasses and a white panama. He kept his habit of working completely naked under the burning sun well into his old age.

On Sevan Aleksandrov worked on various chapters of his joint monograph with Hopf, "*Topologiya*" (Topology) and he helped me to write the German text of my article on the theory of the integral. Besides writing this paper I was busy with ideas about the analytic description of Markov processes with continuous time, the end product of which later became the memoir "*Ob analiticheskikh metodakh v teorii veroyatnostei*" (On analytic methods in the theory of probability).

Given its position Lake Sevan mostly enjoyed sunny weather, but sometimes clouds coming from the East filled up from the mountains, dropped down to the water and then, on contact with it, vanished. We stayed there for about 20 days without leaving the place (apart from excursions to the Erivan monastery under the guidance of our captain). For Aleksandrov the day was approaching on which he had arranged to journey to Gagra, and we set off together for Erevan (where we stayed for some days in a student dormitory). The temperature was 40°, the sky was a hazy blue, and only after sunset did there unexpectedly appear the peak of Ararat, suspended in this blue sky. We visited Echmiadzin (where we decided not to visit Katolikos as we did not have the right clothes). From Echmiadzin we walked to Alagez (spending the first night by the lake, where the physicists who were working on cosmic showers very kindly put us up). After spending one night there (still without suits, wearing only shorts) we climbed the south bank of the Alagez, which did not present any complications (4000 m). From the top there opened up a view of the rocky northern summit (4100 m) separated from the southern summit by a huge ridge of snow, at the very bottom of which could be seen a small lake, its shores frozen and covered with snow. Of course Aleksandrov wanted to go down there and bathe, but I preferred climbing the northern summit.

From Erevan we set out for Tiflis. There we went to the Orbeliani bath house and asked for experienced bath house attendants. They assigned me an old Persian, of very small stature. He worked very energetically on me and at the end of the whole procedure he turned me onto my stomach, jumped onto my back, and began to massage me with his feet. We enjoyed these treatments very much.

In Tiflis Aleksandrov and I separated for a time. He went to Gagra—by train to Batum, then by steamer; I took a short trek on foot in the region of the Upper Terek and Ardon.

My journey began in Kobi, which I reached by bus. From Kobi I set off upwards along the Terek, so that on my right were Kazbek and Gimarai-Khokh, and on my left the mountains of the Glavnyi range, the highest of which was Zil′ga-Khokh (3840 m). I climbed to the Trusovskii pass and then I settled down for the night in the open—I was well equipped for it, as I had a sleeping-bag lined with goat's wool, enclosed in a large waterproof bag. I woke in the morning feeling comfortably warm, and then I discovered that I was covered by a little mound of snow from the snowfall of the night before. But the cloud broke up, and I happily climbed Zil′ga-Khokh. I returned to the Trusovskii pass and then turned towards Zaromak and from there I walked to the Tseiskii glacier. After eating my fill at the tourist centre

Aleksandrov and Kolmogorov (near Moscow in the 60's).

I began to climb on this glacier, and there I spend the next night—it was very clear. This was my first encounter with the land of eternal snows, and it made a great impression on me.

By bus and train through Tuapse and Sochi I arrived in Gagra, where I joined Aleksandrov and a lively group of mathematicians. We all joined in bathing in a fairly strong surf, where simply mastering the exhausting waves requires a planned approach—one has to throw oneself under the wave with the head down. Aleksandrov was a complete master of this art. His enthusiastic expression was worth seeing as he threw himself in to meet the waves of the sea! I was a rather worse swimmer than Aleksandrov, but was on the whole a good enough partner in his cult of active and spontaneous contact with nature, his cult of the sun, the water, and the snow.

I remember well our stay there. On the Gagrian shore he was attracted by a two-story house, which was so situated on the shore that the balcony of the second story projected over the sea and the waves of the sea crashed directly below it. Aleksandrov and I badly wanted to settle down on this balcony. The owner of the house was a former Abkhazian princess, and we did not succeed in getting permission immediately. But when we did settle there with our sleeping bags on the balcony, we were attacked by whole swarms of bed bugs, and immediately lost all desire to stay there. We decided to take other lodgings, although they were a long way from the sea.

I do not remember exactly when we decided to stay together; but once the decision was taken we returned from Gagra to Moscow. In the holiday village of Klyazma we found on Pisarev Street a newly built house which was in two parts, with three rooms in each, and a common kitchen without any conveniences and without running water. We took one half of the house. My aunt Vera Yakovlevna moved in with us and she ran the house. The owners lived in the other half of the house. I can only remember the name and patronymic of the landlady—Vera Arkhipovna. Our landlords had a cow, from which we got milk.

It was fairly late autumn when we settled into this house, but we still managed to run regularly for a bathe in one of our two rivers—Klyazma the nearer one and Ucha the farther one. We took the house for two years, but in 1930–1931 we both spent much time abroad, so that we only lived there for about a year.

In the autumn of 1931 we moved, together with Vera Yakovlevna, into lodgings still in the village of Klyazma, in the house on Nekrasovskaya street, which belonged to Aleksandrov's brother Mikhail Sergeevich Aleksandrov. We lived in strange conditions there: in the summer we occupied the fairly spacious attics, but in winter we moved down to the lower part of the house. There was rather more space here than in Pisarev Street, but this house too was without modern conveniences. Masha Barabanova joined us there as housekeeper—before the Revolution she has been my nanny on our estate near Yaroslavl.

In 1935 we acquired from the heirs of Konstantin Sergeevich Stanislavskiĭ part of an old manor house in the village of Komarovka near Bolshevo (later we bought the whole house). This "house in Komarovka" satisfied all our needs: there was room for a large library and we could put up our guests in separate rooms for several days and even for longer periods.

By the end of the 30's we were both well settled in. As a rule, of the seven days of the week, four were spent in Komarovka, one of which was devoted entirely to physical recreation—skiing, rowing, long excursions on foot (our long ski tours covered on average about 30 kilometers, rising to 50; on sunny March days we went out on skis wearing nothing but shorts, for as much as four hours at a stretch). On the other days, morning exercise was compulsory, supplemented in winter by a 10 kilometer ski run. We were never walruses, bathing every day all the year round; we bathed at any time we felt like it. Especially did we love swimming in the river just as it began to melt, even when there were still snow drifts on the banks. When the frost was not too severe the morning run of about one kilometer was done barefoot and wearing only shorts. I swam only short distances in icy water but Aleksandrov swam much further. It was I however who skied naked for considerably longer distances.

One of our favorite ways of arranging ski-runs was this: we invited young mathematicians to, say, Kalistov, and from there we set out in the direction of Komarovka. Some who did not get as far as Komarovka caught a bus and set off for home. Those who got to Komarovka were offered a shower and then if one felt like it a romp in the snow and then dinner. In the golden years of the Komarovka house the number of guests at the dinner table after skiing could be as many as fifteen.

This was a typical day's programme at Komarovka. Breakfast at 8–9 o'clock. Study from 9 to 2. Second breakfast about 2. Ski run or walk from 3 to 5. When the organization was at its strictest, a pre-dinner nap of 40 minutes. Dinner 5–6 p.m. Then reading, music, discussion of scientific and general topics. And finally a short evening walk, expecially on moonlight nights in winter. Bed between 10 and 11.

There were two cases in which this arrangement could be altered; a) when scientific research became exciting and demanded an unlimited length of time; b) on sunny days in March when skiing was the only occupation.

Chapter III
TRAVEL ABROAD 1930–1931

I had problems with my application for a Rockefeller scholarship and in the end was not awarded one. Instead it was suggested that I go to Germany and France for six months on a mission for Narkompros (People's Commissariat of Education). Since I was already abroad, I wrote to Narkompros saying that the sum allotted to me was enough for a stay of nine months. Narkompros granted my wish and agreed to extend my stay abroad to nine months.

Aleksandrov had an invitation to lecture at the University of Göttingen and in June 1930 we travelled together from Moscow to Berlin. There Aleksandrov stayed with H. Hopf and I stayed in a hotel. On the streets of Berlin the tensions arising from the political and economic situation could be felt. Crowds of unemployed people, often very badly dressed. Many garden sheds were decorated with flags: red for the Communists, yellow, red and black for the Social Democrats, white, red and black for the extreme right parties.

We stayed three days in Berlin and then went on to Göttingen. There Aleksandrov settled down in the home of Neugebauer and I took a furnished room. The adjoining courtyard was occupied by the Student Corps with its rather noisy way of life and not infrequent duels (which in the more serious cases ended with a cut on the cheek, not too dangerous but yet clearly visible). There were few mathematical students in the corps, and they were not very popular. Sometimes those students even took off the insignia of the corps at the entrance to the Mathematical Institute of the University.

The future of Germany was completely in doubt. Courant, whether in jest or in earnest, said that the National Socialists would probably come to power, but not for long, and that then power would pass to the Communists, and he even added, this time really in jest, that in that event Aleksandrov would come from Moscow as the Commissar for the University of Göttingen.

The Institute of Mathematics, the building of which had been finished not long before our arrival, was exceptionally well-arranged for scientific studies. For young academics arriving at the Institute there was a separate small room with a writing desk, two chairs, and small shelves for books and paper of every kind; there was also a key to the library, where one could, without asking anyone, take out any book and take it to one's own study. It was of course assumed that when no longer needed the books would be returned to the library and correctly shelved. Serious students also had a key to the library with the same rights and obligations.

The professors had a fairly big office and a professor's key opened his own office, all the lecture rooms in the Institute, and of course the library. But the Director of the Institute and his Assistenten could, with one key, unlock all the rooms in the Institute. Neugebauer, who had arranged this key-system, was very proud of it.

At that time Göttingen was regarded as the leading mathematical centre of Germany and as a worthy rival to Paris in France and Princeton in the USA. Göttingen had attained this position with only a very limited number of permanent staff. There were four full professors—Hilbert, Courant, Landau, and apparently Bernstein (when he reached the age of 68 Hilbert had to retire, and Hermann Weyl had already been invited to succeed him). There were many of Courant's young collaborators there as Assistenten (Friedrichs, Rellich, Hans Levi, and others). Although she was not a full Professor, Emmy Noether was already being considered

P. S. Aleksandrov, H. Hopf, E. Noether (Göttingen in the 20's).

the head of the school of modern general algebra. Her students, van der Waerden and Deuring, were already there as Assistenten.

At that time the number of mathematicians invited to spend a semester or some other fixed period in Göttingen was unusually large, as was the number going there on their own initiative.

Most of the corpus of Göttingen mathematicians were grouped around Hilbert, Courant, Landau, and Emmy Noether. It was a very friendly group and Aleksandrov was regarded as a full member of the group. It can even be said that he was extremely popular in this circle, and he formed a close friendship with Courant, Neugebauer, and Emmy Noether.

I too had various scientific contacts in Göttingen. First of all with Courant and his students working on limit theorems, where diffusion processes turned out to be the limits of discrete random processes; then with H. Weyl in intuitionistic logic; and lastly with Landau in function theory. With the latter, however, my contact did not progress very successfully. Landau badly wanted to find an answer to this question: can a continuous function have neither a finite nor an infinite derivative at a point (in the well-known examples of Weierstrass and van der Waerden there are infinite derivatives at some points). With great excitement I took up this problem and quickly constructed the necessary example. It proved to be very unwieldy, but I wrote it out in detail and took it to Landau. Landau was very pleased; he told everyone about the success of the young Russian mathematician and asked me to write it up as quickly as possible for publication. But some weeks later I saw to my horror in Fundamenta Mathematicae a paper by Besicovitch in which the same example was constructed and where the methods used by both Besicovitch and myself were extraordinarily similar.

My preoccupation with generalizing the concept of the integral and with measure theory opened up very desirable lines of communication with Carathéodory, and therefore for the two weeks before the end of the summer semester (his stay in

Germany came to an end on 1 August) I left Göttingen for Munich, where Aleksandrov soon followed me after completing his duties in Göttingen. Since we were going on to Paris and we meant to travel on our way there, we forwarded all our best clothes from Göttingen to Paris and so we found ourselves in Munich looking like hikers: I the younger in shorts and Aleksandrov in knickerbockers.

In Munich I met Carathéodory. He liked my work on measure theory and he insisted on having it published as quickly as possible (I also gave him the work I had already prepared on the generalization of the concept of the integral and which seemed to me to be a very great improvement. But Carathéodory, although he recommended it to Mathematische Annalen, reacted rather coldly to it.)

In the summer we were both invited by Fréchet to stay with him on the shores of the Mediterranean, to work with him on probability theory in my case and on set theoretic topology in Aleksandrov's case. Our plan was to hike through Germany and Southern France and then go to Fréchet's—he lived at that time in the small town of Sanary-sur-Mer not far from Toulon—and to go on from there to Paris.

We spent the first two days in the Bavarian Alps in a little inn by the small lake of Spitzensee. From there we did the "ascent of the Rotwand". Along the tourist path to the Rotwand were marching sturdy Bavarians wearing hats with a feather, Bavarian leather shorts, Alpine boots with knitted socks, and with Alpenstocks in their hands. Partly from a desire to tease these respectable Bavarians, we did all the ascent barefoot (Aleksandrov and I generally liked walking barefoot).

Our journey continued to two of the most remarkable Gothic cathedrals, in Ulm and Freiburg. We arrived in both these cities in the afternoon, we took a room in a hotel, and went for a walk round the town looking for different vantage points from which to view the cathedral.

After supper we returned to the cathedral at sunset and we went back again in the morning. Aleksandrov very much wanted to get to know the monuments of Gothic architecture really well, but still he repeated frequently the words of Verlaine: "la mer est plus belle que les cathédrales".

In Ulm we bathed a few times in the Danube. The current near Ulm is very fast, and local youths bathing there showed us that we could have a very pleasant swim if we left our clothes and set off down stream with the current right through the whole town, then ran along the bank to the place from where we started. We did this, leaving our clothes in the care of the local bathers.

In Ulm we found a hotel at a reasonable price. Freiburg was less hospitable from this point of view: the price of a room in the hotel seemed to us excessive. We then went to the outskirts of Freiburg, where we quickly found a little village inn with very attractive tiny rooms. There we spent the night (on beds of striking freshness and whiteness), and in the morning we went back again to the cathedral. Then we went by local train to France, to Mulhouse. Then another short journey across France—partly on foot, partly by the local omnibus. One whole day—from early morning till late evening—we spend on an excursion to the "seven lakes", which lie on the Alpine meadows at a height of about 2000 m. Two days were spent on the banks of the Rhône, and another on the shores of Lake Annecy. Then we had a day in Marseille and from there we went by train to Sanary-sur-Mer, where Fréchet was already waiting for us; he had booked a room for us in the same Pension de la Gorge where he himself was living. The pension was surrounded by orchards, and not far away was a small beach to which we went with Fréchet. We bathed and swam but Fréchet only paddled.

Not far away was Cap Gorge, a picturesque group of red socks. We much enjoyed working on these rocks, leaping from them, and bathing underneath them. As usual Aleksandrov worked in the sun and I in the shade.

At that time Fréchet was studying Markov chains with discrete time and different types and sets of states. We discussed with him all the Markov problems over a wide range. This rather monotonous life continued for about a month, interrupted by occasional short outings.

In September we left Sanary and went by train to Brittany—right across France. Our destination was the small village of Batz, where the grave of Pavel Samuilovich Uryson lies. The deserted granite beaches, against which the huge ocean waves thunder, form a complete contrast to the shores of the Mediterranean. Uryson's grave is well tended because it is looked after by Mademoiselle Cornu in whose house Aleksandrov and Uryson were living at the time of his death. Both the gloomy nature of Brittany and the memory of Uryson inclined us to silent walks along the sea shore. We went to the fishing village of Le Croisic and the little fishing village of La Turballe.

From Batz we finally travelled to Paris. We went straight from the station to a hotel on the Rue Tournefort, where Luzin had always stayed when he visited Paris. The fairly large and rather gloomy rooms were make pleasant by the feeling of complete silence but ... there were a lot of bedbugs there (common in France at that time). With some excitement Aleksandrov began making short work of the bugs, piling up their corpses on a piece of paper. Then we collected our belongings and immediately left. In the foyer the manageress began to question us, uneasily: "What is the matter? Is there something you are not satisfied with?". Aleksandrov answered her majestically: "You will find the explanation in our room". Aleksandrov did not like the next hotel because it was too noisy, and in the third one very sloppily dressed women were sneaking along the corridors. Only by the method of successive approximation did we find in the Latin Quarter a hotel which was not too expensive and which met all Aleksandrov's requirements. There we stayed for the two weeks we spent together in Paris.

It was not Aleksandrov's first visit to Paris. There was much in this city that he liked—most of all that special atmosphere, which casts a spell on all who arrive in this city, right up to the present day. This is more natural for a Russian, since one can acquire from the Russian classics a large amount of information about Paris, its geography, its sights, the street life, and much else.

Aleksandrov already had a fairly wide knowledge of the outstanding works of art kept in the Paris museums. But it must be said that of all the field of fine arts, the ting which was the centre of attraction for him, which outweighed all other impressions, was the Musée Rodin. A photograph of Rodin's sculpture "The bronze age" was always to be found on the wall of his room in Komarovka. It hangs there still.

It was natural that we, like most people visiting Paris, should yield to the irresistible passion for wandering along the streets with no fixed goal in mind. We also visited the Parisian swimming baths (at that time they were interesting and there were still very pronounced social markers: we visited different ones, from aristocratic ones to the largest democratic baths—the Piscine de la Gare).

Aleksandrov left Paris at the end of September and I stayed on until the middle of December. Together with Aleksandrov I had already found a small furnished room in a quiet family house near the University area and the Parc Montsouris.

As I was in Paris, I was naturally interested in seeing how my work was regarded there and in getting some advice about continuing my work from the leaders of the older generation of mathematicians—Borel and Lebesgue. But my contact with them, alas, was limited to a small number of official visits. However, Borel's help was essential if I was to get my French visa extended. Permission was granted immediately on receipt of a letter signed Emile Borel, Ancien Ministre de la Marine.

In mathematical matters I gained much from my contacts with P. Lévy. Time and again he invited me to his home, where we had lengthy serious scientific discussions. For lack of time I did not manage to form any interesting relations with the representatives of the younger generation.

The autumn weather in Paris was extremely unpleasant, especially at the stations and in the tunnels of the Paris métro, where a cold wind burst through the outer doors and turned into clouds of steam. As a result of this the walls of the métro and our clothes were covered with moisture. Apparently it was because of this that I caught a bad cold and arrived in Göttingen a sick man. Aleksandrov immediately put me into an excellent private hospital, where at Christmas they even brought a decorated Christmas tree into my private ward. After convalescing, I was sent by Aleksandrov to the comparatively cheap but well-furnished boarding house Kreiznacher, and only in mid-January was I allowed to move into the room I had seen earlier in the flat of a postman's family. Living conditions there were rather spartan. My postman and his wife, a grown-up daughter, and two boys occupied a fairly large four-room flat, in which however in winter only the kitchen was permanently heated; the other rooms were usually not heated. Since the room was not heated, I thought it wise to keep the window permanently open and to sleep under a duvet and wearing a night cap. My room was very small and cost, if I am not mistaken, 15 marks a month. In addition, I paid extra so that I could study in the drawing room in the mornings near a stove which was heated for me, but my agreement with the landlord laid down that I could use the drawing room only on weekdays, because on Sundays and holidays the drawing room and the dining room were heated only for the sake of visitors, and consequently I could not study there.

I quickly regained my strength and even in the middle of February I was running in the mornings in shorts and sports shirt over distances of approximately one and a half kilometers to the University bath-house Klie.

Chapter IV
EXTRACTS FROM ALEKSANDROV'S LETTERS FROM THE USA

(From a letter dated 20 February 1931)

I have already given two lectures here; from the language point of view the first went very badly, but the second (today) was considerably better. Moreover, I am taking lessons in English from a pleasant elderly lady, a teacher of German at the local grammar school. So far I have had two lessons and am very pleased with them. But this pleasure does not cost just two marks as yours does, but two and a half dollars. But I hope I shall not have to continue these lessons for long; at present they are nevertheless an absolute necessity.

I give a fairly general topology course, ranging from combinatorics (including the laws of duality) to dimension theory; I do it all with closed sets in view; the course is apparently followed with interest, and I think the students attending are infinitely better prepared than anywhere else. Amongst those attending is Alexander; it is difficult to imagine an audience better prepared and more responsive to the teacher. And Alexander himself gives a completely elementary course in topology; just now he is presenting with astonishing elegance the simplest themes from abstract topology—separability axioms, and so on—but later there will be bicompact spaces, and so on. In a word, the present topology season in Princeton could well be taking place in Moscow.

By the way, you have often spoken to me about purely combinatorial topology. Read the recently (this summer) published paper by Alexander: "Combinatorial theory of complexes", Ann. of Math. **31**, no. 2, 292–320. This is one of the latest numbers of the Annals and is in the Lesezimmer (reading room). The paper is completely elementary (it assumes nothing) and is unusually elegant. I am doing a little work on the main part of my book. And now I have to write a review for the Zentralblatt on the topological article for the Enzyklopädie by Tietze and Vietoris (I shall have to do it tomorrow, I fear it will take half a day), the Anhang (Appendix) for Hilbert, footnotes to my paper, as well as a whole lot of other things. Alexander is trying to prove that a two-dimensional complex that consists of all the two-dimensional faces of a six-dimensional simplex cannot be topologically included in a four-dimensional space, and that in general (analogously) the theorem that a k-dimensional space can always fit into E^{2k+1} cannot be improved on.

I swim every day in the swimming pool; it is in fact excellent, many times bigger than the one in Göttingen. I do not know exactly how long it is—here lengths are not measured in metres, but in yards or some similar units—but in any case there is no need for it to be any bigger (the next desirable improvement would be rather different—that it should be in the open). The showers available at the pool are free with an unlimited supply of soap—turn the knob and a handful of very bubbly powder comes out, which, when just damp, immediately gives off a large amount of foam (even without a sponge). So here it is very easy to insist on everyone washing with soap before bathing. And the complete absence of bathing trunks at the swimming pool naturally promotes cleanliness too (and, in spite of Fréchet's misgiving, I have not seen any swimmers with any sort of hernia which should have been hidden under swimming trunks on aesthetic grounds). By the way, the majority of the bathers are, as is to be expected, students and most of them are of excellent physique, so that I think that from the aesthetic point of view the young people swimming in the Princeton swimming pool are considerably better than the public on the French beaches, even if hernias are hidden under swimming trunks.

The Princeton Athletic Field is only two steps away from me; I try to go there to run, and so on. But in America shorts and shirt (sleeveless) are worn for running, since the Athletic Field can be seen from the road and openness in behaviour is less than in Germany! In general it is an amazing fact—Americans do not appreciate the joy of exposing the naked body to the sun's rays, to the wind, in general, to all the natural influences of nature around us. They understand perfectly the satisfaction of doing exercises in the nude—in the various rooms of the Gymnasium one can see young people in the briefest of shorts and often completely naked, throwing a ball against the wall, doing various gymnastic exercises, and so on. But they do not think of transferring all this to the open air (even wearing rather longer shorts). But still, specifically, and from the aesthetic point of view, American University youths are, without doubt, and in the best possible way, some of the best looking young people, so there is no reason for concealing their nakedness so carefully.

I am very glad that you have bought a Trainingsanzug (a track suit). But, if you wear it with nothing under it (for example for running in the morning), it is essential to have two, so that they can be washed fairly often. Ask if they shrink in the wash (beim Waschen eingehen). They probably do. This must be borne in mind and one must buy correspondingly larger sizes. But this combination—the top half of a track suit with a white shirt and tie—would certainly suit you very well—in general loose and simple clothes suit you. Be sure to buy or to order a suit with short trousers; this is fine and practical and can be worn on all ordinary occasions (not on formal ones and not in the evening). Do not fail to buy yourself all these things, especially in case you have to return unexpectedly to Moscow. (Unfortunately I think this cannot be ruled out.) And if happily you are there to welcome me in Germany, I suppose it would be better to buy or order a suit with me. Meanwhile, you should certainly think about my return, what things we should buy both for sports and for holidays; a tent, and so on. What I think is especially desirable is a really good tent, comfortable enough for us to be able to live in it for a long time, and if we are to be able to study in it, there must be good light. We really must have a tent like this so that we can be independent of any Abkhazian princesses and settle down without any bed bugs or other unpleasant things, possibly for a long time and wherever the fancy takes us.

Think carefully about all this and find out where we can best buy such things, get catalogues, find out the prices, and so on. It is doubtful whether America is the best place to buy; such things are more expensive here than in Germany. We can of course buy the things wherever you wish, but it seems better to do it where we thought of buying a boat (Schellenberg, Neugebauer's neighbour, knows the address of a Berg Verlag[1] in Munich where one can find all the information, get catalogues, and so on; try to get in touch directly with him, or ask Neugebauer).

My advice to you is to buy a ski outfit, especially as you seem comfortable in one. To come to the point, all these things are quite inexpensive but their real worth (that is, the benefits derived from them, to use the expression of Aleksandr Yakovlevich) is very great.

You have written very little about your sporting activities, but I should like to have a continuous detailed report: how far, in what clothes, at what temperature were you running? At what time of day? Straight from home or only on the Spielplatz? Did you bathe at Klie? Did you swim in the Schwimmhalle? What gymnastics did you do and where (at home, at Klie)? Also you have not written about how you feel. Are you coughing? Are you hoarse? How is your cold? And the main thing, how do you feel in general? It should be a very good idea for you to buy yourself cream as well as milk. It seems (I heard this from a biochemist) that a small bottle of cream (costing 13 cents) contains more nourishment than a big bottle (more than half a litre) of milk, which here costs 9 cents. By the way, the prices here (for different richness of cream and milk) are fixed almost entirely according to their nutritive value, that is, their chemical analysis (fat content, and so on). And this is confirmed subjectively; the filling effect of cream is very great. By the way, cream is very widely available in America—people buying milk (for

[1] I remember: Berg Verlag Rudolf Rother, Munich (Munich 49, I think), but check with Schellenberg.

example, my Mrs. Gulich) quite often buy cream too, since they know that it is a more concentrated form of that same milk and not some exotic luxury item: in all these matters the American housewife is infinitely more intelligent than her German counterpart: although Grete was a lady of much higher class it took me a long time to convince her that paying a fifth more to get an egg one and a half times larger is sensible, but when it came to saying that paying 13 cents for a small bottle of cream is as sensible as paying 9 cents for a bigger bottle of milk—of that I could not convince her. But my Mrs. Gulich understands all this very well. Therefore in America one can fix the price according to the chemical analysis (and not according to the stupidity of the purchaser). But you should buy both cream and milk (since you drink cream in your coffee, but milk by itself). Of course, the best of all is to drink cream by itself.

It is quite hard for me to convey all this to you: everything here is topsy-turvy; the weight of people is calculated not in kilos, not even in English pounds, but in some special units called "stones". Probably the stone over which Washington's horse stumbled when he was going to be elected First President of the USA! But I am trying to work out the corresponding coefficients and then I'll write to you. I'll send you some money soon. I still have not got the money: it seems that they propose to pay the only due to me monthly: on 5 March, 5 April, May, June (since it is assumed that people invited to give lectures will not arrive with 50 dollars in their pockets). But I will take an advance. I also enjoy unlimited credit (not only with individuals but also, for example, in the University shop, where I just bought myself a typewriter (on credit) for 60 dollars. As well as the usual alphabet it will have the signs $+, -, <, >, =, \varepsilon, \Sigma$, the exponent n (E^n) and the subscripts $_{1,2}$, and of course $_0, ', ''$. Until these signs are added, they have given me an ordinary typewriter to use in the meantime).

From a letter of 22.III.1931

I am definitely falling in love with it, it is even with some emotion that I look at it every day, and it plays a very big role in my life. All around it there is complete silence and even mystery, and it is frankly Undine reincarnated—quiet, transparent, and touching, especially when you see it, lit by the sun's bright rays, gliding between big, old, but not yet bare trees.

And its name is astonishing—Stony Brook; such names are (or more correctly, were) the names of the daughters and brides of the Indian chiefs. Each time I look at it, I say to myself, you are called Stony Brook, and pronouncing this name gives me pleasure. The name fits it perfectly—the water in it is clear, green, and cold, and the bottom is stony.

To get to the place where I swim I have to go a long way through thickets and then suddenly, after climbing a small ridge, there directly in front of you is the river with large trees perched above it, much as they have been for ages, as if out of Turgenev. For a long stretch it flows parallel to the canal, and I swim at the end of this stretch. Between it and the canal the path goes through the thicket; probably only enchanted princes walk along it to Stony Brook in the moonlight, since so far I have not seen even one human being (but near the path I found two empty tins, which shows that enchanted princes occasionally eat tinned beans; I found too some signs of a camp fire—obviously they have a picnic here from time to time).

Be that as it may, so far I have not seen a single human being on this path, and I use it for running before and after my swim.

On the other bank of Stony Brook there is a wooded swamp and further on, a long way further on, some farms, and on the other bank of the canal endless orchards, which give the impression of complete neglect. There is no end to water fowl, so all in all one really gets an impression of mystery—is there not a mysterious illusion of wilderness half an hour's walk from Princeton? It is also a half hour walk via my thickets to the nearest road (without them it would be twenty minutes). And the thickets are very loverly. I can imagine how luxuriant it will be when the spring really comes, when the completely impenetrable thick thorny creeping verdure here turns green. No, definitely much better than any Klie and it can compete with the best spots on our Ucha ...

Some days ago I went for a long four-hour walk and followed the course of the Stony Brook over a distance of 15 kilometers. This walk was often fraught with difficulties, even dangers, since I had not only to penetrate the thickets where there were no paths, but what was much worse—I had to go through private properties, fenced in with wired fences, on which there was every ten paces, instead of barbs, a printed notice stating that any stranger entering here was in breach of the laws of the USA, with all the ensuing consequences! But, thank God, there were not any consequences. In general, as I have written to you time and again, the main thing at Princeton is the sun. Today I woke up at sunrise—through the window I saw the earth, the trees, the houses, lit by the bright red light, and clouds near the horizon, then all at once, without any transition—blue sky. It was a really beautiful sight, which lasted of course only for a few minutes. On the Klyazma you and I were asleep when the sun rose, and it was probably just as beautiful! ...

From a letter of 25.III.1931

"City Lights"
A Romantic Comedy
written and directed
by Charles Chaplin

Yesterday I went to the local cinema. Normally it needs some initiative to drag me to the cinema. This initiative was taken by Alexander and I have no regrets.

For the first time in my life I felt, as I came out of the cinema, moved by a great serious human and universal art. And this overwhelming impression lasted the whole day and, for me at least, this is one of the basic criteria for judging a significant impression in the theatre.

The main thing about Chaplin is that he succeeded in creating a new artistic image, absolutely unique, but definitely new, a figure which simply did not exist before him in human art (although of course it did exist in life—both at the time of King Solomon and in our own time). Please do not take this as comparing standards, as a quantitative estimate; but just as Shakespeare revealed Hamlet to us, and Gogol′ revealed Akakiĭ Akakievich, so Chaplin revealed this figure, which he always plays, of the comic pathetic failure, of the vagabond, who knows not whence he came nor whither he goes, the exact equivalent of Schubert's "organ-grinder"—"willst du zu meinen Liedrn deine Leier drehn?"

Chaplin's artistic discovery does not of course lie in this centuries old context, but in that surprising embodiment he found for himself, moving, and simple, without any rhetoric, chaste. And it was there that I understood the purely material superiority, in a certain aspect, of the cinema to the theatre—this creation of art will indeed survive, we must hope, "for ever and ever". Meanwhile, the work is, in the best sense of the word, contemporary: all this shyness, this fear of phrase-mongering, and many other things seem impossible in the 19th or 18th centuries. I do not know whether you can understand what I mean by this: I am simply saying that contemporary romanticism (and Chaplin is indeed its representative) is, if anything, more romantic than it was a hundred years ago (in everything except music), more romantic in the sense of recognizing that nothing can be conveyed by words.

The plot in all this presentation is of course completely unimportant, if anything something exaggeratedly mechanical—something standard which existed in the "organic periods of the theatre"—for example, in the Italian commedia del arte—it only aggravates the contents of the main characters, more exactly, of the main character. But the expression in his eyes at the end of the film, the gesture with which, as he leaves, he picks up his changeless battered bowler hat, kisses his hand—that is quite unforgettable. It is simply inevitable that all this should take place on the screen—in the real theatre all the beauty would be lost, all would become crude, there would be left only an empty shell of all that.

By the way, I often have doubts about this supposed unity of reality (a question on which we have often touched in our letters) and I am often tempted by the Platonic concept of a dualistic world.

Translated by A. Lofthouse

Newton and Contemporary Mathematical Thought

A. N. Kolmogorov

It is known that Newton suffered from the two phobias "fear of controversy" and "fear of philosophy". In view of this it is especially important in the case of Newton to adhere to the rule we apply in studying the works of most representatives of the mathematical and natural sciences: to study the methodology of the scientist in the first place directly from his scientific works, and not from his methodological pronouncements. What follows is a perhaps imperfect attempt to apply this rule to the mathematical works of Newton.[1]

The fate of his mathematical works is very distinctive. The decisive years of his creativeness on the whole were 1665–1666. In this brief period he made roughly all his fundamental discoveries in mathematics, mechanics, and physics. If we consider that here we mean the creation of mathematical analysis (the differential and integral calculus) and of the mathematical natural sciences,[2] then this case must be acknowledged as unique in the history of science.

It is considered that Newton[3] wrote three fundamental mathematical works in the subsequent years:

Published in the journal *Matematika v Shkole* **1982**, no. 6, 58–64. This article stems from an anniversary lecture given in 1943, and was published in the collection *Moscow University*: *In memory of Isaac Newton*, Izdat. Moskov. Gos. Univ., Moscow, 1946, pp. 27–42.

[1]I owe my interest in the scientific methodology in Newton's mathematical works to the remarkable publications of Alekseĭ Nikolaevich Krylov. In particular, my view of the whole contrast between the sound clarity of Newton's thought and the mathematical mysticism of Leibniz and Euler is taken from Krylov. If I stress the distinction between Newton's "rigor" and modern "set-theoretic rigor" more than Krylov did, then that is due to the natural difference between our generations.

[2]Newton not only made fundamental discoveries in the mathematical natural sciences that we need not go into here, since they are widely known, but he actually was the first to create a mathematics for the natural sciences in the sense of a system for the mathematical investigation of all mechanical, physical, and astronomical phenomena. Before Newton one could say only that this or that individual area of the natural sciences could be studied by mathematical methods. Of course, the ideas of Leibniz about the possibility of a mathematization of the whole of human knowledge were even more universal. But they bore no fruit precisely because of their absolute generality and lack of concreteness. On the subject of universal applicability and at the same time restrictedness see my article "Mathematics" in the *Great Soviet Encyclopedia*.

[3]The quotations below are from the translation of D. D. Mordukhaĭ-Boltovskiĭ. (*Translator's note*: The quotations of Newton below were taken from the translations from the Latin by Andrew Motte, John Stewart, and John Harris, reprinted in *Great books of the western world*, vol. 34, Ency. Britannica, London, 1952, and *The mathematical works of Isaac Newton*, vol. 1, Johnson Reprint Corp., New York, 1964.)

1) *De analysi per aequationes numero terminorum infinitas* in 1665;

2) *Methodus fluxionum et serierum infinitarum*, after *De analysi per aequationes* but before 1671;

3) *De quadratura curvarum*, the main text in 1665–1666, but the final version with an introduction and the concluding Scholium considerably later, apparently in the 1670s after *Methodus fluxionum.*

But these three works were published:

1) in 1711;

2) after Newton's death in 1736;

3) in 1704 as an appendix to *Optics.*

They were all published as not completely consistent fragments, sometimes bearing clear signs of having been written at different times. Even parts of one and the same treatise sometimes bear clear signs of having been written at different times. In the case of *De quadratura curvarum* the lack of agreement is quite deliberate, since the main text is intended to reproduce the state of Newton's ideas in 1665–1666, even though they differ sharply from his views at the time when the introduction and the Scholium[4] were written.

Newton's attitude toward the editing of his famous treatise on mechanics, *Principia mathematica philosophiae naturalis*, and his *Optics* was quite different. In subsequent editions they were subjected to extremely careful (and perhaps at times rather morbidly suspicious) editing and revision. For our purpose the described state of the texts of his mathematical works does make it more difficult for us to give a coherent characterization of his scientific methodology in each period of his work, and yet it also has a certain advantage: it provides an opportunity for penetrating into the laboratory of his scientific thought.

Newton and Leibniz. As is well known, the origin of our modern differential and integral calculus was to a significant extent prepared for by the work of mathematicians in the first half of the seventeenth century: Kepler, Cavalieri, Descartes, Fermat, and others. However, it is not without grounds that the discovery (in the proper sense of the word) of this calculus is ascribed to Newton and Leibniz, since they first reduced the solution of all the diverse problems their predecessors had treated by methods of analysis of infinitesimals, to a systematic use of the two mutually inverse operations of differentiation and integration.

In the sense of a printed publication the priority belongs to Leibniz, who gave a comprehensive account of the fundamental ideas of the new calculus in papers published in *Acta Eruditorum* in 1682–1686.

On the other hand, with regard to the time at which the fundamental results were obtained there is every basis for regarding Newton as having priority: he came upon the fundamental ideas of the differential and integral calculus in the course of 1665 and 1666 and by 1671 had an already complete system of exposition of his theory recorded in *Methodus fluxionum*, while Leibniz began his investigations in the analysis of infinitesimals only in 1673.

We shall not dwell long on the question of the degree of independence of Leibniz, an issue still not completely cleared up. We do know the following.

---

[4]The introduction to *De quadratura curvarum* closes with the words: "From the fluxions to find the fluents is the more difficult problem, and the first step of the solution of it is equivalent to the quadrature of curves; concerning which I have formerly written the following tract."

The first of Newton's works cited above, *De analysi per aequationes*, written in 1665, was given to Barrow and Collins in manuscript form around 1669 and gained some fame among English mathematicians. At the time of his journey to England Leibniz, who would soon begin his work on the analysis of infinitesimals, must undoubtedly have heard something about the contents of Newton's treatise. However, Leibniz became acquainted with the actual manuscript via Collins only in 1676, when on the whole he had already carried out his own investigations. In addition we should bear in mind that in *De analysi per aequationes* Newton does not yet provide an explicit account of his general method of fluxions, confining himself to an exposition of certain elements of it in connection with isolated special problems. In the same year 1676 Newton, in answer to queries of Leibniz conveyed to him by Oldenburg, sent Leibniz a list of his main results in two letters, without completely revealing the method by which they were obtained. These letters could apparently no longer have given Leibniz much that was new.

The converse influence, of Leibniz on Newton, can be seen only in the introduction to *De quadratura curvarum*, which, as the author himself indicated, was written considerably later than the main text of the tract. As for the main text of *De quadratura*, it can be judged from the nature of the exposition to have been written between *De analysi per aequationes* and *Methodus fluxionum*, that is, between 1665 and 1667.

It is considerably more interesting that Newton and Leibniz approached the creation of differential and integral calculus from completely different points of view and with completely opposite methodological systems.

The difference between their approaches is described by A. N. Krylov in a very clear way, though rather simplified and biased in favor of Newton:[5]

> Newton discovered and gave the bases for the calculus of infinitesimals by starting from mechanical and geometric concepts. In his arguments he always applied geometric notions and was absolutely rigorous in them and absolutely precise in language and expressions. Therefore, he first establishes the concept of the limit of a variable quantity, the concept used today, and all his teaching about "fluxions", or "derivatives" in modern terminology, is based on finding the limits of ratios of two infinitesimal quantities that are in a definite mutual dependence and that vary jointly. Posing as the basic problem of integral calculus the determination of a "fluent" from its given "fluxion", that is, of a primitive function from its given derivative, he constantly employs geometric notions, and his tract itself is called *De quadratura curvarum.*
>
> Leibniz proceeded in a different way. He introduced the new term "infinitesimal" instead of an increment of a variable or a function of it that vanishes in the limit. He did not give this concept a precise and rigorous mathematical definition, and in some of his explanations he seemingly did not distinguish between the mathematical concepts of "infinitely small" and "very small", nor between "infinitely large" and "very large", likening for example the one to a speck of dust and the other to the Earth. Moreover, he connected the concept of an infinitesimal with the philosophical

[5]A. N. Krylov, *Leonhard Euler*, Izdat. Akad. Nauk SSSR, Moscow–Leningrad, 1933, p. 16.

> concepts of "finite or infinite divisibility of matter", of an "indivisible atom", of a "monad", and so on, concepts very far from pure mathematics, which has to do not with the quantities themselves but with numbers serving as a measure of them.

In none of his treatises did Newton actually give a completely consistent exposition of the method of fluxions that agrees entirely with Krylov's characterization. Along with the method of "prime and ultimate ratios", that is, the method of limits in our modern terminology, Newton uses the "method of moments", which in essence coincides with the "method of indivisibles" used by his contemporaries and predecessors, who were less demanding with regard to logical rigor.

It is interesting to trace from his works the history of Newton's use of the "method of moments" and the "method of prime and ultimate ratios". In his earliest work, *De analysi per aequationes*, written in 1665, Newton has already mastered a perfectly clear idea of limit, although he does not formulate it as a *definition* but only describes it on one particular occasion when he writes:

> The other thing to be demonstrated, is the literal resolution of affected equations. Namely, that the quotient, when x is sufficiently small, the further it is produced, approaches so much nearer to the truth, so that the defect (p, q, or r, etc.) by which it differs from the full value of y, at length becomes less than any given quantity; and that quotient being produced infinitely is exactly equal to y.

However, where it is convenient to do so Newton uses "moments" without constraint in *De analysi per aequationes*, and his remark, "Neither am I afraid to speak of unity in points, or lines infinitely small, since geometers are wont now to consider proportions even in such a case, when they make use of the methods of indivisibles," hardly helps matters.

In his later treatise *Methodus fluxionum* the main exposition of the method of fluxions is handled quite without moments.

We shall see below that to preserve the advantages of using the method of moments (homogeneity of expressions in studying implicit dependencies between variable quantities) Newton takes here a parametric point of view, regarding all quantities connected with each other as functions of an auxiliary variable, the "time", which does not appear explicitly in the computations. This alone makes one think that getting rid of the method of moments did not just happen but was the implementation of a quite consciously planned program.

Nevertheless, in considering geometric applications Newton again returns to the method of moments in *Methodus fluxionum*.

Newton reveals an even greater inconsistency in *De quadratura curvarum*. In its introduction the letter o no longer stands for an "infinitesimal" moment but an ordinary finite increment. For example, corresponding to the increment o of the variable x is the increment

$$\text{(a)} \qquad nox^{n-1} + \frac{nn-n}{2}oox^{n-2} + \cdots$$

of x^n. By just dividing (a) by o Newton obtains

$$nx^{n-1} + \frac{nn-n}{2}ox^{n-2} + \cdots,$$

and then nx^{n-1} after passing to the limit. Corresponding to the original method of moments would be the assertion that the moment of x^n is simply equal to nox^{n-1}. The introduction concludes with the words:

> By like ways of arguing, and by the method of prime and ultimate ratios, may be gathered the fluxions of lines, whether right or crooked in all cases whatsoever, as also the fluxions of surfaces, angles and other quantities. In finite quantities so to frame a calculus, and thus to investigate the prime and ultimate ratios of nascent or evanescent finite quantities, is agreeable to the geometry of the ancients; and I was willing to shew, that in the method of fluxions there's no need of introducing figures infinitely small into geometry. For this analysis may be performed in any figures whatsoever, whether finite or infinitely small, so they are but imagined to be similar to the evanescent figures; as also in figures which may be reckoned as infinitely small, if you do but proceed cautiously.
>
> From the fluxions to find the fluents is the more difficult problem, and the first step of the solution of it is equivalent to the quadrature of curves; concerning which I have formerly written the following tract.

Examination of the main text that follows in *De quadratura curvarum* at once convinces us that it is not a program for constructing analysis by considering exclusively finite quantities and the theory of limits, but is structured on the extensive use of moments in the sense of infinitesimals. For example, we see already in the first pages that when the time increases by the "moment" o, the fluents x, y, z increase by $o\dot{x}$, $o\dot{y}$, $o\dot{z}$, where $\dot{x}$, $\dot{y}$, $\dot{z}$ are the fluxions (that is, the derivatives).

This is natural if we consider that, as is commonly accepted, the basic text of *De quadratura curvarum* was written soon after *De Analysi per Aequationes*, and much earlier than *Methodus fluxionum*. However this may have been, the program sketched in the introduction of *De quadratura curvarum* thus remained completely unrealized (since in *Methodus fluxionum* it was realized only in part).

It is also very important to note that in the Scholium which concludes *De quadratura curvarum* and which, like the introduction, was apparently written later than the main text and even later than *Methodus fluxionum* Newton takes a very important next step toward the full realization of his program, coming close to the modern definition of differential. This will be discussed below.

Newton included the most definitive exposition of his "method of prime and ultimate ratios" in his *Principia mathematica philosophiae naturalis* (first edition in 1686).

It may be thought that this account of the method of prime and ultimate ratios (that is, the theory of limits) is a carefully thought out presentation of ways of looking at the limit concept that Newton by 1685 regarded as definitive and most modern. We shall return again to this place in *Principia.*

Similarly, Newton did not give a logically complete and at the same time sufficiently full account of the method of fluxions, confining himself to a brief exposition of his final methodological positions on this question in the introduction of *De quadratura curvarum* and in the sixth section of the second book of *Principia.* Not wishing to place his *Principia* under threat of attacks for lack of rigor or lack of precision, Newton preferred to present the basic content of *Principia* without using the method of fluxions.

We shall not analyze the reasons why Newton avoided giving a clear and completely realizable program for a rigorous and intelligible construction of mathematical analysis. The clarity and rigor of exposition in the sense required by Newton (or satisfying such superb applied mathematicians as Krylov in our own day) were undoubtedly still very far from the notions of mathematical rigor which were later put forth by Cauchy or Weierstrass and which are dominant in modern mathematics. However, in comparison to his contemporaries Newton stood at an extremely high level in this respect.

It is a common opinion that in not using infinitesimals in the Leibniz sense, Newton lost the advantages given by the Leibniz algorithm for computing with differentials. We show below that this opinion is not fully justified. The simplified atomic concepts of Leibniz did have the advantage of simplicity, though. It is therefore hard to say to what extent a more systematic development and timely publication by Newton of his system of construction of mathematical analysis might have saved mathematics from being immersed for a century in a period of mystical faith in the unlimited power of mathematical algorithms, even those devoid of any clear meaning.

Indeed, beginning with Leibniz is a period of development of mathematics that reached its culmination in the work of Euler, who is characterized not by a simple disregard for mathematical rigor in the sense of a precise concern that mathematical concepts not stray from the real meaning originally invested in them (toward which Newton strove so diligently), but by an active and militant belief in the benefit and legitimacy of employing mathematical algorithms outside the bounds within which the symbols used in these algorithms have a real meaning.

The mood of this epoch of algorithmic mysticism is reflected even more strikingly with respect to complex numbers or divergent series than in the unconcerned use of actual infinitesimals. Without giving any explanation of the meaning of integration of complex quantities, Leibniz finds integrals of complex functions by decomposing them into real terms and integrating each of these terms separately according to formal rules.

When the result turns out to coincide with what is found without using complex quantities, Leibniz calls this a "miracle", which, however, necessarily follows in his opinion from the predetermined harmony that reigns in the world. Krylov wrote much in the work on Euler cited above about the analogous attitudes of the latter involving his belief that every series naturally arising in analysis (even if divergent and with unboundedly increasing terms) has a quite definite sum,[6] which can be found by various formal transformations.[7]

---

[6]For example,

$$1 - 1 + 1 - 1 + 1 - 1 + \cdots = \frac{1}{2},$$
$$1 - 2 + 4 - 8 + 16 - 32 + \cdots = \frac{1}{3},$$
$$1 - 3 + 9 - 27 + 81 - 243 + \cdots = \frac{1}{4}.$$

[7]It should be noted that Newton confines himself to very general and rough indications with regard to convergence of the series he uses, but he does not *consciously* make use of divergent

This whole direction, so alien to the spirit of Newton, yet had a positive value for the development of mathematics: initially unjustified extensions of old mathematical algorithms were eventually given a rigorous basis, and extreme caution at the time when they first appeared might well have delayed the progress of science.

At the present time, however, this direction should be regarded as finally exhausted: at the contemporary level of logico-mathematical culture any *hypothesis* about the possibility of an extended use of this or that algorithm or about the possibility of supplementing an inventory of previously used mathematical objects by new objects having various desirable properties can be *tested* directly, and if it is true, then the corresponding new definitions of the meaning of the algorithm in the broader domain can be given with complete precision, and the new objects can be constructed.

For example, while it took whole centuries to go from the first idea about the possibility of obtaining correct results by computations with complex numbers (which were then simply symbols for unrealizable operations!) to a clear *construction* of complex numbers (as pairs of real numbers with appropriate definitions of the operations on them), Kummer's idea in the nineteenth century about the possibility of restoring, in a suitable extended domain, unique prime factorization in systems of algebraic integers for which unique factorization does not hold without this extension, almost immediately found its logically irrefutable embodiment as the theory of "ideals".

Newton's fundamental concepts of analysis

Limits. The Newtonian idea of a limit is presented in the most complete form in the first section of the first book of *Principia.* With a view to subsequent applications the stress is on limits of ratios of quantities that are vanishing (that is, tending to zero) or increasing without bound. The first impression of Newton's statements confirm Krylov's opinion that here we have a completely modern rigorous theory of limits. As an example of his statements that sound quite modern we cite the following part of the Scholium to the first section of the first book:[8]

> For those ultimate ratios with which quantities vanish are not truly the ratios of ultimate quantities, but limits towards which the ratios of quantities decreasing without limit do always converge; and to which they approach nearer than by any given difference, but never go beyond, nor in effect attain to, till the quantities are diminished *in infinitum.* This thing will appear more evident in quantities infinitely great. If two quantities, whose difference is given, be augmented *in infinitum*, the ultimate ratio of these quantities will be given, namely, the ratio of equality; but it does not from thence follow, that the ultimate or greatest quantities themselves, whose ratio that is, will be given. Therefore if in what follows, for the sake of being more easily understood, I should happen to mention quantities as least, or evanescent, or ultimate, you are

series, and he repeatedly underscores the necessity of investigating convergence to justify using series expansions.

[8]In his translation Krylov modernizes Newton's terminology somewhat, replacing Newton's "prime and ultimate ratios", "limit ratios", and so on. However, Krylov gives the logical structure of Newton's arguments with sufficient accuracy. (*Translator's note*: see footnote 3.)

> not to suppose that quantities of any determinate magnitude are meant, but such as are conceived to be always diminished without end.

However, it is natural to ask oneself why all this is said only in the concluding Scholium to the whole exposition of the theory of limits. The exposition certainly does not begin with a *definition* of the concept of limit, but with a lemma.

LEMMA 1. *Quantities, and the ratios of quantities, which in any finite time converge continually to equality, and before the end of that time approach nearer to each other than by any given difference, become ultimately equal.*

Since the lemma is given a proof, the concept of limit obviously is regarded as already known in advance. One searches in vain, however, for Newton's definition of this concept. He regards it as not at all necessary to provide such a definition, considering the concept of limit to be one of the fundamental primitive concepts, not subject to definition but only to clarification by examples. This is especially clear from the following arguments of Newton (also given in the Scholium to the section of the *Principia* under consideration):

> Perhaps it may be objected, that there is no ultimate proportion of evanescent quantities; because the proportion, before the quantities have vanished, is not the ultimate, and when they are vanished, is none. But by the same argument it may be alleged that a body arriving at a certain place, and there stopping, has no ultimate velocity; because the velocity, before the body comes to the place, is not its ultimate velocity; when it has arrived, there is none. But the answer is easy; for by the ultimate velocity is meant that with which the body is moved, neither before it arrives at its last place and the motion ceases, nor after, but at the very instant it arrives; that is, that velocity with which the body arrives at its last place, and with which the motion ceases. And in like manner, by the ultimate ratio of evanescent quantities is to be understood the ratio of the quantities not before they vanish, nor afterwards, but with which they vanish.

Here Newton, while defending the consistency of the concept of limit, appeals to the obviousness of the fact that a body can have a definite (nonzero) velocity at the time when its motion ceases.

Summarizing, we can say that *for Newton the concept of limit (like that of velocity) is one of the primal concepts which, because of their primitive nature and intuitive clarity, are not subject to direct definition. However, in all his assertions about the properties of limits and the ways of finding them Newton is quite precise and does not in any way stray from our modern ideas.*

Another misunderstanding can be cleared up in passing. Newton is often accused of inconsistency because he frequently speaks about the limit value of a ratio $\lambda = x : y$ as $x \to 0$, $y \to 0$ as the value to which the ratio tends and which it *attains in the end.* It is often thought that in this he arrives at a definition of the limit ratio in the Leibniz–Euler spirit as a ratio of two actual infinitesimals, or two zeros, in a more vulgar expression. This point is actually explained by the fact that for Newton a function is automatically assumed to be defined "by continuity" at the points where it has a definite limit. For example, setting

$$f(x) = \frac{\sin x}{x}, \tag{1}$$

we say today that the function $f(x)$ is *defined* only at $x \neq 0$. But if we observe that

$$\lim_{x \to 0} f(x) = 1, \tag{2}$$

then we can define the new function

$$\varphi(x) = \begin{cases} \frac{\sin x}{x} & \text{for } x \neq 0, \\ 1 & \text{for } x = 0, \end{cases} \tag{3}$$

which is now defined for all real x. Newton's concept can be formulated as follows in this context: since (2) holds, the function (1) is defined not only where it can be computed directly from the formula (1) but also for $x = 0$, because $f(0) = 1$ in view of (2).

Since for Newton functions (fluents)[9] are continuous by their very definition, there is no logical error in this approach to the matter.

For precisely the same reason, it is no crime from the formally logical point of view to regard the concept of limit as a primitive concept to which no formal definition is given. Such a treatment would not satisfy us today, but it does not conflict with our notions about the necessity of logical consistency and rigor, in contrast, for example, to the Leibniz definitions of derivative and integral (as a ratio or a sum of actual infinitesimals), which from a strictly formal point of view can be regarded only as erroneous or meaningless.

Derivatives. For Newton a "fluxion" (that is, a "derivative" in modern terminology) is always a rate of change of a "fluent" (a function). Analysis of corresponding places in Newton's works compels us to think that to him the concept of velocity is so clear that he did not feel a need to define velocity as the limit of the ratio of the increment of a varying quantity to the time increment Δt as $\Delta t \to 0$. In correspondence to this, the relation

$$\dot{x} = \lim_{\Delta t \to 0} \frac{\Delta x}{\Delta t}$$

is for Newton not the definition of the fluent $\dot{x}$, but only a formula enabling us to find an analytic expression for a fluxion from an analytic expression of the fluent.

Differentials. It is worth dwelling at greater length on the different variants in which Newton's thought comes close to the modern notion of differential.

Today, in regarding variables $x, y, z, \ldots$ as functions of a basic independent variable t we define the differentials $dx, dy, dz, \ldots$ as the principal parts of $\Delta x, \Delta y, \Delta z, \ldots$, that is, as functions of the two variables t and Δt that are linear in Δt and have the property that the differences $\Delta x - dx, \Delta y - dy, \Delta z - dz, \ldots$ are infinitesimal in comparison with Δt. By virtue of this definition,

$$dx = \dot{x}\Delta t, \quad dy = \dot{y}\Delta t, \quad dz = \dot{z}\Delta t,$$

where $\dot{x}$, $\dot{y}$, and $\dot{z}$ are the derivatives.

[9] The Newtonian "fluent" is, in our modern language, always a continuous function $f(x)$ on an interval domain (a, b) with the initial point a included when the limit $\lim_{x \to a} f(x)$ exists, and the terminal point b included when the limit $\lim_{x \to b} f(x)$ exists (the possibility that these limits might not exist was well known to Newton), or on one of the half-lines $(a, +\infty)$ or $(-\infty, b)$ (with a or b included when the corresponding limit exists), or on the whole line $(-\infty, \infty)$.

Applications of differentials in analysis fall into two groups:

1) In the first group the increment Δt can be regarded as an arbitrary *constant.* An advantage of using differentials instead of derivatives in matters of this kind is the formal superiority of the convenient and easily grasped algorithm expressed, for example, in the fact that when working with several variables $x, y, z, \dots$ it is not necessary to choose one of them as the main independent variable, and all the computations can be carried out in *homogeneous form* by writing the differential equations

$$\frac{dy}{dx} = f(x, y, z), \qquad \frac{dz}{dx} = g(x, y, z)$$

as

$$\frac{dx}{P(x, y, z)} = \frac{dy}{Q(x, y, z)} = \frac{dz}{R(x, y, z)},$$

and so on; or in the fact that, in place of the formulas

$$D_x f(\varphi(x)) = f'(\varphi(x))\varphi'(x),$$
$$D_x f^{-1}(x) = \frac{1}{f'(f^{-1}(x))}$$

for the derivative of a function and the inverse function, the simple identities

$$\frac{dz}{dx} = \frac{dz}{dy}\frac{dy}{dx}, \qquad \frac{dy}{dx} = \frac{1}{\frac{dx}{dy}}$$

are obtained when differentials are used.

In many nineteenth-century analysis courses (and even sometimes today) differentials were understood in the following restricted sense: the differential of the independent variable t is by definition an arbitrary constant Δt, while the differentials of the dependent variables x, y, z are *by definition*

$$\dot{x}\Delta t, \ \dot{y}\Delta t, \ \dot{z}\Delta t.$$

2) However, there is a second group of applications of the concept of differential that is based essentially on the relations

$$\Delta x = \dot{x}\Delta t + o(\Delta t),$$
$$\Delta y = \dot{y}\Delta t + o(\Delta t),$$
$$\Delta z = \dot{z}\Delta t + o(\Delta t),$$

where $o(\Delta t)$ denotes a quantity that is infinitesimally small in comparison to Δt. Of course, here Δt must now be regarded as a *variable.* This group of applications includes the definition of the integral as a limit of a sum of differentials, together with all applications of differentials involving geometric considerations of "infinitesimal triangles", "infinitely close normals to a curve", "infinitesimal angles" between such normals, and so on. There are significant difficulties in fully mastering the idea that this whole method of arguing geometrically with infinitesimals, if done on the basis of suitable definitions, is totally in accord with complete logical rigor. For example, I recall how in our own century there was great skepticism (cultivated at the time especially by B. K. Mlodzievskiĭ) toward arguments of this kind in the Moscow school of Luzin.

After these introductory remarks we proceed to Newton himself. We find the following:

a) Chiefly in his early works (1665) he introduces the concept of an actual infinitesimal in the naive sense of Leibniz under the name "moment". He uses arguments of this kind in less crucial places also in his later works. However, one should recall what he writes in the *Principia*:

> Therefore if hereafter I should happen to consider quantities as made up of particles, or should use little curved lines for right ones, I would not be understood to mean indivisibles, but evanescent divisible quantities; not the sums and ratios of determinate parts, but always the limits of sums and ratios; and that the force of such demonstrations always depends on the method laid down in the foregoing lemmas.[10]

b) *Methodus fluxionum* develops in rather unusual form a conception in essence completely equivalent to the modern treatment of differentials with *constant* Δt.

It seems to me that in most works on the history of mathematics there is not enough stress on the fact that Newton's differential calculus, in the form it is given in *Methodus fluxionum*, has all the formal algorithmic advantages of Leibniz' calculus of differentials.

The fact of the matter is that in *Methodus fluxionum* Newton always thinks of fluxions as derivatives with respect to some auxiliary variable t which nowhere appears explicitly in the computations. About this "time" Newton says the following:

> But since we do not consider the time here, any farther than as it is expounded and measured by an equable local motion; and besides whereas things only of the same kind can be compared together, and also their velocities of increase and decrease: therefore in what follows I shall have no regard to time formally considered, but shall suppose some one of the quantities proposed, being of the same kind, to be increased by an equable fluxion, to which the rest may be referred, as it were to time; and therefore by way of analogy it may not improperly receive the name of Time. Whenever therefore the word *Time* occurs in what follows (which for the sake of perspicuity and distinction I have sometimes used), by that word I would not have it understood as if I meant time in its formal acceptation, but only that other quantity by the equable increase or fluxion whereof, Time is expounded and measured.[11]

---

[10]This remark is in the *Principia* after eleven lemmas that include Newton's theory of limits.

[11]In connection with these lines from *Methodus fluxionum* we might also note that in the above characterization Krylov is perhaps too hasty in including Newton among the consistent supporters of giving analysis a *foundation* "by starting from mechanical and geometric concepts." It is true that in appealing to the concept of *velocity* Newton *cannot* do without the concept of *kinematics* in providing a basis for analysis; but now we have seen that he was close to the idea that pure analysis does *not* in essence depend on considerations connected with the introduction of *time* (we might say that in the *real* sense of this word (Newton says: "in its formal sense") the difference is purely a matter of terminology). We hardly need add that, up to the present time, the progressive scientific methodology has been that which clearly keeps in view the origin of abstract concepts from generalizations of concrete experience, but which has also been able to isolate them in pure form by getting rid of everything that is inessential for them.

As for the derivatives

$$\frac{dy}{dx}, \frac{dz}{dx}, \frac{dx}{dy}, \ldots$$

of any one of the explicit variables in the problem with respect to another, they are always expressed in *Methodus fluxionum* as ratios of fluxions:

$$\frac{\dot{y}}{\dot{x}}, \frac{\dot{z}}{\dot{x}}, \frac{\dot{x}}{\dot{y}}, \ldots.$$

Therefore, the fluxions in the computations here play the role of our differentials rather than derivatives.

For example, if he wants to find the maximum of the variable x from the relation

$$x^3 - ax^2 + axy - y^3 = 0, \tag{1}$$

Newton gets from (1) that

$$3x^2\dot{x} - 2ax\dot{x} + ay\dot{x} + ax\dot{y} - 3y^2\dot{y} = 0, \tag{2}$$

and, setting $\dot{x} = 0$ in (2), arrives at

$$ax - 3y^2 = 0, \tag{3}$$

which together with (1) makes it possible to compute the maximum value of x and the corresponding value of y. It is clear that (2) corresponds to the modern notation in working with differentials:

$$3x^2dx - 2ax\,dx + ay\,dx + ax\,dy - 3y^2dy. \tag{2'}$$

We confine ourselves to this very simple example in order not to distract the reader from the discussion of fundamental matters.

However, we remark that in *Methodus fluxionum* Newton solves several very complicated problems (especially if one takes into account that he did not have the general rules for differentiating quotients and functions of functions) in whose solution the advantages of his notation corresponding to our operations with differentials become quite pronounced.

The complete equivalence of the methods used by Newton in *Methodus fluxionum* to the earliest modern approach sketched above to the concept of differential is perfectly natural. If Δt is an arbitrary constant, then it can be set equal to unity, and then

$$dx = \dot{x}\Delta t = \dot{x}, \; dy = \dot{y}\Delta t = \dot{y}, \; \ldots.$$

c) A Scholium was appended to *De quadratura curvarum*, apparently written, like the introduction mentioned earlier, considerably later than the main text. This Scholium begins as follows:

> That the fluxions of flowing quantities may be considered as *first, second, third, fourth* fluxions, etc., hath been said above: and these fluxions are as the *terms of infinitely converging series.* Thus, suppose z^n a flowing quantity, and that by flowing it become $(z+o)^n$, then may it be resolved into this converging series
>
> $$(z+o)^n = z^n + noz^{n-1} + \frac{n^2-n}{2}o^2z^{n-2} + \frac{n^3-3n^2+2n}{6}o^3z^{n-3} + \cdots.$$

> In which series the first term z^n is the flowing quantity itself; the second noz^{n-1} shall be the first *increment* or the first difference, to which considered as just *nascent*, the first fluxion is proportional. The third term
> $$\frac{n^2-n}{2}o^2z^{n-2}$$
> will be the *second increment* or *difference*, to which considered as now *nascent*, the *second fluxion* is proportional. The fourth term
> $$\frac{n^3-3n^2+2n}{6}o^3z^{n-3}$$
> shall be the fluents third *increment* or *difference*, and to which as nascent, the third fluxion is proportional, and so on infinitely.[12]

Here Newton, confining himself to the special case[13]

$$f(z)=z^n,$$

considers the function $f(z)$, expanded in a convergent series of powers of the finite increment o of the independent variable z:

$$f(z+o)=C_0+C_1o+C_2o^2+C_3o^3+\cdots$$

and calls the second, third, fourth, and so on, terms of the series the first, second, third, and so on, *increments* (differences) of $f(z)$. Newton's assertion that these "increments" are proportional to the corresponding fluxions is true in the sense that in the equalities

$$\begin{aligned} C_1o &= of'(z),\\ C_2o^2 &= \frac{o^2}{2}f''(z),\\ C_3o^3 &= \frac{o^3}{6}f'''(z),\\ &\vdots \end{aligned}$$

the coefficients

$$o,\ \frac{o^2}{2},\ \frac{o^3}{6},\ \frac{o^4}{24},\ldots$$

[12]Bernoulli (*Commercium epistolicum Leibnitii et Johannis Bernoulli*, vol. 2, p. 294) found an *error* of Newton here. Or rather one should speak of *unfortunate terminology*. Newton's second, third, fourth, and subsequent increments (differences) are equal, respectively, to

$$\frac{d^2z^n}{2},\ \frac{d^3z^n}{6},\ \frac{d^4z^n}{24},\ldots.$$

It would have been better to call the differentials $d^2z^n, d^3z^n, d^4z^n,\ldots$ themselves the second, third, fourth, and successive increments (differences).

[13]It is known that the series Newton was considering converges for any n (even complex, though Newton supposedly had in view only real n) in our modern sense of the word for sufficiently small o when $z \neq o$. Generally speaking, the question of convergence of series is not always completely clear in Newton's works, but his understanding of it, stemming already from *De analysi per aequationes*, tends to coincide with the modern understanding: a series converges if sufficiently many of its terms give the sum to a previously specified accuracy.

do not depend on the form of the function $f(z)$. The increments of Newton are connected, if we set $dz = o$, with the modern differentials of corresponding orders by the equalities

$$\begin{aligned} C_1 o &= df \\ C_2 o^2 &= \frac{1}{2} d^2 f, \\ C_3 o^3 &= \frac{1}{6} d^3 f, \\ &\vdots \end{aligned}$$

We see that here Newton's "moment" of the independent variable becomes (in contrast to his earlier works) a *variable of finite size*, and the "first increment" of the function is defined as the second term in the expansion of $f(z+o)$ in powers of o, that is, as the principal linear part of the full increment

$$f(z+o) - f(z).$$

This is the modern definition of a differential.

As usual, we remark here that this fragment from *De quadratura curvarum* shows that when he wrote the Scholium to this treatise Newton was very close to discovering the Taylor series (not to say simply that he did discover it!). Unfortunately, this was apparently in the period when in mathematics Newton was mainly interested no longer in further progress, but in defending his priority with regard to his earlier achievements.

Bibliography

I. General list of the main publications by A. N. Kolmogorov

1921

1. *Report to the mathematical circle on covering by squares*, Selected Works of A. N. Kolmogorov, Vol. 3: Information Theory and the Theory of Algorithms, "Nauka", Moscow, 1987, pp. 290–294; English transl., Kluwer, Dordrecht, 1993, pp. 261–265.

1922

1. *On operations on sets.* II, Selected Works of A. N. Kolmogorov, Vol. 3: Information Theory and the Theory of Algorithms, "Nauka", Moscow, 1987, pp. 294–303; English transl., Kluwer, Dordrecht, 1993, pp. 266–274.

1923

1. *Une série de Fourier–Lebesgue divergente presque partout*, Fund. Math. **4** (1923), 324–328.
2. *Sur l'ordre de grandeur des coéfficients de la série de Fourier–Lebesgue*, Bull. Acad. Pol. Sci. Ser. A (1923), 83–86.

1924

1. *Une contribution a l'étude de la convergence des séries de Fourier*, Fund. Math. **5** (1924), 96–97.
2. *Sur la convergence des séries de Fourier*, C. R. Acad. Sci. Paris Sér. Math. **178** (1924), 303–306, jointly with G. A. Seliverstov.

1925

1. *La définition axiomatique de l'intégrale*, C. R. Acad. Sci. Paris Sér. Math. **180** (1925), 110–111.
2. *Sur le bornes de la généralisation de l'intégrale*, Selected Works of A. N. Kolmogorov, Vol. 1: Mathematics and Mechanics, "Nauka", Moscow, 1985, pp. 21–38; English transl., Kluwer, Dordrecht, 1991, pp. 15–32.
3. *Sur la possibilité de la définition générale de la dérivée, de l'intégrale et de la sommation des séries divergentes*, C. R. Acad. Sci. Paris Sér. Math. **180** (1925), 362–364.
4. *Sur les fonctions harmoniques conjuguées et les séries de Fourier*, Fund. Math. **7** (1925), 24–29.
5. *On the* tertium non datur *principle*, Mat. Sb. **32** (1925), 646–667. (Russian)
6. *Über Konvergenz von Reihen, deren Glieder durch den Zufall bestimmt werden*, Mat. Sb. **32** (1925), 668–677, jointly with A. Ya. Khinchin.

In compiling this bibliography of Kolmogorov's works we have used various sources: the journals *Uspekhi Matematicheskikh Nauk* [**8**:3 (1953), **18**:5 (1963), **28**:5 (1973), **38**:4 (1983)], *Teoriya Veroyatnosteĭ i ee Primeneniya* [**8**:2 (1963)], and the books *Selected works of A. N. Kolmogorov, Vol.* 1: *Mechanics and Mathematics* ("Nauka", Moscow, 1985), *A. N. Kolmogorov. Mathematics in its historical evolution* ("Nauka", Moscow, 1991), *Mathematics in the USSR over the* 40*-year Period* 1917–1957, Vol. 2, Bibliography (Fizmatgiz, Moscow, 1959). We also used the card catalogues of several libraries, and Kolmogorov's personal archives. Some of Kolmogorov's works not appearing in the aforementioned lists of publications were communicated to us by A. M. Abramov, A. V. Prokhorov, V. M. Tikhomirov, and V. A. Uspenskiĭ. Most of the work involving final checks, inserting additions, and refinements was done by E. S. Kedrova (Steklov Institute of Mathematics, Russian Academy of Sciences). [A. Shiryaev]

1926

1. *Sur la convergence des séries de Fourier*, Atti Accad. Naz. Lincei Rend. **3** (1926), 307–310, jointly with G. A. Seliverstov.
2. *Une série de Fourier–Lebesgue divergente partout*, C. R. Acad. Sci. Paris Sér. Math. **183** (1926), 1327–1329.

1927

1. *Sur la loi des grands nombres*, C. R. Acad. Sci. Paris Sér. Math. **185** (1927), 917–919.
2. *Sur la convergence des séries de fonctions orthogonales*, Math. Z. **26** (1927), 432–441, jointly with D. E. Men′shov.

1928

1. *On operations on sets*, Mat. Sb. **35** (1928), 415–422. (Russian)
2. *Sur une formule limite de M. A. Khintchine*, C. R. Acad. Sci. Paris Sér. Math. **186** (1928), 824–825.
3. *Sur un procédé d'intégration de M. Denjoy*, Fund. Math. **11** (1928), 27–28.
4. *Über die Summen durch den Zufall bestimmter unabhängiger Grössen*, Math. Ann. **99** (1928), 309–319.

1929

1. *Bemerkungen zu meiner Arbeit "Über die Summen durch den Zufall bestimmter unabhängiger Grössen"*, Math. Ann. **102** (1929), 484–488.
2. *General measure theory and the calculus of probabilities*, Trudy Kommunisticheskoĭ Akad. Razdel Mat. **1** (1929), 8–21. (Russian)
3. *Present-day controversies on the nature of mathematics*, Nauch. Slovo **1929**, no. 6, 41–54. (Russian)
4. *Über das Gesetz des iterierten Logarithmus*, Math. Ann. **101** (1929), 126–135.
5. *Sur la loi des grands nombres*, Atti Accad. Naz. Lincei Rend. **9** (1929), 470–474.

1930

1. *Sur la loi forte des grands nombres*, C. R. Acad. Sci. Paris Sér. Math. **191** (1930), 910–912.
2. *Zur topologisch-gruppentheoretischen Begründung der Geometrie*, Nachr. Ges. Wiss. Göttingen. Fachgr. 1 (Mathematik) **8** (1930), 208–210.
3. *Untersuchungen über den Integralbegriff*, Math. Ann. **103** (1930), 654–696.
4. *Sur la notion de la moyenne*, Atti Accad. Naz. Lincei Rend. **12** (1930), 388–391.

1931

1. *Über die analytischen Methoden in der Wahrscheinlichkeitsrechnung*, Math. Ann. **104** (1931), 415–458.
2. *Sur la problème d'attente*, Mat. Sb. **38** (1931), 101–106.
3. *The method of medians in the theory of errors*, Mat. Sb. **38** (1931), no. 3/4, 47–50. (Russian)
4. *Eine Verallgemeinerung des Laplace–Liapounoffschen Satzes*, Izv. Akad. Nauk SSSR OMEN **1931**, no. 7, 959–962.
5. *Über Kompaktheit der Funktionenmengen bei der Konvergenz im Mittel*, Nachr. Ges. Wiss. Göttingen **1** (1931), 60–63.

1932

1. *The theory of functions of a real variable*, Fifteen Years of Mathematics in the Soviet Union, GTTI, Moscow–Leningrad, 1932, pp. 37–48. (Russian)
2. *Sulla forma generale di un processo stocastico omogeneo (Un problēma di Bruno de Finetti)*, Atti Accad. Naz. Lincei Rend. **15** (1932), 805–808.
3. *Ancora sulla forma generale di un processo stocastico omogeneo*, Atti Accad. Naz. Lincei Rend. **15** (1932), 866–869.
4. *Zur Deutung der intuitionistischen Logik*, Math. Z. **35** (1932), 58–65.
5. *Zur Begründung der projektiven Geometrie*, Ann. Math. **33** (1932), 175–176.
6. *Introduction to the theory of functions of a real variable*, GTTI, Moscow–Leningrad, 1932, jointly with P. S. Aleksandrov. (Russian)

1933

1. *Introduction to the theory of functions of a real variable*, 2nd ed., GTTI, Moscow–Leningrad, 1933, jointly with P. S. Aleksandrov. (Russian)

2. *Grundbegriffe der Wahrscheinlichkeitsrechnung*, Springer, Berlin, 1933.
3. *Beiträge zur Maßtheorie*, Math. Ann. **107** (1933), 351–366.
4. *Zur Berechnung der mittleren Brownschen Fläche*, Phys. Z. Sow. **4** (1933), no. 1, 1–13, jointly with M. A. Leontovich.
5. *Sulla determinazione empirica di una legge di distribuzione*, Giorn. Inst. Ital. Attuari **4** (1933), 83–91.
6. *Über die Grenzwertsätze der Wahrscheinlichkeitsrechnung*, Izv. Akad. Nauk SSSR OMEN **1933**, 363–372.
7. *Zur Theorie der stetigen zufälligen Prozesse*, Math. Ann. **108** (1933), 149–160.
8. *Sur la détermination empirique d'une loi de distribution*, Uch. Zapiski Moskov. Gos. Univ. **1** (1933), 9–10.
9. *On the question of suitability of forecast formulas found statistically*, Zh. Geofiz. **3** (1933), 78–82. (Russian)

1934

1. *On points of discontinuity of functions of two variables*, Dokl. Akad. Nauk SSSR **1** (1934), no. 3, 105–106, jointly with I. Ya. Verchenko. (Russian)
2. *Zur Normierbarkeit eines allgemeinen topologischen linearen Raumes*, Studia Math. **5** (1934), 29–33.
3. *Continuation of investigations on points of discontinuity of functions of two variables*, Dokl. Akad. Nauk SSSR **4** (1934), 361–362, jointly with I. Ya. Verchenko. (Russian)
4. *On the convergence of series in orthogonal polynomials*, Dokl. Akad. Nauk SSSR **1** (1934), 291–294. (Russian)
5. *Quelques remarques sur l'approximation des fonctions continues*, Mat. Sb. **41** (1934), 99–103. (Russian)
6. *On some new trends in probability theory*, Abstracts of Reports in Bulletin of the 2nd All-Union Math. Conf. (Leningrad, 24–30 June 1934), Izdat. Akad. Nauk SSSR, Leningrad, 1934, p. 8. (Russian)
7. *Contemporary mathematics*, Front Nauki i Tekhniki, no. 5/6 (1934), 25–28. (Russian)
8. *The Institute of Mathematics and Mechanics of Moscow University*, Front Nauki i Tekhniki, no. 5/6 (1934), 75–78.
9. *The Second All-Union Mathematics Congress*, Sotsialistich. Rekonstruktsiya i Nauka, no. 7 (1934), 142–145. (Russian)
10. *Zufällige Bewegungen (Zur Theorie der Brownschen Bewegung)*, Ann. Math. **35** (1934), 116–117.

1935

1. *On some contemporary trends in probability theory*, Proc. 2nd All-Union Math. Conf. (Leningrad, 24–30 June 1934), vol. 1, Izdat. Akad. Nauk SSSR, Moscow–Leningrad, 1935, pp. 349–358. (Russian)
2. *Deviations from Hardy's formulas for partial isolation*, Dokl. Akad. Nauk SSSR **3** (1935), 129–132. (Russian)
3. *La transformation de Laplace dans les espaces linéaires*, C. R. Acad. Sci. Paris Sér. Math. **200** (1935), 1717–1718.
4. *Zur Grössenordnung des restgliedes Fourierschen Reihen differenzierbarer Funktionen*, Ann. Math. **36** (1935), 521–526.

1936

1. *Über die beste Annäherung von Funktionen einer gegebenen Funktionenklasse*, Ann. Math. **37** (1936), 107–110.
2. *Foundations of the theory of probability*, ONTI, Moscow–Leningrad, 1936; Russian transl. of the 1933 German original [2].
3. *Equation*, Great Sov. Encycl., Vol. 56, 1936, pp. 163–165. (Russian)
4. *Contemporary mathematics*, Collection of Articles on the Philosophy of Mathematics, ONTI, Moscow, 1936, pp. 7–13. (Russian)
5. *Theory and practice in mathematics*, Front Nauki i Tekhniki, no. 5 (1936), 39–42. (Russian)
6. *Über die Dualität im Aufbau der kombinatorische Topologie*, Mat. Sb. **1** (1936), 97–102.
7. *Anfangsgründe der Theorie der Markoffschen Ketten mit unendlich vielen möglichen Zuständen*, Mat. Sb. **1** (1936), 607–610.

8. *Homologierung des Komplexes und des lokalbikompakten Räumes*, Mat. Sb. **1** (1936), 701–706.
9. *Zur Theorie der Markoffschen Ketten*, Math. Ann. **112** (1936), 155–160.
10. *On Plesner's condition for the law of large numbers*, Mat. Sb. **1** (1936), 847–849. (Russian)
11. *Endliche Überdeckungen topologischer Räume*, Fund. Math. **26** (1936), 267–271, jointly with P. S. Aleksandrov.
12. *Les groupes de Betti des espaces localement bicompacts*, C. R. Acad. Sci. Paris Sér. Math. **202** (1936), 1144–1147.
13. *Propriétés des groupes de Betti des espaces localement bicompacts*, C. R. Acad. Sci. Paris Sér. Math. **202** (1936), 1325–1327.
14. *Les groupes de Betti des espaces métriques*, C. R. Acad. Sci. Paris Sér. Math. **202** (1936), 1558–1560.
15. *Cycles rélatifs. Théorème de dualité de M. Alexandre*, C. R. Acad. Sci. Paris Sér. Math. **202** (1936), 1641–1643.
16. *Foreword*, Mathematische Grundlagenforschung. Intuitionismus. Beweistheorie, by A. Heyting (translated from German to Russian), ONTI, Moscow–Leningrad, 1936, pp. 3–4.
17. *Sulla teoria di Volterra della lotta per l'esistenza*, Giorn. Inst. Ital. Attuar. **7** (1936), 74–80.

1937

1. *Über offene Abbildungen*, Ann. Math. **38** (1937), 36–38.
2. *Skew-symmetric quantities and topological invariants*, Proc. Seminar on Vector and Tensor Analysis with Applications to Geometry, Mechanics, and Physics, Vol. 4, GONTI, Moscow–Leningrad, 1937, pp. 345–347. (Russian)
3. *Markov chains with a denumerable number of possible states*, Byull. Moskov. Univ. Mat. i Mekh. **1** (1937), no. 3, 1–16. (Russian)
4. *The investigation of a diffusion equation connected with an increasing amount of substance and its application to a biological problem*, Byull. Moskov. Univ. Mat. i Mekh. **1** (1937), no. 6, 1–26, jointly with I. G. Petrovskiĭ and N. S. Piskunov. (Russian)
5. *On the statistical theory of crystallization of metals*, Izv. Akad. Nauk SSSR Ser. Mat. **3** (1937), 353–359. (Russian)
6. *Ein vereinfachter Beweis der Birkhoff–Khintchineschen Ergodensatzes*, Mat. Sb. **2** (1937), 367–368.
7. *Zur Umkehrbarkeit der statistischen Naturgesetze*, Math. Ann. **113** (1937), 766–772.
8. *Foreword* and *Appendix* (editor), Mengenlehre, by F. Hausdorff (translated from German to Russian), GONTI, Moscow–Leningrad, 1937, pp. 3–4, 266–290, jointly with P. S. Aleksandrov.
9. *Continuum*, Great Sov. Encycl., Vol. 34, 1937, pp. 139–140. (Russian)

1938

1. *Introduction to the theory of functions of a real variable*, 3rd rev. ed., GONTI, Moscow–Leningrad, 1938, jointly with P. S. Aleksandrov.
2. *Andreĭ Andreevich Markov*, Great Sov. Encycl., Vol. 38, 1938, pp. 152–153. (Russian)
3. *Mathematics*, Great Sov. Encycl., Vol. 38, 1938, pp. 359–402. (Russian)
4. *Mathematical induction*, Great Sov. Encycl., Vol. 38, 1938, pp. 405–406. (Russian)
5. *Measure*, Great Sov. Encycl., Vol. 38, 1938, pp. 831–832. (Russian)
6. *Multidimensional space*, Great Sov. Encycl., Vol. 39, 1938, pp. 577–578. (Russian)
7. *Probability theory and its applications*, Mathematics and the Natural Sciences in the USSR, GONTI, Moscow–Leningrad, 1938, pp. 51–61. (Russian)
8. *On the information section in the 1st volume of "Uspekhi Matematicheskikh Nauk"*, Uspekhi Mat. Nauk **4** (1938), 326–327. (Russian)
9. *A remark on the fondations of geometry* (*On the question of the new translation of "Foundations of Geometry" by D. Hilbert*), Uspekhi Mat. Nauk **4** (1938), 347–348. (Russian)
10. *From the editorial board* (*a series of articles on the theory of stochastic processes*), Uspekhi Mat. Nauk **5** (1938), 3–4. (Russian)
11. *On analytic methods in probability theory*, Uspekhi Mat. Nauk **5** (1938), 5–41, transl. of the 1931 paper [1] into Russian.
12. *A simplified proof of the Birkhoff–Khinchin ergodic theorem*, Uspekhi Mat. Nauk **5** (1938), 52–56, transl. of the 1937 paper [6] into Russian.
13. *Some problems in the theory of functions of a real variable*, Uspekhi Mat. Nauk **5** (1938), 232–234, jointly with G. M. Fikhtengol'ts and I. M. Gel'fand. (Russian)

14. *On the solution of a biological problem*, Izv. Nauchn. Issled. Inst. Mat. i Mekh. Tomsk. Univ. **2** (1938), no. 1, 7–12. (Russian)
15. *Une géréralisation de l'inégalité de M. J. Hadamard entre les bornes supérieures des dérivées successives d'une fonction*, C. R. Acad. Sci. Paris Sér. Math. **207** (1938), 764–765.
16. *Foreword* (editor), Sur le mesure des grandeurs, by H. Lebesgue (translated from French to Russian), Uchpedgiz, Moscow, 1938, pp. 7–16.

1939

1. *Algebra, Part* 1, Uchpedgiz, Moscow, 1939, jointly with P. S. Aleksandrov. (Russian)
2. *Orientation*, Great Sov. Encycl., Vol. 43, 1939, pp. 342–344. (Russian)
3. *On inequalities between upper bounds of the successive derivatives of an arbitrary function on an infinite interval*, Uchen. Zap. Moskov. Univ. Mat. **30** (1939), 3–16. (Russian)
4. *On rings of continuous functions on topological spaces*, Dokl. Akad. Nauk SSSR **22** (1939), 11–15, jointly with I. M. Gel′fand. (Russian)
5. *Sur l'interpolation et extrapolation des suites stationnaires*, C. R. Acad. Sci. Paris Sér. Math. **208** (1939), 2043–2045.

1940

1. *Surface*, Great Sov. Encycl., Vol. 45, 1940, pp. 746–748. (Russian)
2. *Curves in Hilbert space invariant under a one-parameter group of motions*, Dokl. Akad. Nauk SSSR **26** (1940), 6–9. (Russian)
3. *Wiener's spiral and some other interesting curves in Hilbert space*, Dokl. Akad. Nauk SSSR **26** (1940), 115–118. (Russian)
4. *On the sixtieth birthday of Sergeĭ Natanovich Bernshteĭn*, Izv. Akad. Nauk SSSR Ser. Mat. **4** (1940), 249–260, jointly with V. L. Goncharov. (Russian)
5. *Valeriĭ Ivanovich Glivenko* (1897–1940) [obituary], Uspekhi Mat. Nauk **8** (1940), 379–383. (Russian)
6. *Review of book: Mathematical Statistics, by V. I. Romanovskiĭ*, Uspekhi Mat. Nauk **7** (1940), 327–329. (Russian)
7. *On a new confirmation of Mendel's laws*, Dokl. Akad. Nauk SSSR **27** (1940), 38–42. (Russian)

1941

1. *Stationary sequences in Hilbert space*, Byull. Moskov. Univ. Mat. **2** (1941), no. 6, 1–40. (Russian)
2. *Interpolation and extrapolation of stationary random sequences*, Izv. Akad. Nauk SSSR **5** (1941), 3–14. (Russian)
3. *Points of local topological character of countably multiple open mappings of compact spaces*, Dokl. Akad. Nauk SSSR **30** (1941), 477–479. (Russian)
4. *Local structure of turbulence in an incompressible viscous fluid for very large Reynolds numbers*, Dokl. Akad. Nauk SSSR **30** (1941), 299–303; English transl., Proc. Roy. Soc. London Ser. A **434** (1991), no. 1890, 9–13.
5. *On the lognormal distribution law of the particle size under grinding*, Dokl. Akad. Nauk SSSR **31** (1941), 99–101. (Russian)
6. *On degeneration of isotropic turbulence in an incompressible viscous fluid*, Dokl. Akad. Nauk SSSR **31** (1941), 538–541. (Russian)
7. *Dissipation of energy in locally isotropic turbulence*, Dokl. Akad. Nauk SSSR **32** (1941), 19–21. (Russian)
8. *Properties of inequalities and the concept of approximate computations*, Mat. v Shkole **1941**, no. 2, 1–12, jointly with P. S. Aleksandrov. (Russian)
9. *Irrational numbers*, Mat. v Shkole **1941**, no. 3, 1–15, jointly with P. S. Aleksandrov. (Russian)
10. *Confidence limits for an unknown distribution function*, Ann. Math. Statist. **12** (1941), 461–463.
11. *Introduction to the theory of functions of a real variable*, "Radyans′ka Shkola", Kiev, 1941, jointly with P. S. Aleksandrov. (Ukrainian)

1942

1. *Determination of the center of scattering and measure of accuracy from a limited number of observations*, Izv. Akad. Nauk SSSR Ser. Mat. **6** (1942), 3–32. (Russian)
2. *Equations of turbulent motion of an incompressible fluid*, Izv. Akad. Nauk SSSR Ser. Fiz. **6** (1942), 56–58. (Russian)

1943

1. *Nikolaĭ Ivanovich Lobachevskiĭ*, Gostekhizdat, Moscow–Leningrad, 1943, jointly with P. S. Aleksandrov. (Russian)

1945

1. *The number of hits in several shots and general principles for estimating the effectiveness of gunnery systems*, Trudy Mat. Inst. Steklov. **12** (1954), 7–25. (Russian)
2. *Artificial scattering in the case of hitting with a single shot and scattering in a single measurement*, Trudy Mat. Inst. Steklov. **12** (1945), 26–45. (Russian)

1946

1. *On the law of resistance in turbulent flow in smooth tubes*, Dokl. Akad. Nauk SSSR **52** (1946), 669–671. (Russian)
2. *Justification of the method of least squares*, Uspekhi Mat. Nauk **1** (1946), no. 1, 57–70. (Russian)
3. *Justification of the theory of real numbers*, Uspekhi Mat. Nauk **1** (1946), no. 1, 217–219. (Russian)
4. *Newton and contemporary mathematical thought*, Moscow University, in Memory of Isaac Newton 1643–1943, Izdat. Moskov. Gos. Univ., Moscow, 1946, pp. 27–43; English transl. in this volume.
5. *On the kinematics of a fluid of variable turbidity* (Discussion of the article "Transport of deposits suspended in a turbulent stream", by M. A. Velikanov, Corresponding Member of the USSR Academy of Sciences), Izv. Otd. Tekh. Nauk **5** (1946), 781–784. (Russian)

1947

1. *The development of mathematics in the Soviet Union*, Great Sov. Encycl., "USSR" volume, 1947, pp. 1318–1323. (Russian)
2. *Average values*, Great Sov. Encycl., Vol. 52, 1947, pp. 508–509. (Russian)
3. *The formula of Gauss in the theory of least squares*, Izv. Akad. Nauk SSSR Ser. Mat. **11** (1947), 561–566, jointly with A. A. Petrov and Yu. M. Smirnov. (Russian)
4. *Branching stochastic processes*, Dokl. Akad. Nauk SSSR **56** (1947), 7–10, jointly with N. A. Dmitriev. (Russian)
5. *Computation of the final probabilities for branching stochastic processes*, Dokl. Akad. Nauk SSSR **56** (1947), 783–786, jointly with B. A. Sevast′yanov. (Russian)
6. *The statistical theory of oscillations with a continuous spectrum*, Collection dedicated to the Thirtieth Anniversary of the Great October Socialist Revolution, Part 1, Izdat. Akad. Nauk SSSR, Moscow, 1947, pp. 242–252. (Russian)
7. *The role of Russian science in the development of probability theory*, The Role of Russian Science in the Development of World Science and Culture, Vol. 1, Book 1, Izdat. Moskov. Gos. Univ., Moscow, 1947, pp. 53–64. (Russian)
8. *Foreword* (editor), Random Variables and Probability Distributions, by Harald Cramér (translated from English to Russian), Inostran. Lit., Moscow, 1947, p. 8.

1948

1. *The statistical theory of oscillations with a continuous spectrum*, Joint Meeting of the USSR Academy of Sciences Dedicated to the Thirtieth Anniversary of the Great October Socialist Revolution, Izdat. Akad. Nauk SSSR, Moscow–Leningrad, 1948, pp. 465–472. (Russian)
2. *Probability theory*, Thirty Years of Mathematics in the Soviet Union, 1917–1947, Gostekhizdat, Moscow–Leningrad, 1948, pp. 701–727, jointly with B. V. Gnedenko. (Russian)
3. *On two theorems concerning probabilities*: *Comments*, Complete Collected Works of P. L. Chebyshev, Vol. 3: Mathematical Analysis, Gostekhizdat, Moscow–Leningrad, 1948, pp. 404–409. (Russian)
4. *The structure of complete metric Boolean algebras*, Uspekhi Mat. Nauk **3** (1948), no. 1, 212. (Russian)
5. *A remark on Chebyshev polynomials deviating least from a given function*, Uspekhi Mat. Nauk **3** (1948), no. 1, 216–221. (Russian)
6. *Evgeniĭ Evgen′evich Slutskiĭ*, Uspekhi Mat. Nauk **3** (1948), no. 4, 143–151. (Russian)
7. *Foreword* (editor), Mathematical Methods of Statistics, by H. Cramér (translated from English to Russian), Inostran. Lit., Moscow, 1948, pp. 5–8.

8. *Algèbras de Boole métriques complètes*, Sixth Polish Congress of Mathematicians, Warsaw, 1948, pp. 22–30.
9. *Foreword*, Introduction to Set Theory and the Theory of Functions, Part I, by P. S. Aleksandrov, Gostekhizdat, Moscow–Leningrad, 1948, pp. 8–12, jointly with P. S. Aleksandrov. (Russian)

1949

1. *Limit distributions for sums of independent random variables*, Gostekhizdat, Moscow–Leningrad, 1949, jointly with B. V. Gnedenko.
2. *On "geometric screening" of crystals*, Dokl. Akad. Nauk SSSR **65** (1949), 681–684. (Russian)
3. *Solution of a probability problem related to the mechanism of stratification*, Dokl. Akad. Nauk SSSR **65** (1949), 793–796. (Russian)
4. *On sums of a random number of terms*, Uspekhi Mat. Nauk **4** (1949), no. 4, 168–172, jointly with Yu. V. Prokhorov. (Russian)
5. *A local limit theorem for classical Markov chains*, Izv. Akad. Nauk SSSR Ser. Mat. **13** (1949), 281–300. (Russian)
6. *Fundamental problems of theoretical statistics*, Proc. 2nd All-Union Conference on Mathematical Statistics (Tashkent, Sept. 27–Oct. 2, 1948), Uzbekgosizdat, Tashkent, 1949, pp. 216–220. (Russian)
7. *The real meaning of the results of analysis of variance*, Proc. 2nd All-Union Conference on Mathematical Statistics (Tashkent, Sept. 27–Oct. 2, 1948), Uzbekhizdat, Tashkent, 1949, pp. 240–268. (Russian)
8. *On the breakdown of drops in a turbulent flow*, Dokl. Akad. Nauk SSSR **66** (1949), 825–828. (Russian)
9. *Absolute value*, Great Sov. Encycl., 2nd ed., Vol. 1, 1949, p. 32. (Russian)
10. *Absolute geometry*, Great Sov. Encycl., 2nd ed., Vol. 1, 1949, p. 33.
11. *Hadamard, Jacques*, Great Sov. Encycl., 2nd ed., Vol. 1, 1949, p. 388. (Russian)
12. *Additive quantities*, Great Sov. Encycl., 2nd ed., Vol. 1, 1949, p. 394. (Russian)
13. *Axiom*, Great Sov. Encycl., 2nd ed., Vol. 1, 1949, pp. 613–616. (Russian)
14. *Perspective geometry*, Great Sov. Encycl., 2nd ed., Vol. 1, 1949, p. 617. (Russian)
15. *On the inequalities between upper bounds of the successive derivatives of an arbitrary function on an infinite interval*, Amer. Math. Soc. Transl. (1949), no. 4.
16. *Research session of the Probability Theory Department of the Steklov Institute of Mathematics and of the Probability Theory Department of Moscow State University*, Uspekhi Mat. Nauk **4** (1949), no. 4, 189–190. (Russian)

1950

1. *Unbiased estimates*, Izv. Akad. Nauk SSSR Ser. Mat. **14** (1950), 303–326.
2. *On determining the thermal conductivity of soil*, Izv. Akad. Nauk SSSR Ser. Geogr. Geofiz. **14** (1950), 97–98. (Russian)
3. *Algebra in secondary school*, Great Sov. Encycl., 2nd ed., Vol. 2, 1950, pp. 61–62. (Russian)
4. *Algebraic expression*, Great Sov. Encycl., 2nd ed., Vol. 2, 1950, p. 64. (Russian)
5. *Algorithm*, Great Sov. Encycl., 2nd ed., Vol. 2, 1950, p. 65. (Russian)
6. *Euclidean algorithm*, Great Sov. Encycl., Vol. 2, 1950, pp. 65–67. (Russian)
7. *Aleksandrov, Aleksandr Danilovich*, Great Sov. Encycl., 2nd ed., Vol. 2, 1950, p. 83. (Russian)
8. *Aleksandrov, Pavel Sergeevich*, Great Sov. Encycl., 2nd ed., Vol. 2, 1950, p. 84. (Russian)
9. *Mathematical analysis*, Great Sov. Encycl., 2nd ed., Vol. 2, 1950, pp. 325–326. (Russian)
10. *Asymptote*, Great Sov. Encycl., 2nd ed., Vol. 3, 1950, pp. 238–239. (Russian)
11. *Asymptotic expressions*, Great Sov. Encycl., 2nd ed., Vol. 3, 1950, p. 239. (Russian)
12. *Akhiezer, Naum Il'ich*, Great Sov. Encycl., 2nd ed., Vol. 3, 1950, p. 565. (Russian)
13. *Banach, Stefan*, Great Sov. Encycl., 2nd ed., Vol. 4, 1950, p. 183. (Russian)
14. *Bari, Nina Karlovna*, Great Sov. Encycl., 2nd ed., Vol. 4, 1950, p. 245. (Russian)
15. *Bernshteĭn, Sergeĭ Natanovich*, Great Sov. Encycl., 2nd ed., Vol. 5, 1950, p. 52. (Russian)
16. *Infinitely large values*, Great Sov. Encycl., 2nd ed., Vol. 5, 1950, pp. 66–67. (Russian)
17. *Infinitesimals*, Great Sov. Encycl., 2nd ed., Vol. 5, 1950, pp. 67–71, jointly with V. F. Kagan. (Russian)
18. *Infinitely remote elements*, Great Sov. Encycl., 2nd ed., Vol. 5, 1950, pp. 71–72, jointly with B. N. Delone. (Russian)
19. *Infinity (in mathematics)*, Great Sov. Encycl., 2nd ed., Vol. 5, 1950, pp. 73–74. (Russian)

20. *Biharmonic functions*, Great Sov. Encycl., 2nd ed., Vol. 5, 1950, p. 159. (Russian)
21. *Bilinear form*, Great Sov. Encycl., 2nd ed., Vol. 5, 1950, p. 167. (Russian)
22. *The law of large numbers*, Great Sov. Encycl., 2nd ed., Vol. 5, 1950, pp. 538–540. (Russian)
23. *A theorem on convergence of conditional mathematical expectations and some of its applications*, Report of 2 Sept. 1950, Proc. First Congress of Hungarian Mathematicians, Akad. Kiadó, Budapest, 1950, pp. 367–376. (Russian)
24. *Exposition of the fundamentals of Lebesgue measure*, Uspekhi Mat. Nauk **5** (1950), no. 1, 211–213. (Russian)
25. *Lazar′ Aronovich Lyusternik (On his fiftieth birthday)*, Uspekhi Mat. Nauk **5** (1950), no. 1, 234–235. (Russian)
26. *On the distribution of a target*, no. 12, Izdat. Voennogo Ministerstva SSSR, Moscow, 1950, pp. 12–19. (Russian)

1951

1. *On the differentiability of transition functions in time-homogeneous Markov processes with a denumerable number of states*, Uchen. Zap. Moskov. Univ. **148** (1951), no. 4, 53–59. (Russian)
2. *Generalization of Poisson's formula to the case of sampling from a finite population*, Uspekhi Mat. Nauk **6** (1951), no. 3, 133–134. (Russian)
3. *Ivan Georgievich Petrovskiĭ*, Uspekhi Mat. Nauk **6** (1951), no. 3, 160–164, jointly with P. S. Aleksandrov. (Russian)
4. *Statistical acceptance sampling when the admissible number of defective items is zero*, All-Union Society for the Propagation of Political and Scientific Knowledge, Leningrad, House of Scientific and Technological Propaganda, Leningrad, 1951, pp. 1–24. (Russian)
5. *Brouwer, Luitzen E. J.*, Great Sov. Encycl., 2nd ed., Vol. 6, 1951, pp. 62–63, jointly with S. A. Yanovskaya. (Russian)
6. *Order statistics (variational series)*, Great Sov. Encycl., 2nd ed., Vol. 6, 1951, p. 641. (Russian)
7. *Weyl, Hermann*, Great Sov. Encycl., 2nd ed., Vol. 7, 1951, pp. 106–107, jointly with S. A. Yanovskaya. (Russian)
8. *Quantity*, Great Sov. Encycl., 2nd ed., Vol. 7, 1951, pp. 340–341. (Russian)
9. *Probable deviation*, Great Sov. Encycl., 2nd ed., Vol. 7, 1951, p. 507. (Russian)
10. *Probability*, Great Sov. Encycl., 2nd ed., Vol. 7, 1951, pp. 508–510. (Russian)
11. *Sampling procedure*, Great Sov. Encycl., 2nd ed., Vol. 9, 1951, pp. 417–418, jointly with T. I. Kozlov. (Russian)
12. *Introductory article* and *comments*, Complete Collected Works of N. I. Lobachevskiĭ, Vol. 5, Gostekhizdat, Moscow–Leningrad, 1951, pp. 329–332, 342–348, jointly with A. N. Khovanskiĭ. (Russian)
13. *Solution of a probability problem related to the question of formation of strata*, Amer. Math. Soc. Transl. (1) **53** (1951).
14. *The work of N. V. Smirnov on properties of an order statistic and nonparametric problems of mathematical statistics*, Uspekhi Mat. Nauk **6** (1951), no. 4, 190–192, jointly with A. Ya. Khinchin. (Russian)
15. *The work of I. M. Gel′fand on algebraic questions in functional analysis*, Uspekhi Mat. Nauk **6** (1951), no. 4, 184–186. (Russian)

1952

1. *On the velocity profile drag in turbulent flow in tubes*, Dokl. Akad. Nauk SSSR **84** (1952), 29–30. (Russian)
2. *Gaussian distribution*, Great Sov. Encycl., 2nd ed., Vol. 10, 1952, p. 275. (Russian)
3. *Geodesic curvature*, Great Sov. Encycl., 2nd ed., Vol. 10, 1952, p. 481. (Russian)
4. *Hilbert, David*, Great Sov. Encycl., 2nd ed., Vol. 11, 1952, pp. 370–371. (Russian)
5. *Histogram*, Great Sov. Encycl., 2nd ed., Vol. 11, 1952, p. 447. (Russian)
6. *Gnedenko, Boris Vladimirovich*, Great Sov. Encycl., 2nd ed., Vol. 11, 1952, p. 545. (Russian)
7. *Homeomorphism*, Great Sov. Encycl., 2nd ed., Vol. 12, 1952, p. 21. (Russian)
8. *Homotopy*, Great Sov. Encycl., 2nd ed., Vol. 12, 1952, p. 35. (Russian)
9. *Motion (in geometry)*, Great Sov. Encycl., 2nd ed., Vol. 13, 1952, pp. 447–448. (Russian)
10. *Binomial*, Great Sov. Encycl., 2nd ed., Vol. 13, 1952, p. 518. (Russian)
11. *Real numbers*, Great Sov. Encycl., 2nd ed., Vol. 13, 1952, pp. 570–571. (Russian)
12. *Division*, Great Sov. Encycl., 2nd ed., Vol. 13, 1952, p. 628. (Russian)

13. *Discreteness*, Great Sov. Encycl., 2nd ed., Vol. 14, 1952, p. 425. (Russian)
14. *Variance*, Great Sov. Encycl., 2nd ed., Vol. 14, 1952, pp. 438–439. (Russian)
15. *Distributivity*, Great Sov. Encycl., 2nd ed., Vol. 14, 1952, p. 479. (Russian)
16. *Distributive operator*, Great Sov. Encycl., 2nd ed., Vol. 14, 1952, p. 479. (Russian)
17. *Differential*, Great Sov. Encycl., 2nd ed., Vol. 14, 1952, p. 497. (Russian)
18. *Differential equations*, Great Sov. Encycl., 2nd ed., Vol. 14, 1952, pp. 520–526, jointly with B. P. Demidovich and V. V. Nemytskiĭ. (Russian)
19. *Confidence probability*, Great Sov. Encycl., 2nd ed., Vol. 14, 1952, pp. 616–617. (Russian)
20. *Confidence limits*, Great Sov. Encycl., 2nd ed., Vol. 14, 1952, p. 617. (Russian)
21. *Mathematical signs*, Great Sov. Encycl., 2nd ed., Vol. 17, 1952, pp. 115–119, jointly with I. G. Bashmakova and A. P. Yushkevich. (Russian)
22. *Significant digits*, Great Sov. Encycl., 2nd ed., Vol. 17, 1952, p. 135. (Russian)
23. *Isomorphism*, Great Sov. Encycl., 2nd ed., Vol. 17, 1952, pp. 478–479, jointly with V. I. Bityutskov. (Russian)
24. *Isotropic lines*, Great Sov. Encycl., 2nd ed., Vol. 17, 1952, p. 509. (Russian)
25. *Concrete number*, Great Sov. Encycl., 2nd ed., Vol. 17, 1952, p. 557. (Russian)
26. *Imshenetskiĭ, Vasiliĭ Grigor′evich*, Great Sov. Encycl., 2nd ed., Vol. 17, 1952, p. 607. (Russian)
27. *Mathematics as a profession: A guide to matriculants of institutions of higher learning*, "Sov. Nauka", Moscow, 1952. (Russian)
28. *Geodesic coordinates*, Great Sov. Encycl., 2nd ed., Vol. 10, 1952, p. 486. (Russian)
29. *Graph*, Great Sov. Encycl., 2nd ed., Vol. 12, 1952, pp. 453–454. (Russian)
30. *Address at a conference on algebra and number theory*, Uspekhi Mat. Nauk **7** (1952), no. 3, 168–170. (Russian)
31. *Foreword* (editor), Introduction to probability theory and its applications, by W. Feller (translated from English to Russian), Inostran. Lit., Moscow, 1952, pp. 3–6.
32. *Measure theory and the Lebesgue integral*, mimeographed lecture notes, 1952. (Russian)
33. *Matematika, középiskolai oktatókáderek és ált. továbbképzésben résztvevőközépiskolái nevelők tananyaga*. I (*Pedagógus továbbképzés*), collotyped edition, 1952. (Hungarian)
34. *Foreword*, Bevetés a halmazelméletbe és a függvénytanba. I, Hungarian transl. of 1948, [9], Budapest, 1952, jointly with P. S. Aleksandrov.

1953

1. *On the concept of an algorithm*, Uspekhi Mat. Nauk **8** (1953), no. 4, 175–176. (Russian)
2. *Some recent work in the area of limit theorems in probability theory*, Vestnik Moskov. Univ. Ser. I Mat. Mekh. **10** (1953), 29–38. (Russian)
3. *On dynamical systems with an integral invariant on a torus*, Dokl. Akad. Nauk SSSR **93** (1953), 763–766. (Russian)
4. *Mathematical induction*, Great Sov. Encycl., 2nd ed., Vol. 18, 1953, pp. 146–147. (Russian)
5. *Integral*, Great Sov. Encycl., 2nd ed., Vol. 18, 1953, pp. 250–253, jointly with V. I. Glivenko. (Russian)
6. *Probability integral*, Great Sov. Encycl., 2nd ed., Vol. 18, 1953, p. 253. (Russian)
7. *Interpolation*, Great Sov. Encycl., 2nd ed., Vol. 18, 1953, pp. 304–305. (Russian)
8. *Intuitionism*, Great Sov. Encycl., 2nd ed., Vol. 18, 1953, p. 319. (Russian)
9. *Elimination of unknowns*, Great Sov. Encycl., 2nd ed., Vol. 18, 1953, p. 483. (Russian)
10. *Trial*, Great Sov. Encycl., 2nd ed., Vol. 18, 1953, p. 604. (Russian)
11. *Method of exhaustion*, Great Sov. Encycl., 2nd ed., Vol. 19, 1953, pp. 50–51. (Russian)
12. *Quadrant*, Great Sov. Encycl., 2nd ed., Vol. 20, 1953, p. 434. (Russian)
13. *Compact space*, Great Sov. Encycl., 2nd ed., Vol. 22, 1953, p. 282. (Russian)
14. *Constant*, Great Sov. Encycl., 2nd ed., Vol. 22, 1953, p. 416. (Russian)
15. *Continuum*, Great Sov. Encycl., 2nd ed., Vol. 22, 1953, pp. 454–455. (Russian)
16. *Coordinates*, Great Sov. Encycl., 2nd ed., Vol. 22, 1953, pp. 524–525. (Russian)
17. *Correlation*, Great Sov. Encycl., 2nd ed., Vol. 23, 1953, pp. 55–58. (Russian)
18. *Unbiased estimates*, Amer. Math. Soc. Transl. (1953), no. 98 (English transl. of 1950, [1]).

1954

1. *On preservation of conditionally periodic motions under a small change in the Hamiltonian functions*, Dokl. Akad. Nauk SSSR **98** (1954), 527–530. (Russian)

2. *The general theory of dynamical systems and classical mechanics*, Proc. Internat. Congress Math., Vol. 1, Amsterdam, 1954, pp. 315–333. (Russian)
3. *Curve*, Great Sov. Encycl., 2nd ed., Vol. 25, 1954, pp. 167–170. (Russian)
4. *Law of small numbers*, Great Sov. Encycl., 2nd ed., Vol. 26, 1954, p. 169. (Russian)
5. *Markov, Andreĭ Andreevich*, Great Sov. Encycl., 2nd ed., Vol. 26, 1954, pp. 294–295. (Russian)
6. *Mathematics*, Great Sov. Encycl., 2nd ed., Vol. 26, 1954, p. 464–483. (Russian)
7. *Mathematical statistics*, Great Sov. Encycl., 2nd ed., Vol. 26, 1954, pp. 485–490. (Russian)
8. *Mathematical physics*, Great Sov. Encycl., 2nd ed., Vol. 26, 1954, p. 490. (Russian)
9. *Mises, Richard von*, Great Sov. Encycl., 2nd ed., Vol. 27, 1954, p. 414. (Russian)
10. *Multidimensional space*, Great Sov. Encycl., 2nd ed., Vol. 27, 1954, pp. 660–661. (Russian)
11. *Set theory*, Great Sov. Encycl., 2nd ed., Vol. 28, 1954, pp. 14–17, jointly with P. S. Aleksandrov. (Russian)
12. *Elements of the theory of functions and functional analysis*: *A lecture course*, Vol. 1. *Metric and normed spaces*, Izdat. Moskov. Gos. Univ., Moscow, 1954, jointly with S. V. Fomin.
13. *Foreword* (editor), Rekursive Funktionen, by R. Péter (translated from German to Russian), Inostran. Lit., Moscow, 1954, pp. 3–10.
14. *A new variant of M. A. Velikanov's gravitational theory of motion of suspended sediment*, Vestnik Moskov. Gos. Univ. **1954**, no. 3, 41–45. (Russian)
15. *Limit distributions for sums of independent random variables*, English transl. of 1949, [1], Addison–Wesley, Reading, MA, 1954, jointly with B. V. Gnedenko.
16. *Grösse. Axiom*, Teubner, Leipzig, 1954, German transl. of 1951, [8] and 1949, [13].

1955

1. *Bounds for the smallest number of elements of ε-nets in various classes of functions and their application to the question of representing functions of several variables as superpositions of functions of a fewer number of variables*, Uspekhi Mat. Nauk **10** (1955), no. 1, 192–194. (Russian)
2. *Orientation*, Great Sov. Encycl., 2nd ed., Vol. 31, 1955, pp. 188–189. (Russian)
3. *Foundations of geometry*, Great Sov. Encycl., 2nd ed., Vol. 31, 1955, pp. 296–297. (Russian)
4. *Surface*, Great Sov. Encycl., 2nd ed., Vol. 33, 1955, pp. 346–347, jointly with L. A. Skornyakov. (Russian)
5. *Ordinal numbers*, Great Sov. Encycl., 2nd ed., Vol. 34, 1955, p. 227. (Russian)
6. *Statistical acceptance sampling*, Great Sov. Encycl., 2nd ed., Vol. 34, 1955, pp. 498–499. (Russian)
7. *O matematyce*, PAN, Warsaw, 1955. (Polish)

1956

1. *On Skorokhod convergence*, Teor. Veroyatnost. i Primenen. **1** (1956), 239–247; English transl. in Theory Probab. Appl. **1** (1956).
2. *Two uniform limit theorems for sums of independent terms*, Teor. Veroyatnost. i Primenen. **1** (1956), 426–436; English transl. in Theory Probab. Appl. **1** (1956).
3. *Zufällige Funktionen und Grenzverteilungssätze*, Bericht über die Tagung Wahrscheinlichkeitsrechnung und mathematische Statistik, Berlin, 1956, pp. 113–126, jointly with Yu. V. Prokhorov.
4. *Some fundamental questions of representing a function of one or several variables approximately and exactly*, Proc. 3rd All-Union Math. Conf., Vol. 2, Izdat. Moskov. Gos. Univ., Moscow, 1956, pp. 28–29. (Russian)
5. *On the Shannon theory of information transmission in the case of continuous signals*, IEEE Trans. Inform. Theory **IT-2** (1956), 102–108.
6. *On the representation of continuous functions of several variables as superpositions of continuous functions of fewer variables*, Dokl. Akad. Nauk SSSR **108** (1956), 179–182. (Russian)
7. *On some asymptotic characteristics of totally bounded metric spaces*, Dokl. Akad. Nauk SSSR **108** (1956), 385–388. (Russian)
8. *On a general definition of amount of information*, Dokl. Akad. Nauk SSSR **111** (1956), 745–748, jointly with I. M. Gel′fand and and A. M. Yaglom. (Russian)
9. *Probability theory*, Mathematics, its Content, Methods, and Meaning, Vol. 2, Izdat. Akad. Nauk SSSR, Moscow, 1956, pp. 252–284. (Russian)
10. *Slutskiĭ, Evgeniĭ Borisovich*, Great Sov. Encycl., 2nd ed., Vol. 39, 1956, p. 378. (Russian)
11. *Smirnov, Nikolaĭ Vasil′evich*, Great Sov. Encycl., 2nd ed., Vol. 39, 1956, p. 406. (Russian)

12. *Sergeĭ Mikhaĭlovich Nikol′skiĭ (On his fiftieth birthday)*, Uspekhi Mat. Nauk **11** (1956), no. 2, 239–244, jointly with S. B. Stechkin. (Russian)
13. Editor, *Mathematics, its Content, Methods, and Meaning,* Vols. 1–3, Izdat. Akad. Nauk SSSR, Moscow, 1956, jointly with A. D. Aleksandrov and M. A. Lavrent′ev. (Russian)
14. *Foundations of the theory of probability*, Chelsea, New York, 1956, English transl. of 1936, [2].

1957

1. *The theory of information transmission*, Session of the USSR Academy of Sciences on Scientific Problems in Automatization of Production, October 15–20, 1956: Plenary Meetings, Izdat. Akad. Nauk SSSR, Moscow, 1957, pp. 66–69. (Russian)
2. *On representing continuous functions of several variables as a composition of continuous functions of a single variable and addition*, Dokl. Akad. Nauk SSSR **114** (1957), 953–956. (Russian)
3. *On a justification of the theory of real numbers*, Mat. Prosveshchenie **2** (1957), 169–173. (Russian)
4. *Elements of the theory of functions and functional analysis: A lecture course,* Vol. 1. *Metric and normed spaces*, English transl. of 1954, [12], Graylock, Rochester, New York, 1957, jointly with S. V. Fomin.
5. *Rozklady graniczne sum zmiennych nezależnych*, Polish transl. of 1949, [1], PAN, Warsaw, 1957, jointly with B. V. Gnedenko.

1958

1. *Mathematics as a profession*, 2nd ed., Izdat. Moskov. Gos. Univ., Moscow, 1958. (Russian)
2. *The amount of information and entropy for continuous distributions*, Proc. 3rd All-Union Math. Conf., Izdat. Akad. Nauk SSSR, Moscow, 1958, pp. 300–320, jointly with I. M. Gel′fand and A. M. Yaglom. (Russian)
3. *Sufficient statistics*, Great Sov. Encycl., 2nd ed., Vol. 51, 1958, p. 106. (Russian)
4. *Information*, Great Sov. Encycl., 2nd ed., Vol. 51, 1958, pp. 129–130. (Russian)
5. *Cybernetics*, Great Sov. Encycl., 2nd ed., Vol. 51, 1958, pp. 149–151. (Russian)
6. *A new metric invariant of transitive dynamical systems and of automorphisms of Lebesgue spaces*, Dokl. Akad. Nauk SSSR **119** (1958), 861–864. (Russian)
7. *Sur les propriétés des fonctions de concentrations de M. P. Lévy*, Ann. Inst. H. Poincaré **16** (1958), 27–34.
8. *On the linear dimension of topological vector spaces*, Dokl. Akad. Nauk SSSR **120** (1958), 239–241. (Russian)
9. *On the definition of algorithm*, Uspekhi Mat. Nauk **13** (1958), no. 4, 3–28. (Russian)
10. *Spectra of dynamical systems generated by stationary random processes*, Teor. Veroyatnost. i Primenen. **3** (1958), 214–215; English transl. in Theory Probab. Appl. **3** (1958).
11. *Ergodic stationary random processes with discrete spectrum*, Teor. Veroyatnost. i Primenen. **3** (1958), 212–213; English transl. in Theory Probab. Appl. **3** (1958).

1959

1. *On entropy per unit time as a metric invariant of automorphisms*, Dokl. Akad. Nauk SSSR **124** (1959), 754–755. (Russian)
2. *ε-entropy and ε-capacity of sets in a function space*, Uspekhi Mat. Nauk **14** (1959), no. 2, 3–86, jointly with V. M. Tikhomirov. (Russian)
3. *The transition of branching processes into diffusion processes and associated problems of genetics*, Teor. Veroyatnost. i Primenen. **4** (1959), 233–236; English transl. in Theory Probab. Appl. **4** (1959).
4. *Remarks on papers of R. A. Minlos and V. V. Sazonov*, Teor. Veroyatnost. i Primenen. **4** (1959), 237–239; English transl. in Theory Probab. Appl. **4** (1959).
5. *Probability theory*, Forty Years of Mathematics in the Soviet Union, Vol. 1, Fizmatgiz, Moscow, 1959, pp. 781–795. (Russian)
6. *Foreword*, An introduction to cybernetics, by W. Ross Ashby (translated from English to Russian), Inostran. Lit., Moscow, 1959, pp. 5–8.
7. *Grenzverteilungen von Summen unabhängiger Zufallsgrössen*, German transl. of 1949, [1], Akad. Verlag, Berlin, 1959, jointly with B. V. Gnedenko.

1960

1. *The work of S. N. Bernshteĭn on the theory of probability*, Teor. Veroyatnost. i Primenen. **5** (1960), 215–221, jointly with O. V. Sarmanov; English transl. in Theory Probab. Appl. **5** (1960).
2. *On the classes $\Phi^{(n)}$ of Fortet and Blanc-Lapierre*, Teor. Veroyatnost. i Primenen. **5** (1960), 373; English transl. in Theory Probab. Appl. **5** (1960).
3. *Random functions of several variables whose realizations are almost all periodic*, Teor. Veroyatnost. i Primenen. **5** (1960), 374; English transl. in Theory Probab. Appl. **5** (1960).
4. *On the work of N. V. Smirnov in mathematical statistics: On his sixtieth birthday*, Teor. Veroyatnost. i Primenen. **5** (1960), 436–440, jointly with B. V. Gnedenko, Yu. V. Prokhorov, and O. V. Sarmanov; English transl. in Theory Probab. Appl. **5** (1960).
5. *On strong mixing conditions for stationary Gaussian processes*, Teor. Veroyatnost. i Primenen. **5** (1960), 222–227; English transl. in Theory Probab. Appl. **5** (1960).
6. *Academician Aleksandr Yakovlevich Khinchin* [obituary], Uspekhi Mat. Nauk **15** (1960), no. 4, 97–110; English transl. in Russian Math. Surveys **15** (1960).
7. *Mathematics as a profession*, 3rd ed., Izdat. Moskov. Gos. Univ., Moscow, 1960. (Russian)
8. *Elements of the theory of functions and functional analysis*, Vol. 2: *Measure, Lebesgue integral, Hilbert space*, Izdat. Moskov. Gos. Univ., Moscow, 1960, jointly with S. V. Fomin.
9. *Foreword* (editor), Sur le mesure des grandeurs, 2nd ed., by H. Lebesgue (translated from French to Russian), Uchpedgiz, Moscow, 1960, pp. 7–16.

1961

1. *Automata and life: notes for the report at the methodology seminar of the Mechanics and Mathematics Department of Moscow University, April 5, 1961*, Mashin. Perevod i Priklad. Lingvistika **6** (1961), 3–8. (Russian)
2. *Automata and life*, Tekhnika—Molodezhi **1961**, no. 10, 16–19; no. 11, 30–33. (Russian)
3. *Remark on the report of V. K. Lezerson*, Teor. Veroyatnost. i Primenen. **6** (1961), 367; English transl. in Theory Probab. Appl. **6** (1961).
4. *Properties of inequalities and the idea of approximate computations. Irrational numbers*, Questions on Teaching Mathematics in Secondary Schools, Uchpedgiz, Moscow, 1961, jointly with P. S. Aleksandrov. (Russian)
5. *Elements of the theory of functions and functional analysis*, Vol. 2: *Measure, Lebesgue integral, Hilbert space*, English transl. of 1960, [8], Graylock, Albany, New York, 1961, jointly with S. V. Fomin.
6. *ε-entropy and ε-capacity of sets in a function space*, Amer. Math. Soc. Transl. (2) **17** (1961), 277–304, jointly with V. M. Tikhomirov.
7. *On the representation of continuous functions of several variables as superpositions of continuous functions of fewer variables*, Amer. Math. Soc. Transl. (2) **17** (1961), 369–373, English transl. of 1956, [6].
8. *Letter to the editor (in connection with S. N. Mergelyan's paper, "Uniform approximations of functions of a complex variable"*, Uspekhi Mat. Nauk **7**:2 (1952)), Izv. Vyssh. Uchebn. Zaved. Mat. **20** (1961), 177. (Russian)
9. *Remarks on a report of M. S. Pinsker*, Teor. Veroyatnost. i Primenen. **6** (1961), 366; English transl. in Theory Probab. Appl. **6** (1961).
10. *Remarks on a report of B. A. Sevast'yanov*, Teor. Veroyatnost. i Primenen. **6** (1961), 367; English transl. in Theory Probab. Appl. **6** (1961).

1962

1. *Estimation of the parameters of a stationary Gaussian Markov process*, Dokl. Akad. Nauk SSSR **146** (1962), 747–750, jointly with M. Aramo and Ya. G. Sinaĭ; English transl. in Soviet Math. Dokl. **3** (1962).
2. *The rhythmics in Mayakovskiĭ's poems*, Vopr. Yazykoznaniya **1962**, no. 3, 62–74, jointly with A. M. Kondrat'ev. (Russian)
3. *A probability problem of optimal control*, Dokl. Akad. Nauk SSSR **145** (1962), 993–995, jointly with L. S. Pontryagin and E. F. Mishchenko; English transl. in Soviet Math. Dokl. **3** (1962).
4. *B. V. Gnedenko's work in probability theory*, Teor. Veroyatnost. i Primenen. **7** (1962), 323–329; English transl. in Theory Probab. Appl. **7** (1962).

5. *A refinement of ideas about the local structure of turbulence in a viscous incompressible fluid at high Reynolds number*, Mécanique de la turbulence (Colloq. Internat. CNRS, Marseilles, August–September 1961), Paris, 1962, pp. 447–458. (French and Russian)
6. *A refinement of previous hypotheses concerning the local structure of a viscous incompressible fluid at high Reynolds number*, J. Fluid Mech. **13** (1962), 82–85.
7. *A local limit theorem for classical Markov chains*, Select. Transl. Math. Statist. Probab. **2** (1962), 109–129.
8. *Review of the book "Mathematical statistics in engineering"*, 3rd ed., by A. M. Dlin, Teor. Veroyatnost. i Primenen. **7** (1962), 243–248, jointly with S. A. Aĭvazyan and L. D. Meshalkin; English transl. in Theory Probab. Appl. **7** (1962).
9. *Life and thought from the point of view of cybernetics*, Life and its relation to other forms of motion of matter, by A. I. Oparin, Izdat. Akad. Nauk SSSR, Moscow, 1962, pp. 1–11. (Russian)
10. *Boris Vladimirovich Gnedenko (On his fiftieth birthday)*, Uspekhi Mat. Nauk **17** (1962), no. 4, 191–200; English transl. in Russian Math. Surveys **17** (1962).

1963

1. *Approximation of distributions of sums of independent terms by infinitely divisible distributions*, Trudy Moskov. Mat. Obshch. **12** (1963), 437–451; English transl. in Trans. Moscow Math. Soc. **1963** (1965).
2. *Discrete automata and finite algorithms*, Proc. 4th All-Union Math. Conf., Vol. 1, Izdat. Leningrad. Gos. Univ., Leningrad, 1963, p. 120. (Russian)
3. *Various approaches to estimating the difficulty of an approximate representation and computation of functions*, Proc. Internat. Congress of Mathematicians (Stockholm), 1963, pp. 351–356. (Russian)
4. *On tables of random numbers*, Indian J. Statist. Ser. A **25** (1963), 369–376.
5. *On a study of Mayakovskiĭ's rhythmics*, Vopr. Yazykoznaniya **1963**, no. 4, 64–71. (Russian)
6. *On the dolnik mode in contemporary Russian poetry: general characteristics*, Vopr. Yazykoznaniya **6** (1963), 84–95, jointly with A. V. Prokhorov.
7. *Statistics and probability theory in the study of Russian versification*, Abstracts and Annotations, Sympos. on the Comprehensive Study of Artistic Creativity, "Nauka", Leningrad, 1963, p. 23, jointly with A. V. Prokhorov. (Russian)
8. *How I became a mathematician*, Ogonek **48** (1963), 12–13. (Russian)
9. *Foreword*, Works on information theory and cybernetics by C. Shannon (translated from English to Russian), Inostran. Lit., Moscow, 1963, pp. 5–6.
10. *Probability theory*, Mathematics: Its Content, Methods, and Meaning, Vol. 1, Part 4, English transl. of 1956, [9], Amer. Math. Soc., Providence, RI, 1963, pp. 33–71.
11. *On representing continuous functions of several variables as a composition of continuous functions of a single variable and addition*, Amer. Math. Soc. Transl. (2) **28** (1963), 55–59 (English transl. of 1957, [2]).
12. *On the definition of an algorithm*, Amer. Math. Soc. Transl. (2) **29** (1963), 217–245, jointly with V. A. Uspenskiĭ.
13. *On the approximation of distributions of sums of independent summands by infinitely divisible distributions*, Sankhyā Ser. A **25** (1963), no. 1, 159–174.
14. *Automata and life*, The possible and impossible in cybernetics: Collection under the editorship of Academicians A. Berg and E. Kol'man, Izdat. Akad. Nauk SSSR, Moscow, 1963, pp. 10–29. (Russian)

1964

1. *On the dolnik in contemporary Russian poetry: a statistical characteristic of the dolnik of Mayakovskiĭ, Bagritskiĭ, and Akhmatova*, Vopr. Yazykoznaniya **1964**, no. 1, 75–94, jointly with A. V. Prokhorov. (Russian)
2. *Foreword*, An Introduction to Probability Theory and its Applications, 2nd ed., by W. Feller (translated from English to Russian), "Mir", Moscow, 1964, pp. 5–6.
3. *On the approximation of distributions of sums of independent summands by infinitely divisible distributions*, Contributions to Statistics, 1964, pp. 159–174.
4. *Izrail' Moiseevich Gel'fand (On his fiftieth birthday)*, Uspekhi Mat. Nauk **19** (1964), no. 3, 187–205, jointly with S. V. Fomin, G. E. Shilov, and M. I. Vishik; English transl. in Russian Math. Surveys **19** (1964).

5. *Lev Abramovich Tumarkin (Topologist. On his fiftieth birthday)*, Uspekhi Mat. Nauk **19** (1964), no. 4, 219–221, jointly with P. S. Aleksandrov; English transl. in Russian Math. Surveys **19** (1964).

1965

1. *Three approaches to the definition of the amount of information*, Problemy Peredachi Informatsii **1** (1965), 3–11; English transl. in Problems Inform. Transmission **3** (1965).
2. *The body of mathematical knowledge for eight-year schools* (Committee on Mathematical Education of the Mathematics Section of the USSR Academy of Sciences), Mat. v Shkole **1965**, no. 2, 21–24. (Russian)
3. *Geometric transformations in a school geometry course*, Mat. v Shkole **1965**, no. 2, 24–29. (Russian)
4. *On the content of a school mathematics course*, Mat. v Shkole **1965**, no. 4, 53–62, jointly with I. M. Yaglom. (Russian)
5. *Functions, graphs, and continuous functions*, Mat. v Shkole **1965**, no. 6, 12–21. (Russian)
6. *Remarks on an analysis of the rhythm in Mayakovskiĭ's "Poem about a Soviet passport"*, Vopr. Yazykoznaniya **1965**, no. 3, 70–75. (Russian)
7. *Natural numbers and positive scalar quantities*, Mathematics School. Lectures and Problems, Vol. IV–V, Izdat. Moskov. Gos. Univ., Moscow, 1965, pp. 19–35. (Russian)
8. *Sergeĭ Mikhaĭlovich Nikol'skiĭ (On his sixtieth birthday)*, Uspekhi Mat. Nauk **20** (1965), no. 5, 275–287, jointly with V. K. Dzyadyk and L. D. Kudryavtsev; English transl. in Russian Math. Surveys **20** (1965).

1966

1. *On the meter in Pushkin's "Songs of the Western Slavs"*, Russk. Literatura **1966**, no. 1, 98–111. (Russian)
2. *An introduction to analysis*, Izdat. Moskov. Gos. Univ., Moscow, 1966. (Russian)
3. *P. S. Aleksandrov and the theory of δS-operations*, Uspekhi Mat. Nauk **21** (1966), no. 4, 275–278. (Russian)
4. *On textbooks for the* 1966–67 *academic year*, Mat. v Shkole **1966**, no. 3, 26–30. (Russian)
5. *On a school definition of identity*, Mat. v Shkole **1966**, no. 2, 33–35. (Russian)
6. *Introduction to S. B. Suvorova's article "An experiment on introducing the fundamentals of differential calculus at an early stage"*, Mat. v Shkole **1966**, no. 4, 23. (Russian)
7. *On textbooks for the* 1966–1967 *academic year*, Mat. v Shkole **1966**, no. 6, 31–37. (Russian)
8. *Geometry on the sphere and geology*, Nauka i Zhizn' **1966**, no. 2, 32. (Russian)
9. *A problem in the theory of curves*, Mathematics School. Lectures and Problems, Vol. VIII, Izdat. Moskov. Gos. Univ., Moscow, 1966, p. 35. (Russian)
10. Editor, *The advanced theory of statistics.* Vol 1: *Distribution theory,* by M. G. Kendall and A. Stuart (translated from English to Russian), "Nauka", Moscow, 1966.
11. *Foreword* (editor), Mathématiques nouvelles, by R. Faure and A. Kaufmann (translated from French to Russian), "Mir", Moscow, 1966, pp. 5–6.
12. *The "stochastic theory of teaching" project of N. A. Romanov*, Priroda **1966**, no. 7, 88–89. (Russian)
13. *Pavel Sergeevich Aleksandrov (On his seventieth birthday and fifty years of research)*, Uspekhi Mat. Nauk **21** (1966), no. 4, 4–7, jointly with L. A. Lyusternik, Yu. M. Smirnov, A. N. Tikhonov, et al.; English transl. in Russian Math. Surveys **21** (1966).

1967

1. *On the realization of networks in three-dimensional space*, Problemy Kibernet. **19** (1967), 261–268, jointly with J. M. Barzdin. (Russian)
2. *Curriculum project for secondary schools in mathematics*, Mat. v Shkole **1967**, no. 1, 4–23, jointly with A. I. Markushevich and I. M. Yaglom. (Russian)
3. *On textbooks for the* 1966–1967 *academic year. "Algebra and elementary functions", Part* II, *by E. S. Kochetkov and E. S. Kochetkova*, Mat. v Shkole **1967**, no. 1, 43–48. (Russian)
4. *On elective studies in mathematics*, Mat. v Shkole **1967**, no. 2, 2–3. (Russian)
5. *New curricula and some basic questions on improving the program of mathematics in secondary schools*, Mat. v Shkole **1967**, no. 2, 4–13. (Russian)
6. *The content of elective studies in mathematics for the academic years of* 1967–1968 *and* 1968–1969, Mat. v Shkole **1967**, no. 2, 33–38. (Russian)
7. *Curricula of special courses in mathematics*, Mat. v Shkole **1967**, no. 3, 73–75. (Russian)

8. *Curricula of special courses in mathematics*, Mat. v Shkole **1967**, no. 4, 58–59. (Russian)
9. *On changes in the text of A. N. Barsukov's algebra book for grades* 6–9, Mat. v Shkole **1967**, no. 6, 22–24. (Russian)
10. *Igor' Vladimirovich Girsanov* [obituary], Teor. Veroyatnost. i Primenen. **12** (1967), 532–535, jointly with E. B. Dynkin, B. T. Polyak, and M. I. Freĭdlin; English transl in Theory Probab. Appl. **12** (1967).
11. *Foreword* (editor), Information Theory and Statistics, by S. Kullback (translated from English to Russian), "Mir", Moscow, 1966, pp. 5–6.

1968

1. *Some theorems on algorithmic entropy and algorithmic amount of information*, Uspekhi Mat. Nauk **23** (1968), no. 2, 201. (Russian)
2. *A generalization of the concept of a power and the exponential function*, Mat. v Shkole **1968**, no. 1, 24–32. (Russian)
3. *A mathematics curriculum for secondary schools*, Mat. v Shkole **1968**, no. 2, 5–20. (Russian)
4. *On the study of the exponential function and logarithms in eight-year schools*, Mat. v Shkole **1968**, no. 2, 23–25. (Russian)
5. *On new curricula in mathematics*, Mat. v Shkole **1968**, no. 2, 21–22. (Russian)
6. *Introduction to probability theory and combinatorial analysis*, Mat. v Shkole **1968**, no. 2, 63–72. (Russian)
7. *A supplement to Yu. A. Shikhanovich's review (of A. A. Stolyar's book)*, Mat. v Shkole **1968**, no. 3, 92. (Russian)
8. *Elements of the theory of functions and functional analysis*, 2nd rev. aug. ed., Fizmatgiz, Moscow, 1968, jointly with S. V. Fomin. (Russian)
9. *On the roots of the classical Russian metrics*, Cooperation of the Sciences and the Mysteries of Creativity, "Iskusstvo", Moscow, 1968, pp. 397–432, jointly with A. V. Prokhorov. (Russian)
10. *An example of the study of meter and its metric variants*, The Theory of Poetry, "Nauka", Leningrad, 1968, pp. 145–167. (Russian)
11. *Three approaches to the quantitative definition of information*, Internat. J. Comput. Math. **2** (1968), no. 2, 157–168.
12. *Limit distributions for sums of independent random variables*, rev. ed., Addison-Wesley, Reading, MA, 1968, jointly with B. V. Gnedenko.
13. *Logical basis for information theory and probability theory*, IEEE Trans. Inform. Theory **IT-14** (1968), 662–664.

1969

1. *On the logical foundations of information theory and probability theory*, Problemy Peredachi Informatsii **5** (1969), no. 3, 3–7; English transl. in Problems Inform. Transmission **5** (1969).
2. *Sergeĭ Natanovich Bernshteĭn*, Uspekhi Mat. Nauk **24** (1969), no. 3, 211–218, jointly with P. S. Aleksandrov, N. I. Akhiezer, and B. V. Gnedenko; English transl. in Russian Math. Surveys **24** (1969).
3. *Letter to the editor* (about inaccuracies in B. E. Veĭts' article), Mat. v Shkole **1969**, no. 2, 93. (Russian)
4. *Scientific foundations for a school course in mathematics. First lecture: Contemporary views on the nature of mathematics*, Mat. v Shkole **1969**, no. 3, 12–17. (Russian)
5. *Scientific foundations for a school course in mathematics. Second lecture: Natural numbers*, Mat. v Shkole **1969**, no. 5, 8–17. (Russian)
6. *New developments in school mathematics*, Nauka i Zhizn' **1969**, no. 3, 62–66. (Russian)
7. *Sergeĭ Natanovich Bernshteĭn* [obituary], Teor. Veroyatnost. i Primenen. **14** (1969), 113–121; English transl. in Theory Probab. Appl. **14** (1969).
8. Editor, *Algebra and the elements of analysis,* by B. E. Veĭts and I. T. Demidov, Trial textbook for 9th grade of secondary school, Prosveshchenie, Moscow, 1969. (Russian)

1970

1. *Viktor Nikolaevich Zasukhin: In memory of mathematicians killed in the Second World War*, Uspekhi Mat. Nauk **25** (1970), no. 3, 243; English transl. in Russian Math. Surveys **25** (1970).
2. *Gleb Aleksandrovich Seliverstov* [obituary], Uspekhi Mat. Nauk **25** (1970), no. 3, 244–245; English transl. in Russian Math. Surveys **25** (1970).
3. *Scientific foundations for a school course in mathematics. Third lecture: Generalizing the notion of number. Nonnegative rational numbers*, Mat. v Shkole **1970**, no. 2, 27–32. (Russian)

4. *On a trial geometry book for 6th grade*, Mat. v Shkole **1970**, no. 4, 21–34, jointly with A. F. Semenovich. (Russian)
5. *Educational materials on geometry for 5th grade*, Mat. v Shkole **1970**, no. 5, 30–45, jointly with A. F. Semenovich and R. S. Cherkasov. (Russian)
6. *What is a function?*, Kvant **1970**, no. 1, 27–36. (Russian)
7. *Problem M3*, Kvant **1970**, no. 1, 52–53. (Russian)
8. *What is the graph of a function?*, Kvant **1970**, no. 2, 3–13. (Russian)
9. *A parquet of regular polygons*, Kvant **1970**, no. 3, 24–27. (Russian)
10. *On the solution of Hilbert's tenth problem*, Kvant **1970**, no. 7, 39–44, jointly with F. L. Varpakhovskiĭ. (Russian)
11. *Foreword to the article*: *The epistemology of Lenin and the concepts of mathematics, by V. G. Boltyanskiĭ and N. Kh. Rozov*, Kvant **1970**, no. 7, 2. (Russian)
12. *The statistical hydrodynamics of the ocean*, Uspekhi Mat. Nauk **25** (1970), no. 4, 167. (Russian)
13. *Geometry for 6th grade*, Trial textbook, Prosveshchenie, Moscow, 1970, jointly with A. Semenovich, R. Cherkasov, and F. Nagibin. (Russian)
14. *Geometry for 6th grade*, Methods manual, Prosveshchenie, Moscow, 1970, jointly with A. F. Semenovich, F. F. Nagibin, and R. S. Cherkasov. (Russian)

1971

1. *Quantity*, Great Sov. Encycl., 3rd ed., Vol. 4, 1971, pp. 456–457. (Russian)
2. *Wiener, Norbert*, Great Sov. Encycl., 3rd ed., Vol. 5, 1971, p. 72. (Russian)
3. *Hilbert, David*, Great Sov. Encycl., 3rd ed., Vol. 6, 1971, p. 519. (Russian)
4. *On a system of basic concepts and notation for a school course in mathematics*, Mat. v Shkole **1971**, no. 2, 17–22. (Russian)
5. *A trial geometry book for 7th grade*, Mat. v Shkole **1971**, no. 3, 9–17, jointly with A. F. Semenovich, F. F. Nagibin, and R. S. Cherkasov. (Russian)
6. *Summary of the report "Elements of logic in secondary school"* (to the Committee on Mathematics of the Academic Methods Council of the USSR Ministry of Education), Mat. v Shkole **1971**, no. 3, 91–92. (Russian)
7. *On a new edition of the trial geometry book for 6th grade*, Mat. v Shkole **1971**, no. 4, 23–35, jointly with A. F. Semenovich, F. F. Nagibin, and R. S. Cherkasov. (Russian)
8. *On a new edition of the trial geometry book for 6th grade*, Mat. v Shkole **1971**, no. 5, 25–38, jointly with A. F. Semenovich, F. F. Nagibin, and R. S. Cherkasov. (Russian)
9. *From a new geometry textbook for 6th grade* (*Geometric constructions*), Mat. v Shkole **1971**, no. 6, 13–21, jointly with A. F. Semenovich, F. F. Nagibin, and R. S. Cherkasov. (Russian)
10. *Contemporary mathematics, and mathematics in contemporary schools*, Mat. v Shkole **1971**, no. 6, 2–3. (Russian)
11. *A mathematics curriculum for mathematics and physics schools*, Izdat. Moskov. Gos. Univ., Moscow, 1971, jointly with V. A. Gusev, A. V. Sosinskiĭ, and A. A. Shershevskiĭ. (Russian)
12. *The summer school on Rubskoe Lake*, Prosveshchenie, Moscow, 1971. (Russian)
13. *Geometry for 7th grade*, Trial textbook, Prosveshchenie, Moscow, 1971, jointly with A. F. Semenovich, R. S. Cherkasov, and F. F. Nagibin. (Russian)
14. *Naum Il'ich Akhiezer* (*On his seventieth birthday*), Uspekhi Mat. Nauk **26** (1971), no. 6, 257–261, jointly with Yu. M. Berezanskiĭ, M. G. Kreĭn, et al.; English transl. in Russian Math. Surveys **26** (1971).
15. Editor, *Algebra and the elements of analysis,* by B. E. Veĭts and I. T. Demidov, Trial textbook for 10th grade, Prosveshchenie, Moscow, 1971. (Russian)
16. *Ivan Geogrievich Petrovskiĭ* (*On his seventieth birthday*), Uspekhi Mat. Nauk **26** (1971), no. 1, 3–24, jointly with P. S. Aleksandrov, O. A. Oleĭnik, A. N. Tikhonov, and M. I. Vishik; English transl. in Russian Math. Surveys **26** (1971).
17. *Geometry for 6th grade*, Educational manual. Project for discussion (A. N. Kolmogorov, ed.), NIIS, MO Akad. Pedagog. Nauk SSSR, MGONO, MGIUU, Moscow, 1971, jointly with A. F. Semenovich, R. S. Cherkasov, and F. F. Nagibin. (Russian)
18. *Ivan Georgievich Petrovskiĭ* (*On his seventieth birthday*), Uspekhi Mat. Nauk **26** (1971), no. 2, 3–24, jointly with P. S. Aleksandrov, V. I. Arnol'ld, I. M. Gel'fand, S. P. Novikov, and O. A. Oleĭnik; English transl. in Russian Math. Surveys **26** (1971).

1972

1. *Complexity of determining and complexity of constructing mathematical objects*, Uspekhi Mat. Nauk **27** (1972), no. 2, 159. (Russian)
2. *Integral*, Great Sov. Encycl., 3rd ed., Vol. 10, 1972, pp. 300–302. (Russian)
3. *Method of exhaustion*, Great Sov. Encycl., 3rd ed., Vol. 10 (1972), 586. (Russian)
4. *Elements of the theory of functions and functional analysis*, 3rd rev. ed., Fizmatgiz, Moscow, 1972, jointly with S. V. Fomin. (Russian)
5. *The science instructor*, The Joy of Discovery, by L. Neĭman, Detskaya Lit., Moscow, 1972. (Russian)
6. *A qualitative study of mathematical models of population dynamics*, Problemy Kibernet. **25** (1972), 101–106. (Russian)
7. *From a new geometry textbook for 6th grade*, Mat. v Shkole **1972**, no. 1, 22–31, jointly with A. F. Semenovich, F. F. Nagibin, and R. S. Cherkasov. (Russian)
8. *Boris Vladimirovich Gnedenko*, Mat. v Shkole **1972**, no. 1, 85–86, jointly with R. S. Cherkasov. (Russian)
9. *On V. Ya. Vilenkin's letter* (the letter "Equality or congruence", in same issue), Mat. v Shkole **1972**, no. 6, 34–35. (Russian)
10. *Coordination of the teaching of mathematics and physics*, All-Union Scientific and Practical Conf. on the Problem of Teaching and Educational Work in Schools and in Classes with Intensive Study of Individual Subjects, Abstracts of Reports, SRI, SIMO, Akad. Pedagog. Nauk SSSR, Moscow, 1972. (Russian)
11. *Elementos de la teoria de funciones y del analisis funcional*, Spanish transl. of 1972, [4], "Mir", Moscow, 1972, jointly with S. V. Fomin.
12. *Georgiĭ Fedorovich Rybkin* [obituary], Uspekhi Mat. Nauk **27** (1972), no. 5, 223–225, jointly with P. S. Aleksandrov, B. V. Gnedenko, A. I. Markushevich, V. B. Orlov, A. T. Tsvetkov, and A. P. Yushkevich; English transl. in Russian Math. Surveys **27** (1972).
13. *Leonid Vital'evich Kantorovich (On his sixtieth birthday)*, Uspekhi Mat. Nauk **27** (1972), no. 3, 221–227, jointly with B. Z. Vulikh, M. K. Gavurin, Yu. V. Linnik, V. L. Makarov, B. S. Mityagin, et al.; English transl. in Russian Math. Surveys **27** (1972).
14. Editor, *Probabilistic methods of investigation*, collection, vyp. 41, Izdat. Moskov. Gos. Univ., Moscow, 1972. (Russian)
15. *Dmitriĭ Evgen'evich Men'shov (On his eightieth birthday)*, Uspekhi Mat. Nauk **27** (1972), no. 2, 185–195, jointly with P. S. Aleksandrov and P. L. Ul'yanov; English transl. in Russian Math. Surveys **27** (1972).
16. *Geometry for 8th grade*, Trial textbook, Prosveshchenie, Moscow, 1972, jointly with A. Semenovich, R. Cherkasov, and V. Gusev.
17. *Boris Vladimirovich Gnedenko (On his sixtieth birthday)*, Uspekhi Mat. Nauk **27** (1972), no. 2, 197–202, jointly with Yu. K. Belyaev and A. D. Solov'ev; English transl. in Russian Math. Surveys **27** (1972).
18. *Georgiĭ Pavlovich Tolstov (On his sixtieth birthday)*, Uspekhi Mat. Nauk **27** (1972), no. 1, 255–264, jointly with I. Ya. Verchenko and V. Ya. Kozlov; English transl. in Russian Math. Surveys **27** (1972).
19. *Measurement*, Great Sov. Encycl., 3rd ed., Vol. 10, 1972, pp. 220–221. (Russian)
20. *Geometry for 6th grade*, 2nd rev. ed., textbook (A. N. Kolmogorov, ed.), Prosveshchenie, Moscow, 1972, jointly with A. F. Semenovich, F. F. Nagibin, and R. S. Cherkasov. (Russian; published also in Azerbaijani, Armenian, Bashkir, Belorussian, Hungarian, Georgian, Kazakh, Karakalpak, Kirghiz, Latvian, Lithuanian, Moldavian, Tadjik, Tatar, Tuvinian, Turkmen, Uzbek, and Uigur)
21. Editor, *Geometry in 8th grade: Methods manual for the textbook "Geometry"*, by V. A. Gusev, G. G. Maslov, A. F. Semenovich, and R. S. Cherkasov, Prosveshchenie, Moscow, 1972. (Russian)
22. *A letter to P. L. Kapitsa*, Voprosy Filosofii **1972**, no. 9, 127–128. (Russian)
23. *Geometry*, Trial textbook for 7th grade, Tbilisi, 1972, jointly with A. F. Semenovich, F. F. Nagibin, and R. S. Cherkasov. (Georgian; published also in Belorussian and Tatar)
24. Editor, *Geometry in 6th grade: Methods manual (teacher's aid)*, by V. A. Gusev, G. G. Maslov, F. F. Nagibin, et al., Prosveshchenie, Moscow, 1972. (Russian)

1973

1. *Continuum*, Great Sov. Encycl., 3rd ed., Vol. 13, 1973, p. 64. (Russian)

2. *Semilogarithmic and logarithmic coordinate scales*, Kvant **1973**, no. 3, 2–7. (Russian)
3. *Mathematics as a profession*, Kvant **1973**, no. 4, 12–18. (Russian)
4. *On procedures for studying "parallel translation" in a 7th-grade geometry course*, Mat. v Shkole **1973**, no. 1, 24–29, jointly with A. F. Semenovich and R. S. Cherkasov. (Russian)
5. *On the structure of the new geometry book for 7th grade*, Mat. v Shkole **1973**, no. 2, 17–29, jointly with A. F. Semenovich and R. S. Cherkasov. (Russian)
6. *Ivan Georgievich Petrovskiĭ*, Mat. v Shkole **1973**, no. 4, 81–86, jointly with P. S. Aleksandrov and O. A. Oleĭnik. (Russian)
7. *Methodological comments on the 9th-grade trial textbook "Algebra and the elements of analysis"*, Mat. v Shkole **1973**, no. 5, 64, jointly with B. E. Veĭts and I. T. Demidov. (Russian)
8. *Scientific foundations for a school curriculum in mathematics*, Curricula for Pedagogical Institutes, Prosveshchenie, Moscow, 1973. (Russian)
9. *Materials for discussion in the Committee on School Terminology and Notation*, Acad. Methods Council, USSR Ministry of Education, 1973. (Russian)
10. *Geometry for 7th grade*, 2nd rev. ed., Prosveshchenie, Moscow, 1973, jointly with A. Semenovich, R. Cherkasov, and F. Nagibin. (Russian; published also in Belorussian, Moldavian, and Tatar)
11. *Foreword* (editor), The advanced theory of statistics. Vol 2: Statistical inference and relationship, by M. G. Kendall and A. Stuart (translated from English to Russian), "Nauka", Moscow, 1973.
12. Editor, *Geometry in 8th grade: Methods manual for the textbook "Geometry",* by V. A. Gusev, G. G. Maslov, A. F. Semenovich, and R. S. Cherkasov, Tallin, 1973. (Russian)
13. Editor, *Geometry in 6th grade: Methods manual (teacher's aid),* by V. A. Gusev, G. G. Maslov, F. F. Nagibin, et al., "Luis", Erevan, 1973. (Armenian; also published in Georgian, Belorussian, Kazakh, Latvian, Lithuanian, Moldavian, Tadjik, and Ukrainian)
14. Editor, *Algebra and the elements of analysis,* by B. E. Veĭts and I. T. Demidov, 2nd rev. ed., Trial textbook for 9th grade, Prosveshchenie, Moscow, 1973. (Russian)
15. *Geometry*, 3rd. ed., textbook for 6th grade of secondary school (A. N. Kolmogorov, ed.), Prosveshchenie, Moscow, 1973, jointly with A. F. Semenovich, F. F. Nagibin, and R. S. Cherkasov. (Russian; published also in Azerbaijani, Armenian, Bashkir, Hungarian, Georgian, Kazakh, Karakalpak, Kirghiz, Latvian, Lithuanian, Tadjik, Tuvinian, Turkmen, Uzbek, Uigur, and Ukrainian)
16. Editor, *Geometry in 7th grade: Methods manual for teachers,* by V. A. Gusev, G. G. Maslov, F. F. Nagibin, et al., Prosveshchenie, Moscow, 1973. (Russian)

1974

1. *Foundations of the theory of probability*, 2nd ed., "Nauka", Moscow, 1974. (Russian)
2. *In memory of Ivan Georgievich Petrovskiĭ (January 18, 1901–January 15, 1973)*, Trudy Moskov. Mat. Obshch. **31** (1974), 5–16, jointly with P. S. Aleksandrov and O. A. Oleĭnik; English transl. in Trans. Moscow Math. Soc. **31** (1976).
3. *Ivan Georgievich Petrovskiĭ* [obituary], Uspekhi Mat. Nauk **29** (1974), no. 2, 3–5; English transl. in Russian Math. Surveys **29**.
4. *Andreĭ Andreevich Markov*, Great Sov. Encycl., 3rd ed., Vol. 15, 1974, p. 379. (Russian)
5. *Mathematics*, Great Sov. Encycl., 3rd ed., Vol. 15, 1974, pp. 467–478. (Russian)
6. *Mathematical statistics*, Great Sov. Encycl., 3rd ed., Vol. 15, 1974, pp. 480–484, jointly with Yu. V. Prokhorov. (Russian)
7. *Multidimensional space*, Great Sov. Encycl., 3rd ed., Vol. 16, 1974, p. 372. (Russian)
8. *Orientation*, Great Sov. Encycl., 3rd ed., Vol. 18, 1974, pp. 509–510. (Russian)
9. *On ensuring worthy reserves*, Vestnik Vyssh. Shkoly **1974**, no. 6, 26–33, jointly with I. T. Tropin and K. V. Chernyshev. (Russian)
10. *New curricula: specialized schools*, Mathematical Education Today, no. 6, Znanie, Moscow, 1974, pp. 5–12. (Russian)
11. *The boarding school of the university. What is it*? (in the section "Ten years of the physics and mathematics school of Moscow State University", Mat. v Shkole **1974**, no. 2, 58–60. (Russian)
12. *Anna Maksimilianovna Fisher* [obituary], Mat. v Shkole **1974**, no. 2, 87, jointly with A. F. Semenovich. (Russian)
13. *The sieve of Eratosthenes*, Kvant **1974**, no. 1, 77. (Russian)

14. Editor, *Algebra and the elements of analysis,* by B. E. Veĭts and I. T. Demidov, 2nd rev. ed., Trial textbook for 10th grade, Prosveshchenie, Moscow, 1974. (Russian)
15. *Eléménts de la théorie des fonctions et l'analysis fonctionnelle*, French transl. of 1972, [4], "Mir", Moscow, 1974, jointly with S. V. Fomin.
16. *Geometry for 8th grade*, 2nd ed., textbook, Prosveshchenie, Moscow, 1974, jointly with A. F. Semenovich, R. S. Cherkasov, and V. A. Gusev. (Russian; published also in Belorussian, Moldavian, and Tatar)
17. *Complexity of algorithms and an objective definition of randomness*, Uspekhi Mat. Nauk **29** (1974), no. 4, 155. (Russian)
18. Editor, *Geometry in 8th grade: Manual for teachers,* by V. A. Gusev, G. G. Maslov, A. F. Semenovich, and R. S. Cherkasov, Prosveshchenie, Moscow, 1974. (Russian)
19. *Geometry*, 4th ed., textbook for 6th grade of secondary school (A. N. Kolmogorov, ed.), Prosveshchenie, Moscow, 1974, jointly with A. F. Semenovich, F. F. Nagibin, and R. S. Cherkasov. (Russian; published also in Polish)
20. *Geometry*, 3rd ed., textbook for 7th grade in secondary school, Prosveshchenie, Moscow, 1974, jointly with A. F. Semenovich, F. F. Nagibin, and R. S. Cherkasov. (Russian; published also in Azerbaijani, Armenian, Bashkir, Hungarian, Georgian, Kazakh, Karakalpak, Kirghiz, Latvian, Lithuanian, Moldavian, Polish, Tadjik, Tuvinian, Turkmen, and Uzbek)

1975

1. *Statistical acceptance sampling*, Great Sov. Encycl., 3rd ed., Vol. 20, 1975, pp. 572–573, jointly with Yu. K. Belyaev. (Russian)
2. *Algebra and the elements of analysis. Mathematical induction*, Mat. v Shkole **1975**, no. 1, 8–14, jointly with S. I. Shvartsburd. (Russian)
3. *The elements of combinatorial analysis*, Mat. v Shkole **1975**, no. 2, 16–25. (Russian)
4. *Real numbers, infinite sequences, and their limits*, Mat. v Shkole **1975**, no. 2, 25–35, jointly with O. S. Ivashev-Musatov. (Russian)
5. *Algebra and the elements of analysis*, textbook for 9th grade of secondary school, Prosveshchenie, Moscow, 1975, jointly with B. E. Veĭts, I. T. Demidov, O. S. Ivashev-Musatov, and S. I. Shvartsburd. (Russian; published also in Belorussian, Moldavian and Ukrainian)
6. *Elementos de la teoria de funciones y del analisis funcional*, 2nd ed., "Mir", Moscow, 1975, jointly with S. V. Fomin.
7. *Sergeĭ Mikhaĭlovich Nikol'skiĭ (On his seventieth birthday)*, Uspekhi Mat. Nauk **30** (1975), no. 4, 271–280, jointly with V. K. Dzyadyk and L. D. Kudryavtsev; English transl. in Russian Math. Surveys **30** (1975).
8. *His path in science (memories of the mathematician A. I. Mal'tsev*, 1909–1967), Nauka i Zhizn' **1975**, no. 7, 112–115. (Russian)
9. *Ol'ga Arsen'evna Oleĭnik (mathematician)*, Vestnik Moskov. Univ. Ser. I Mat. Mekh. **1975**, no. 4, 118–124, jointly with P. S. Aleksandrov and S. L. Sobolev. (Russian)
10. *Foreword* (editor), Mathematical Methods of Statistics, by H. Cramér (translated from English to Russian), "Mir", Moscow, 1975; Foreword to the second Russian edition, p. 5; Foreword to the first Russian edition, pp. 6–8.
11. *Geometry*, 5th ed., textbook for 6th grade of secondary school (A. N. Kolmogorov, ed.), Prosveshchenie, Moscow, 1975, jointly with A. F. Semenovich, F. F. Nagibin, and R. S. Cherkasov. (Russian; published also in Bashkir, Belorussian, Moldavian, and Ukrainian)
12. *Geometry*, 4th ed., textbook for 7th grade of secondary school (A. N. Kolmogorov, ed.), Prosveshchenie, Moscow, 1975, jointly with A. F. Semenovich, F. F. Nagibin, and R. S. Cherkasov. (Russian; published also in Kirghiz, Lithuanian, Moldavian, Polish, and Tuvinian)
13. *Geometry*, 3rd ed., textbook for 8th grade of secondary school (A. N. Kolmogorov, ed.), Prosveshchenie, Moscow, 1975, jointly with A. F. Semenovich, F. F. Nagibin, and R. S. Cherkasov. (Russian; published also in Armenian, Bashkir, Hungarian, Kazakh, Karakalpak, Kirghiz, Latvian, Moldavian, Tadjik, Turkmen, Uzbek, and Uigur)

1976

1. *The trigonometric functions, their graphs and derivatives, in a 10th grade textbook*, Mat. v Shkole **1976**, no. 1, 11–25, jointly with S. I. Shvartsburd. (Russian)
2. *Thirty-eighth Moscow Mathematics Olympiad (February–March* 1975), Mat. v Shkole **1976**, no. 4, 68–72, jointly with G. A. Gal'perin. (Russian)
3. *The integral in a 10th-grade textbook*, Mat. v Shkole **1976**, no. 6, 15–17. (Russian)

4. *Groups of transformations*, Kvant **1976**, no. 10, 2–5. (Russian)
5. *Pavel Sergeevich Aleksandrov (On his eightieth birthday)*, Uspekhi Mat. Nauk **31** (1976), no. 5, 3–14, jointly with A. V. Arkhangel′skiĭ, A. A. Mal′tsev, and O. A. Oleĭnik; English transl. in Russian Math. Surveys **31** (1976).
6. Editor, *Multivariate statistical analysis and time series,* by M. G. Kendall and A. Stuart (translated from English to Russian), "Nauka", Moscow, 1976, jointly with Yu. V. Prokhorov.
7. *Memories of Sergeĭ Vasil′evich Fomin*, Uspekhi Mat. Nauk **31** (1976), no. 4, 199–212, jointly with P. S. Aleksandrov; English transl. in Russian Math. Surveys **31** (1976).
8. *Tashmukhamed Alievich Sarymsakov (On his sixtieth birthday)*, Uspekhi Mat. Nauk **31** (1976), no. 2, 241–246, jointly with P. S. Aleksandrov, B. V. Gnedenko, and Yu. V. Prokhorov; English transl. in Russian Math. Surveys **31** (1976).
9. Editor, *Geometry in 6th grade (teacher's aid),* by V. A. Gusev, G. G. Maslov, F. F. Nagibin, et al., Mektep, Alma-Ata, 1976. (Kazakh)
10. *Algebra and the elements of analysis*, textbook for 10th grade of secondary school (A. N. Kolmogorov, ed.), Prosveshchenie, Moscow, 1976, jointly with O. S. Ivashev-Musatov, B. M. Ivlev, and S. I. Shvartsburd. (Russian; published also in Belorussian, Lithuanian, and Ukrainian)
11. *Algebra and the elements of analysis*, 2nd ed., textbook for 9th grade of secondary school (A. N. Kolmogorov, ed.), Prosveshchenie, Moscow, 1976, jointly with B. E. Veĭts, I. T. Demidov, O. S. Ivashev-Musatov, and S. I. Shvartsburd. (Russian; published also in Azerbaijani, Armenian, Hungarian, Kazakh, Kirghiz, Latvian, Lithuanian, Moldavian, Tadjik, Tatar, Turkmen, Uzbek, and Uigur)
12. *Georgiĭ Evgen′evich Shilov* [obituary], Uspekhi Mat. Nauk **31** (1976), no. 1, 212–228, jointly with P. S. Aleksandrov, S. V. Fomin, I. M. Gel′fand, et al.; English transl. in Russian Math. Surveys **31** (1976).
13. *It is worth thinking about "blockhead"* (Introduction to an article by G. Shilov and V. Berman about a word game), Nauka i Zhizn′ **1976**, no. 5, 138. (Russian)
14. *Geometry*, 5th ed., textbook for 7th grade of secondary school (A. N. Kolmogorov, ed.), Prosveshchenie, Moscow, 1976, jointly with A. F. Semenovich, F. F. Nagibin, and R. S. Cherkasov. (Russian; published also in Latvian and Polish)
15. *Elements of the theory of functions and functional analysis*, 4th rev. ed., "Nauka", Moscow, 1976, jointly with S. V. Fomin. (Russian)
16. *Geometry*, 6th ed., textbook for 6th grade of secondary school (A. N. Kolmogorov, ed.), Prosveshchenie, Moscow, 1976, jointly with A. F. Semenovich, F. F. Nagibin, and R. S. Cherkasov. (Russian; published also in Hungarian)
17. *Geometry*, 4th ed., textbook for 8th grade of secondary school (A. N. Kolmogorov, ed.), Prosveshchenie, Moscow, 1976, jointly with A. F. Semenovich, F. F. Nagibin, and R. S. Cherkasov. (Russian; published also in Latvian, Polish, and Ukrainian)

1977

1. *Infinity*, Math. Encyclopedia, Vol. 1, 1977, pp. 455–458. (Russian)
2. *Quantity*, Math. Encyclopedia, Vol. 1, 1977, pp. 651–653. (Russian)
3. *Probability*, Math. Encyclopedia, Vol. 1, 1977, pp. 667–669. (Russian)
4. *The physics-mathematics school of Moscow State University*, Kvant **1977**, no. 1, 56–57, jointly with V. V. Vavilov. (Russian)
5. *Adol′f Pavlovich Yushkevich (On his seventieth birthday)*, Uspekhi Mat. Nauk **32** (1977), no. 3, 197–202, jointly with I. G. Bashmakova, A. T. Grigor′yan, A. I. Markushevich, F. A. Medvedev, and B. A. Rozenfe′ld; English transl. in Russian Math. Surveys **32** (1977).
6. *Eléménts de la théorie des fonctions et l'analyse fonctionnelle*, 2nd ed., "Mir", Moscow, 1977, jointly with S. V. Fomin.
7. *Algebra and the elements of analysis*, 3nd ed., textbook for 9th grade of secondary school (A. N. Kolmogorov, ed.), Prosveshchenie, Moscow, 1977, jointly with B. E. Veĭts, I. T. Demidov, O. S. Ivashev-Musatov, and S. I. Shvartsburd. (Russian; published also in Azerbaijani, Latvian, Lithuanian, and Polish)
8. *Algebra and the elements of analysis*, textbook for 10th grade of secondary school (A. N. Kolmogorov, ed.), Prosveshchenie, Moscow, 1977, jointly with O. S. Ivashev-Musatov, B. M. Ivlev, and and S. I. Shvartsburd. (Russian; published also in Azerbaijani, Armenian, Hungarian, Georgian, Kazakh, Karakalpak, Kirghiz, Lithuanian, Moldavian, Polish, Tadjik, Tatar, Turkmen, Uzbek, and Uigur)

9. *Geometry*, 7th ed., textbook for 6th grade of secondary school (A. N. Kolmogorov, ed.), Prosveshchenie, Moscow, 1977, jointly with A. F. Semenovich, F. F. Nagibin, and R. S. Cherkasov. (Russian)
10. *Geometry*, 6th ed., textbook for 7th grade of secondary school (A. N. Kolmogorov, ed.), Prosveshchenie, Moscow, 1977, jointly with A. F. Semenovich, F. F. Nagibin, and R. S. Cherkasov. (Russian; published also in Hungarian)
11. *Geometry*, 5th ed., textbook for 8th grade of secondary school (A. N. Kolmogorov, ed.), Prosveshchenie, Moscow, 1977, jointly with A. F. Semenovich, V. A. Gusev, and R. S. Cherkasov. (Russian; published also in Latvian, Polish, and Lithuanian)

1978

1. *What is a function?*, Mat. v Shkole **1978**, no. 2, 27–29 (in connection with G. V. Dorofeev's article in the same issue: "The concept of function in mathematics and in school"). (Russian)
2. *On cultivating a dialectic-materialistic world view in mathematics and physics lessons*, Mat. v Shkole **1978**, no. 3, 6–9. (Russian)
3. *Draft of a mathematics curriculum for secondary schools*, Mat. v Shkole **1978**, no. 4, 7–32. (Russian)
4. *Sergeĭ L'vovich Sobolev*, Mat. v Shkole **1978**, no. 6, 67–73, jointly with O. A. Oleĭnik. (Russian)
5. *The new curricula in French secondary schools*, Mat. v Shkole **1978**, no. 6, 74–78, jointly with A. M. Abramov. (Russian)
6. *How I became a mathematician. What is mathematics? Science in your profession*, Znanie **1978**, no. 11, 5–9. (Russian)
7. *Bounds for the spectral functions of stochastic processes*, Abstracts, Eleventh European Conf. on Statistics (Oslo, August 14–18, 1978), Oslo, 1978, p. 36, jointly with I. G. Zhurbenko. (Russian)
8. *On the formation of a dialectic-materialistic ideology in school children through mathematics and physics lessons*, The Role of Educational Literature in the Formation of Ideology in School Children, Pedagogika, Moscow, 1978, pp. 69–74. (Russian)
9. Editor, *Dimension theory and related topics: Articles of a general nature*, Selected Works of P. S. Aleksandrov, "Nauka", Moscow, 1978, jointly with A. A. Mal'tsev, L. S. Pontryagin, and A. N. Tikhonov. (Russian)
10. Editor, *The theory of functions of a real variable and the theory of topological spaces*, Selected Works of P. S. Aleksandrov, "Nauka", Moscow, 1978, jointly with A. A. Mal'tsev, L. S. Pontryagin, and A. N. Tikhonov. (Russian)
11. *Foreword* (editor), Mathematics of the Nineteenth Century, "Nauka", Moscow, 1978, pp. 7–10, jointly with A. P. Yushkevich. (Russian)
12. *Mark Grigor'evich Kreĭn (On his seventieth birthday)*, Uspekhi Mat. Nauk **33** (1978), no. 3, 197–203, jointly with V. M. Adamyan, Yu. M. Berezanskiĭ, N. N. Bogolyubov, M. A. Lavrent'ev, and Yu. A. Mitropol'skiĭ; English transl. in Russian Math. Surveys **33** (1978).
13. *Elementos de la teoria de funciones y del analisis funcional*, 3rd ed., "Mir", Moscow, 1978, jointly with S. V. Fomin.
14. *Memories of Boris Pavlovich Demidovich* [obituary], Uspekhi Mat. Nauk **33** (1978), no. 2, 169–174; English transl. in Russian Math. Surveys **33** (1978).
15. *Remarks on statistical solutions of the Navier-Stokes equation*, Uspekhi Mat. Nauk **33** (1978), no. 3, 124. (Russian)
16. *Algebra and the elements of analysis*, 4th ed., textbook for 9th grade of secondary school (A. N. Kolmogorov, ed.), Prosveshchenie, Moscow, 1978, jointly with B. E. Veĭts, I. T. Demidov, O. S. Ivashev-Musatov, and S. I. Shvartsburd. (Russian)
17. *Algebra and the elements of analysis*, 3rd ed., textbook for 10th grade of secondary school (A. N. Kolmogorov, ed.), Prosveshchenie, Moscow, 1978, jointly with O. S. Ivashev-Musatov, B. M. Ivlev, and S. I. Shvartsburd. (Russian; published also in Polish)
18. *Geometry*, 8th ed., textbook for 6th grade of secondary school (A. N. Kolmogorov, ed.), Prosveshchenie, Moscow, 1978, jointly with A. F. Semenovich, F. F. Nagibin, and R. S. Cherkasov. (Russian; published also in Lithuanian and Polish)
19. *Foreword* (editor), Géométrie euclidienne plane, by A. Doneddu (translated from French to Russian), "Nauka", Moscow, 1978.

20. *Geometry*, 7th ed., textbook for 7th grade of secondary school (A. N. Kolmogorov, ed.), Prosveshchenie, Moscow, 1978, jointly with A. F. Semenovich, F. F. Nagibin, and R. S. Cherkasov. (Russian)
21. *Geometry*, 6th ed., textbook for 8th grade of secondary school (A. N. Kolmogorov, ed.), Prosveshchenie, Moscow, 1978, jointly with A. F. Semenovich, F. F. Nagibin, and R. S. Cherkasov. (Russian; published also in Hungarian)

1979

1. *Linear sampling estimates of sums*, Teor. Veroyatnost. i Primenen. **24** (1979), 241–251, jointly with A. V. Bulinskiĭ; English transl. in Theory Probab. Appl. **24** (1979).
2. *The physics and mathematics school of Moscow State University after* 15 *years*, Kvant **1979**, no. 1, 55–57, jointly with V. V. Vavilov and I. T. Tropin. (Russian)
3. *On the textbook "Geometry* 6–8*"*, Mat. v Shkole **1979**, no. 3, 38–42, jointly with A. F. Semenovich and R. S. Cherkasov. (Russian)
4. *Alekseĭ Ivanovich Markushevich* [obituary], Mat. v Shkole **1979**, no. 5, 77–78, jointly with V. D. Belousov, V. G. Boltyanskiĭ, et al. (Russian)
5. *The exponential and logarithmic functions*, Mat. v Shkole **1979**, no. 6, 22–27, jointly with A. M. Abramov, O. S. Ivashev-Musatov, B. M. Ivlev, and S. I. Shvartsburd. (Russian)
6. Editor, *General theory of homology*, Selected Works of P. S. Aleksandrov, "Nauka", Moscow, 1979, jointly with A. A. Mal'tsev, L. S. Pontryagin, and A. N. Tikhonov. (Russian)
7. *Geometry for* 6*th*–8*th grades*, textbook (A. N. Kolmogorov, ed.), Prosveshchenie, Moscow, 1979, jointly with A. F. Semenovich and R. S. Cherkasov. (Russian; published also in Hungarian, Belorussian, and Ukrainian)
8. Editor, *Theory of sets and functions*, Selected Works of P. S. Novikov, "Nauka", Moscow, 1979. (Russian)
9. *Algebra and the elements of analysis*, 5th ed., textbook for 9th grade of secondary school (A. N. Kolmogorov, ed.), Prosveshchenie, Moscow, 1979, jointly with B. E. Veĭts, I. T. Demidov, O. S. Ivashev-Musatov, and S. I. Shvartsburd. (Russian; published also in Hungarian, Belorussian, and Ukrainian)
10. *Algebra and the elements of analysis*, 4th ed., textbook for 10th grade of secondary school (A. N. Kolmogorov, ed.), Prosveshchenie, Moscow, 1979, jointly with O. S. Ivashev-Musatov, B. M. Ivlev, and S. I. Shvartsburd. (Russian)
11. *Geometry*, 8th ed., textbook for 7th grade of secondary school (A. N. Kolmogorov, ed.), Prosveshchenie, Moscow, 1979, jointly with A. F. Semenovich, F. F. Nagibin, and R. S. Cherkasov. (Russian)
12. *Geometry*, 7th ed., textbook for 8th grade of secondary school (A. N. Kolmogorov, ed.), Prosveshchenie, Moscow, 1979, jointly with A. F. Semenovich, V. A. Gusev, and R. S. Cherkasov. (Russian)
13. *Geometry*, textbook for 6th grade of secondary school, Lumina, Kishinev, 1979, jointly with A. F. Semenovich and R. S. Cherkasov. (Moldavian)
14. Editor, *Geometry in* 6*th grade* (*teacher's aid*), 2nd ed., by V. A. Gusev, G. G. Maslov, F. F. Nagibin, et al., Mektep, Alma-Ata, 1979. (Kazakh)

1980

1. *The dialectic-materialistic world view in school mathematics and physics courses*, Kvant **1980**, no. 4, 15–18. (Russian)
2. *On the textbook "Algebra and the elements of analysis* 9–10*"*, Mat. v Shkole **1980**, no. 4, 18–21, jointly with A. M. Abramov, O. S. Ivashev-Musatov, B. M. Ivlev, and S. I. Shvartsburd. (Russian)
3. *Algebra and the elements of analysis*, textbook for 9th and 10th grades of secondary school, Prosveshchenie, Moscow, 1980, jointly with A. M. Abramov, B. E. Veĭts, O. S. Ivashev-Musatov, B. M. Ivlev, and S. I. Shvartsburd. (Russian; published also in Belorussian and Polish)
4. *The mathematician and the historian of mathematics: The works of A. I. Markushevich* (1908–1979), Voprosy Istorii Estestvoznaniya i Tekhniki **1980**, no. 2, 96–100. (Russian)
5. *Geometry*, 2nd ed., textbook for 6th–8th grades of secondary school, Prosveshchenie, Moscow, 1980, jointly with A. F. Semenovich and R. S. Cherkasov. (Russian; published also in Bashkir, Hungarian, Georgian, Kazakh, Kirghiz, Latvian, Lithuanian, Polish, Tadjik, Tatar, Turkmen, Uzbek)

6. *Elementi di teoria della funzion e di analisi funzionale*, Italian transl. of 1976, [15], "Mir", Moscow, 1980, jointly with S. V. Fomin.
7. *Geometria*, 2nd, ed., Pomoc. naukowa dla klas 8–9 szkoly sredniej (A. N. Kolmogorov, ed.), Sviesa, Kaunas, 1980, jointly with A. F. Semenovich, V. A. Gusev, and R. S. Cherkasov. (Polish)
8. *Algebra and the elements of analysis*, 5th ed., textbook for 10th grade of secondary school (A. N. Kolmogorov, ed.), Prosveshchenie, Moscow, 1980, jointly with O. S. Ivashev-Musatov, B. M. Ivlev, and S. I. Shvartsburd. (Russian; published also in Karakalpak)
9. *Geometry*, 8th ed., textbook for 8th grade of secondary school (A. N. Kolmogorov, ed.), Prosveshchenie, Moscow, 1980, jointly with A. F. Semenovich, V. A. Gusev, and R. S. Cherkasov. (Russian; published also in Uzbek)
10. *Geometry*, textbook for 6th grade of secondary school, Luĭs, Erevan, 1980, jointly with A. F. Semenovich and R. S. Cherkasov. (Armenian; published also in Uigur)
11. *Geometry*, textbook for 7th grade of secondary school, Lumina, Kishinev, 1980, jointly with A. F. Semenovich and R. S. Cherkasov. (Moldavian)

1981

1. *Elements of the theory of functions and functional analysis*, 5th rev. ed., "Nauka", Moscow, 1981, jointly with S. V. Fomin. (Russian)
2. *The physics and mathematics school of Moscow State University* (Matematika, kibernetika, no. 5), Izdat. Moskov. Gos. Univ., Moscow, 1981, jointly with V. V. Vavilov and I. T. Tropin. (Russian)
3. *Geometry for 6th–8th grades*, 3rd ed., textbook (A. N. Kolmogorov, ed.), Prosveshchenie, Moscow, 1981, jointly with A. F. Semenovich and R. S. Cherkasov. (Russian)
4. *On the concept of vector in the secondary school curriculum*, Mat. v Shkole **1981**, no. 3, 7–8. (Russian)
5. *On giving the first lessons on the topic "Vectors"*, Mat. v Shkole **1981**, no. 3, 8–11, jointly with A. M. Abramov. (Russian)
6. *Review of L. S. Pontryagin's book "The analysis of infinitesimals"*, Mat. v Shkole **1981**, no. 5, 73–74. (Russian)
7. *Izabella Grigor′evna Bashmakova* (*On her sixtieth birthday*), Mat. v Shkole **1981**, no. 1, 73–74, jointly with P. S. Aleksandrov, B. V. Gnedenko, S. S. Demidov, S. S. Petrov, K. A. Rybnikov, and A. P. Yushkevich. (Russian)
8. *On sampling estimates of sums of random variables*, Trudy Inst. Prikl. Geofiziki **46** (1981), 73–77, jointly with A. V. Bulinskiĭ. (Russian)
9. Editor, *Mathematics of the nineteenth century*, "Nauka", Moscow, 1981, jointly with A. P. Yushkevich. (Russian)
10. *Sagdy Khasanovich Sirazhdinov* (*On his sixtieth birthday*), Uspekhi Mat. Nauk **36** (1981), no. 6, 237–242, jointly with T. A. Azlarov, B. V. Gnedenko, Yu. V. Prokhorov, T. A. Sarymsakov, and V. A. Statulyavichus; English transl. in Russian Math. Surveys **36** (1981).
11. Editor, *Some questions in mathematics and mechanics*, Izdat. Moskov. Gos. Univ., Moscow, 1981. (Russian)
12. *Aleksandr Filippovich Timan* (*On his sixtieth birthday*), Uspekhi Mat. Nauk **36** (1981), no. 2, 221–225, jointly with V. F. Vlasov, V. G. Ponomarenko, V. N. Trofimov, and F. B. Yudin; English transl. in Russian Math. Surveys **36** (1981).
13. *Foreword*, Kalejdoskop Matematyczny, by H. Steinhaus (translated from Polish to Russian), "Kvant" Library series, no. 8, "Nauka", Moscow, 1981, pp. 3–4.
14. *To the memory of Anatoliĭ Illarionovich Shirshov*, Uspekhi Mat. Nauk **36** (1981), no. 5, 153–157, jointly with L. A. Bokut′, Yu. L. Ershov, A. I. Kostrikin, E. N. Kuz′min, V. N. Latyshev, S. L. Sobolev, and I. P. Shestakov; English transl. in Russian Math. Surveys **36** (1981).
15. *Naum Il′ich Akhiezer* [obituary], Uspekhi Mat. Nauk **36** (1981), no. 4, 183–184, jointly with M. G. Kreĭn, B. Ya. Levin, Yu. I. Lyubich, V. A. Marchenko, I. V. Ostrovskiĭ, A. Ya. Povzner, and A. V. Pogorelov; English transl. in Russian Math. Surveys **36** (1981).
16. *To the memory of Mikhail Alekseevich Lavrent′ev*, Uspekhi Mat. Nauk **36** (1981), no. 2, 3–10, jointly with P. S. Aleksandrov, N. N. Bogolyubov, L. A. Lyusternik, G. I. Marchuk, S. L. Sobolev, and B. V. Shabat; English transl. in Russian Math. Surveys **36** (1981).
17. *Vladimir Mikhaĭlovich Alekseev* [obituary], Uspekhi Mat. Nauk **36** (1981), no. 4, 177–182, jointly with D. V. Anosov, V. I. Arnol′d, M. I. Zelikin, et al.; English transl. in Russian Math. Surveys **36** (1981).

18. *A függvényelmélet és a funkcionálanalízis elemei*, Hungarian transl. of 1976, [15], Muszaki Konuvkiadó, Budapest, 1981, jointly with S. V. Fomin.
19. *Algebra and the elements of analysis*, 2nd ed., textbook for 9th and 10th grades of secondary school (A. N. Kolmogorov, ed.), Prosveshchenie, Moscow, 1981, jointly with A. M. Abramov, B. E. Veĭts, O. S. Ivashev-Musatov, B. M. Ivlev, and S. I. Shvartsburd. (Russian; published also in Azerbaijani, Armenian, Belorussian, Georgian, Kazakh, Kirghiz, Latvian, Lithuanian, Tatar, Uzbek, Uigur, and Ukrainian)
20. *Geometry*, 3rd rev. ed., textbook for 6th–8th grades of secondary school (A. N. Kolmogorov, ed.), Prosveshchenie, Moscow, 1981, jointly with A. F. Semenovich and P. S. Cherkasov. (Russian; published also in Azerbaijani, Armenian, Lithuanian, Moldavian, Uigur, and Tuvinian)

1982

1. *Introduction to mathematical logic*, Izdat. Moskov. Gos. Univ., Moscow, 1982, jointly with A. G. Dragalin. (Russian)
2. *Introduction to probability theory*, "Kvant" Library series, no. 23, "Nauka", Moscow, 1982, jointly with I. G. Zhurbenko and A. V. Prokhorov. (Russian)
3. *Mathematics*, Math. Encyclopedia, Vol. 3, 1982, pp. 560–564, jointly with Yu. V. Prokhorov. (Russian)
4. *Boris Vladimirovich Gnedenko*, Mat. v Shkole **1982**, no. 1, 72–73, jointly with R. S. Cherkasov. (Russian)
5. *Leonid Vital'evich Kantorovich (On his seventieth birthday)*, Mat. v Shkole **1982**, no. 2, 77–78, jointly with V. A. Zalgaller. (Russian)
6. *On the concept of limit in schools of general education*, Mat. v Shkole **1982**, no. 5, 56. (Russian)
7. *Newton and contemporary mathematical thought*, Mat. v Shkole **1982**, no. 5, 58–64.
8. *Mathematical statistics*, Math. Encyclopedia, Vol. 3, 1982, pp. 576–581, jointly with Yu. V. Prokhorov. (Russian)
9. *On tables of random numbers*, Semiotika i Informatika, Akad. Nauk SSSR, VINITI **18** (1982), 3–13. (Russian)
10. *Dmitriĭ Evgen'evich Men'shov (On his ninetieth birthday)*, Uspekhi Mat. Nauk **37** (1982), no. 5, 209–219, jointly with S. M. Nikol'skiĭ, S. A. Skvortsov, and P. L. Ul'yanov; English transl. in Russian Math. Surveys **37** (1982).
11. *Boris Vladimirovich Gnedenko (On his seventieth birthday)*, Uspekhi Mat. Nauk **37** (1982), no. 6, 243–248, jointly with Yu. K. Belyaev and A. D. Solov'ev; English transl. in Russian Math. Surveys **37** (1981).
12. *Elementos da teoria das funcões e de análise funcional*, Portugese transl. of 1981, [1], "Mir", Moscow, 1982, jointly with S. V. Fomin.
13. *Basic concepts of the theory of probability*, Bernoulli, Laplace, and Kolmogorov in probability, "Nauka i Izkustvo", Sofia, 1982, pp. 155–249. (Bulgarian)
14. *A valószinűségszámitás alapfogalmai*, Hungarian transl. of 1974, [1], Gondolat, Budapest, 1982.
15. *Algebra and the elements of analysis*, 3rd ed., textbook for 9th and 10th grades of secondary school (A. N. Kolmogorov, ed.), Prosveshchenie, Moscow, 1982, jointly with A. M. Abramov, B. E. Veĭts, O. S. Ivashev-Musatov, B. M. Ivlev, and S. I. Shvartsburd. (Russian; published also in Belorussian, Hungarian, Polish, and Tadjik)
16. *Geometry*, 4th rev. ed., textbook for 6th–8th grades of secondary school (A. N. Kolmogorov, ed.), Prosveshchenie, Moscow, 1982, jointly with A. F. Semenovich and P. S. Cherkasov. (Russian)

1983

1. *Algebra and the elements of analysis*, 4th ed., textbook for 9th and 10th grades of secondary school, Prosveshchenie, Moscow, 1983, jointly with A. M. Abramov, B. E. Veĭts, O. S. Ivashev-Musatov, B. M. Ivlev, and S. I. Shvartsburd. (Russian)
2. *Pavel Sergeevich Aleksandrov*, Mat. v Shkole **1983**, no. 1, 47–48, jointly with B. V. Gnedenko. (Russian)
3. *On the textbook "Geometry", by A. V. Pogorelov*, Mat. v Shkole **1983**, no. 2, 45–46. (Russian)
4. *Combinatorial foundations of information theory and the calculus of probabilities*, Uspekhi Mat. Nauk **38** (83), no. 4, 27–36; English transl. in Russian Math. Surveys **38** (1983).
5. *On the logical foundations of probability*, Lecture Notes in Math. **1021** (1983), 1–5.

6. Editor, *Probabilistic-statistical methods of investigation*, Izdat. Moskov. Gos. Univ., Moscow, 1983, jointly with I. G. Zhurbenko. (Russian)

1984

1. *Remarks on the concept of a set in school mathematics courses*, Mat. v Shkole **1984**, no. 1, 52–53. (Russian)
2. *S. L. Sobolev and contemporary mathematics*, Mat. v Shkole **1984**, no. 1, 73–77, jointly with O. A. Oleĭnik. (Russian)
3. *Analysis of the metric structure of A. S. Pushkin's poem "Arion"*, Problems in the Theory of Poetry, "Nauka", Leningrad, 1984, pp. 118–124. (Russian)
4. *Mathematical logic: Supplementary chapters*, Izdat. Moskov. Gos. Univ., Moscow, 1984, jointly with A. G. Dragalin. (Russian)

1985

1. *A model of the rhythmic structure of Russian speech suitable for studying the metrics of classical Russian verse*, Russian versification. Tradition and Problems of Evolution, "Nauka", Moscow, 1985, pp. 113–134. (Russian)
2. *Selected works. Mathematics and mechanics*, "Nauka", Moscow, 1985. (Russian)
3. *A new metric invariant of transitive dynamical systems and automorphisms of Lebesgue spaces. New revision*, Trudy Mat. Inst. Steklov. **169** (1985), 94–98; English transl. in Proc. Steklov Inst. Math. **169** (1986).
4. *A student on his teacher* (interview of Kolmogorov by V. A. Uspenskiĭ about N. N. Luzin), Uspekhi Mat. Nauk **40** (1985), no. 3, 7–8; English transl. in Russian Math. Surveys **40** (1985).
5. *Foreword* (editor), Works on Set Theory by G. Cantor, "Nauka", Moscow, 1985, pp. 373–381, jointly with A. P. Yushkevich. (Russian)
6. *Guriĭ Ivanovich Marchuk (On his sixtieth birthday)*, Uspekhi Mat. Nauk **40** (1985), no. 5, 3–17, jointly with N. N. Bogolyubov and V. S. Vladimirov; English transl. in Russian Math. Surveys **40** (1985).
7. *Sergeĭ Mikhaĭlovich Nikol'skiĭ (On his eightieth birthday)*, Uspekhi Mat. Nauk **40** (1985), no. 5, 3–17, jointly with V. K. Dzyadyk, L. D. Kudryavtsev, and S. L. Sobolev; English transl. in Russian Math. Surveys **40** (1985).
8. *Algebra and the elements of analysis*, 5th ed., textbook for 9th and 10th grades of secondary school (A. N. Kolmogorov, ed.), Prosveshchenie, Moscow, 1985, jointly with A. M. Abramov, B. E. Veĭts, O. S. Ivashev-Musatov, B. M. Ivlev, and S. I. Shvartsburd. (Russian; published also in Karakalpak)

1986

1. *On scalar quantities*, Mat. v Shkole **1986**, no. 3, 32–33. (Russian)
2. *Memories of P. S. Aleksandrov*, Uspekhi Mat. Nauk **41** (1986), no. 6, 187–203; English transl. in Russian Math. Surveys **41** (1986).
3. *Selected works. Probability theory and mathematical statistics*, "Nauka", Moscow, 1986. (Russian)
4. *Foreword*, On the Law of Large Numbers, by Jakob Bernoulli, "Nauka", Moscow, 1986, pp. 3–5. (Russian)
5. *Problems of cybernetics today*, A time to search, Molodaya Gvardiya, Moscow, 1986, pp. 99–110, the paper "Automata and life", dated 1961. (Russian)
6. *Algebra and the elements of analysis*, 6th ed., textbook for 9th and 10th grades of secondary school (A. N. Kolmogorov, ed.), Prosveshchenie, Moscow, 1986, jointly with A. M. Abramov, B. E. Veĭts, O. S. Ivashev-Musatov, B. M. Ivlev, and S. I. Shvartsburd. (Russian; published also in Ukrainian)
7. *Foreword* (editor), The Moscow Mathematics Olympiads, by G. A. Gal'perin and A. K. Tolpygo, Prosveshchenie, Moscow, 1986, pp. 3–4. (Russian)

1987

1. *Selected works. Information theory and the theory of algorithms*, "Nauka", Moscow, 1987. (Russian)
2. *Welcoming speech to the participants of the 1st World Congress of the Bernoulli Society*, Teor. Veroyatnost. i Primenen. **32** (1987), 218; English transl. in Theory Probab. Appl. **32** (1987).
3. *Algorithms and randomness*, Teor. Veroyatnost. i Primenen. **32** (1987), 425–455; English transl. in Theory Probab. Appl. **32** (1987).

4. *Foreword* (editor), Mathematics of the Nineteenth Century, "Nauka", Moscow, 1987, pp. 7–8, jointly with A. P. Yushkevich. (Russian)
5. Editor, *John von Neumann. Selected works on functional analysis*, Vols. 1, 2, "Nauka", Moscow, 1987. (Russian)
6. *John von Neumann*, John von Neumann, Selected Works on Functional Analysis, Vol. 1, "Nauka", Moscow, 1987, pp. 337–351, jointly with A. M. Vershik and Ya. G. Sinaĭ. (Russian)
7. *Algorithms and randomness*, Proc. 1st World Congress of the Bernoulli Society (Tashkent, 8–14 September 1986) (Yu. V. Prokhorov and V. V. Sazonov, eds.), Vol. 1, VNU Science Press, Utrecht, 1987, pp. 3–56, jointly with V. A. Uspenskiĭ.
8. *Algebra and the elements of analysis*, 7th ed., textbook for 9th and 10th grades of secondary school (A. N. Kolmogorov, ed.), Prosveshchenie, Moscow, 1987, jointly with A. M. Abramov, B. E. Veĭts, O. S. Ivashev-Musatov, B. M. Ivlev, and S. I. Shvartsburd. (Russian; published also in Latvian, Lithuanian, and Turkmen)
9. *Boris Vladimirovich Gnedenko (On his seventy-fifth birthday)*, Mat. v Shkole **1987**, no. 2, 62–63, jointly with I. G. Bashmakova, Yu. K. Belyaev, S. S. Petrova, A. D. Solov′ev, and R. S. Cherkasov. (Russian)

1988

1. *Mathematics: science and profession*, "Kvant" Library series, no. 64, "Nauka", Moscow, 1988, jointly with G. A. Gal′perin. (Russian)
2. *Probabilistic-statistical methods of observing spontaneously arising effects*, Trudy Mat. Inst. Steklov. **182** (1988), 4–23; English transl. in Proc. Steklov Inst. Math. **1990**, no. 1.
3. *Letters of A. N. Kolmogorov to (the Dutch mathematician) A. Heyting*, Uspekhi Mat. Nauk **43** (1988), no. 6, 75–77; English transl. in Russian Math. Surveys **43** (1988).
4. *Algebra and the elements of analysis*, 8th ed., textbook for 9th and 10th grades of secondary school (A. N. Kolmogorov, ed.), Prosveshchenie, Moscow, 1988, jointly with A. M. Abramov, B. E. Veĭts, O. S. Ivashev-Musatov, B. M. Ivlev, and S. I. Shvartsburd. (Russian; published also in Armenian, Belorussian, Georgian, Kazakh, Kirghiz, Latvian, Moldavian, Lithuanian, Tadjik, Uzbek, and Ukrainian)
5. *Elements of the theory of functions and functional analysis*, Vol. 1, Dari transl. of 1954, [12], "Mir", Moscow, 1988, jointly with S. V. Fomin.
6. *Elements of the theory of functions and functional analysis*, Classical Arabic transl. of 1981, [1], "Mir", Moscow, 1988, jointly with S. V. Fomin.

1989

1. *Elements of the theory of functions and functional analysis*, 6th rev. ed., "Nauka", Moscow, 1989, jointly with S. V. Fomin. (Russian)
2. *Algebra i początki analizy*, Pomoc naukowa dla klas 10–12 Szkoly średniej (A. N. Kolmogorov, ed.), Šviesa, Kaunas, 1989, jointly with A. M. Abramov, B. E. Veĭts, O. S. Ivashev-Musatov, B. M. Ivlev, and S. I. Shvartsburd. (Polish; published also in Azerbaijani, Armenian, Georgian, Karakalpak, Latvian, Lithuanian, Tadjik, and Ukrainian)

1990

1. *Algebra and the elements of analysis*, textbook for 10th and 11th grades of secondary school (A. N. Kolmogorov, ed.), Prosveshchenie, Moscow, 1990, jointly with A. M. Abramov, Yu. P. Dudnitsyn, B. M. Ivlev, and S. I. Shvartsburd. (Russian; published also in Azerbaijani, Armenian, Belorussian, Georgian, Kazakh, Uzbek, and Ukrainian)

1991

1. *Algebra and the elements of analysis*, 2nd ed., textbook for 10th and 11th grades of secondary school (A. N. Kolmogorov, ed.), Prosveshchenie, Moscow, 1991, jointly with A. M. Abramov, Yu. P. Dudnitsyn, B. M. Ivlev, and S. I. Shvartsburd. (Russian; published also in Belorussian, Georgian, Moldavian, Kazakh, Kirghiz, Turkmen, Ukrainian, and Uigur)
2. *Algorithm, information, complexity*, Thematic collection of (6) papers completed in the 1950's and 1960's, Znanie, Moscow, 1991. (Russian)
3. V. M. Tikhomirov (ed.), *Selected works of A. N. Kolmogorov*. Vol. 1: *Mathematics and mechanics*, Kluwer, Dordrecht, 1991.

1992

1. A. N. Shiryaev (ed.), *Selected works of A. N. Kolmogorov*. Vol. 2: *Probability theory and mathematical statistics*, Kluwer, Dordrecht, 1992.

2. *Algebra and the elements of analysis*, textbook for 10th and 11th grades of secondary school (A. N. Kolmogorov, ed.), "Radyans'ka Shkola", Kiev, 1992. (Ukrainian; published also in Uzbek and Turkmen)

1993

1. A. N. Shiryaev (ed.), *Selected works of A. N. Kolmogorov*. Vol. 3: *Information theory and the theory of algorithms*, Kluwer, Dordrecht, 1993.
2. *Problems in probability theory* (*theses of reports at the Session of the Moscow Mathematical Society* 11 *December* 1944, Teor. Veroyatnost. i Primenen. **38** (1993); English transl. in Theory Probab. Appl. **38** (1993), 177–178.
3. *Algebra and the elements of analysis*, 3rd ed., textbook for 10th and 11th grades of secondary school (A. N. Kolmogorov, ed.), Prosveshchenie, Moscow, 1993, jointly with A. M. Abramov, Yu. P. Dudnitsyn, B. M. Ivlev, and S. I. Shvartsburd. (Russian)

II. Encyclopedia articles by Kolmogorov

Equation, Great Sov. Encycl., Vol. 56, 1936, pp. 163–165.

Continuum, Great Sov. Encycl., Vol. 34, 1937, pp. 139–140.

Markov, Andreĭ Andreevich , Great Sov. Encycl., Vol. 38, 1938, pp. 152–153.

Mathematics, Great Sov. Encycl., Vol. 38, 1938, pp. 359–402.

Mathematical induction, Great Sov. Encycl., Vol. 38, 1938, pp. 405–406.

Measure, Great Sov. Encycl., Vol. 38, 1938, pp. 831– 832.

Multidimensional space, Great Sov. Encycl., Vol. 39, 1938, pp. 577–578.

Orientation, Great Sov. Encycl., Vol. 43, 1939, pp. 342–344.

Surface, Great Sov. Encycl., Vol. 45, 1940, pp. 746–748.

The development of mathematics in the USSR, Great Sov. Encycl., "USSR" Vol., 1947, pp. 1318–1323.

Average values, Great Sov. Encycl., Vol. 52, 1947, p. 508–509.

Absolute value, Great Sov. Encycl., 2nd ed., Vol. 1, 1949, p. 32.

Absolute geometry, Great Sov. Encycl., 2nd ed., Vol. 1, 1949, p. 33.

Hadamard, Jacques, Great Sov. Encycl., 2nd ed., Vol. 1, 1949, p. 388.

Additive quantities, Great Sov. Encycl., 2nd ed., Vol. 1, 1949, p. 394.

Axiom, Great Sov. Encycl., 2nd ed., Vol. 1, 1949, pp. 613–616.

Perspective geometry, Great Sov. Encycl., 2nd ed., Vol. 1, 1949, p. 617.

Algebra in secondary school, Great Sov. Encycl., 2nd ed., Vol. 2, 1950, pp. 61–62.

Algebraic expression, Great Sov. Encycl., 2nd ed., Vol. 2, 1950, p. 64.

Algorithm, Great Sov. Encycl., 2nd ed., Vol. 2, 1950, p. 65.

Euclidean algorithm, Great Sov. Encycl., 2nd ed., Vol. 2, 1950, pp. 65–67.

Aleksandrov, Aleksandr Danilovich, Great Sov. Encycl., 2nd ed., Vol. 2, 1950, p. 83

Aleksandrov, Pavel Sergeevich, Great Sov. Encycl., 2nd ed., Vol. 2, 1950, p. 84.

Mathematical analysis, Great Sov. Encycl., 2nd ed., Vol. 2, 1950, pp. 325–326.

Asymptote, Great Sov. Encycl., 2nd ed., Vol. 3, 1950, pp. 238–239.

Asymptotic expressions, Great Sov. Encycl., 2nd ed., Vol. 3, 1950, p. 239.

Naum Il'ich Akhiezer, Great Sov. Encycl., 2nd ed., Vol. 3, 1950, p. 565.

Banach, Stefan, Great Sov. Encycl., 2nd ed., Vol. 4, 1950, p. 183.

Bari, Nina Karlovna, Great Sov. Encycl., 2nd ed., Vol. 4, 1950, p. 245.

Bernshteĭn, Sergeĭ Natanovich, Great Sov. Encycl., 2nd ed., Vol. 5, 1950, p. 52.

Infinitely large values, Great Sov. Encycl., 2nd ed., Vol. 5, 1950, p. 66–67.

Infinitesimals, Great Sov. Encycl., 2nd ed., Vol. 5, 1950 , pp. 67–71, jointly with V. F. Kagan.

Infinitely remote elements, Great Sov. Encycl., 2nd ed., Vol. 5, 1950, pp. 71–72, jointly with B. N. Delone.

Infinity (in mathematics), Great Sov. Encycl., 2nd ed., Vol. 5, 1950, pp. 73–74.

Biharmonic functions, Great Sov. Encycl., 2nd ed., Vol. 5, 1950, p. 159.

Bilinear form, Great Sov. Encycl., 2nd ed., Vol. 5, 1950, p. 167.

The law of large numbers, Great Sov. Encycl., 2nd ed., Vol. 5, 1950, pp. 538-540.

Brouwer, Luitzen E. J., Great Sov. Encycl., 2nd ed., Vol. 6, 1951, pp. 62–63, jointly with S. A. Yanovskaya.

Order statistics (variational series), Great Sov. Encycl., 2nd ed., Vol. 6, 1951, p. 641.

Weyl, Hermann, Great Sov. Encycl., 2nd ed., Vol. 7, 1951, pp. 106–107, jointly with S. A. Yanovskaya.

Quantity, Great Sov. Encycl., 2nd ed., Vol. 7, 1951, pp. 340–341.

Probable deviation, Great Sov. Encycl., 2nd ed., Vol. 7, 1951, pp. 507–508.

Probability, Great Sov. Encycl., 2nd ed., Vol. 7, 1951, pp. 508–510.

Sampling procedure, Great Sov. Encycl., 2nd ed., Vol. 9, 1951, pp. 417–418, jointly with T. I. Kozlov.

Gaussian distribution, Great Sov. Encycl., 2nd ed., Vol. 10, 1952, p. 275.

Geodesic curvature, Great Sov. Encycl., 2nd ed., Vol. 10, 1952, p. 481.

Geodesic coordinates, Great Sov. Encycl., 2nd ed., Vol. 10, 1952, p. 486.
Hilbert, David, Great Sov. Encycl., 2nd ed., Vol. 11, 1952, pp. 370–371.
Histogram, Great Sov. Encycl., 2nd ed., Vol. 11, 1952, p. 447.
Gnedenko, Boris Vladimirovich, Great Sov. Encycl., 2nd ed., Vol. 11, 1952, p. 545.
Homeomorphism, Great Sov. Encycl., 2nd ed., Vol. 12, 1952, p. 21.
Homotopy, Great Sov. Encycl., 2nd ed., Vol. 12, 1952, p. 35.
Graph, Great Sov. Encycl., 2nd ed., Vol. 12, 1952, pp. 453–454.
Motion (in geometry), Great Sov. Encycl., 2nd ed., Vol. 13, 1952, pp. 447–448.
Binomial, Great Sov. Encycl., 2nd ed., Vol. 13, 1952, p. 518.
Real numbers, Great Sov. Encycl., 2nd ed., Vol. 13, 1952, pp. 570–571.
Division, Great Sov. Encycl., 2nd ed., Vol. 13, 1952, p. 628.
Discreteness, Great Sov. Encycl., 2nd ed., Vol. 14, 1952, p. 425.
Variance, Great Sov. Encycl., 2nd ed., Vol. 14, 1952, pp. 438–439.
Distributivity, Great Sov. Encycl., 2nd ed., Vol. 14, 1952, p. 479.
Distributive operator, Great Sov. Encycl., 2nd ed., Vol. 14, 1952, p. 479.
Differential, Great Sov. Encycl., 2nd ed., Vol. 14, 1952, p. 497.
Differential equations, Great Sov. Encycl., 2nd ed., Vol. 14, 1952, pp. 520–526, jointly with B. P. Demidovich and V. V. Nemytskiĭ.
Confidence probability, Great Sov. Encycl., 2nd ed., Vol. 14, 1952, pp. 616–617.
Confidence bounds, Great Sov. Encycl., 2nd ed., Vol. 14, 1952, p. 617.
Mathematical signs, Great Sov. Encycl., 2nd ed., Vol. 17, 1952, pp. 115–119, jointly with I. G. Bashmakova and A. P. Yushkevich.
Significant digits, Great Sov. Encycl., 2nd ed., Vol. 17, 1952, p. 135.
Isomorphism, Great Sov. Encycl., 2nd ed., Vol. 17, 1952, pp. 478–479, jointly with V. I. Bityutskov.
Isotropic lines, Great Sov. Encycl., 2nd ed., Vol. 17, 1952, p. 509.
Concrete number, Great Sov. Encycl., 2nd ed., Vol. 17, 1952, p. 557.
Mathematical induction, Great Sov. Encycl., 2nd ed., Vol. 18, 1953, pp. 146–147.
Integral, Great Sov. Encycl., 2nd ed., Vol. 18, 1953, pp. 250–253, jointly with V. I. Glivenko.
Probability integral, Great Sov. Encycl., 2nd ed., Vol. 18, 1953, p. 253.
Interpolation, Great Sov. Encycl., 2nd ed., Vol. 18, 1953, pp. 304–305.
Intuitionism, Great Sov. Encycl., 2nd ed., Vol. 18, 1953, p. 319.
Elimination of unknowns, Great Sov. Encycl., 2nd ed., Vol. 18, 1953, p. 483.
Trial, Great Sov. Encycl., 2nd ed., Vol. 18, 1953, p. 604.
Method of exhaustion, Great Sov. Encycl., 2nd ed., Vol. 19, 1953, pp. 50–51.
Quadrant, Great Sov. Encycl., 2nd ed., Vol. 20, 1953, p. 434.
Compact space, Great Sov. Encycl., 2nd ed., Vol. 22, 1953, p. 282.
Constant, Great Sov. Encycl., 2nd ed., Vol. 22, 1953, p. 416.
Continuum, Great Sov. Encycl., 2nd ed., Vol. 22, 1953, pp. 454–455.
Coordinates, Great Sov. Encycl., 2nd ed., Vol. 22, 1953, pp. 524–525.
Correlation, Great Sov. Encycl., Vol. 23, 2nd ed., 1953, pp. 55–58.
Curve, Great Sov. Encycl., 2nd ed., Vol. 25, 1954, pp. 167–170.
The law of small numbers, Great Sov. Encycl., 2nd ed., Vol. 26, 1954, p. 169.
Markov, Andreĭ Andreevich, Great Sov. Encycl., 2nd ed., Vol. 26, 1954, pp. 294–295.
Mathematics, Great Sov. Encycl., 2nd ed., Vol. 26, 1954, pp. 464–483.
Mathematical statistics, Great Sov. Encycl., 2nd ed., Vol. 26, 1954, pp. 485–490.
Mathematical physics, Great Sov. Encycl., 2nd ed., Vol. 26, 1954, p. 490.
Mises, Richard von, Great Sov. Encycl., 2nd ed., Vol. 27, 1954, p. 414.
Multidimensional space, Great Sov. Encycl., 2nd ed., Vol. 27, 1954, pp. 660–661.
Set theory, Great Sov. Encycl., 2nd ed., Vol. 28, 1954, pp. 14–17, jointly with P. S. Aleksandrov.
Orientation, Great Sov. Encycl., 2nd ed., Vol. 31, 1955, pp. 188–189.
Foundations of geometry, Great Sov. Encycl., 2nd ed., Vol. 31, 1955, pp. 296–297.
Surface, Great Sov. Encycl., 2nd ed., Vol. 33, 1955, pp. 346-347, jointly with L. A. Skornyakov.

Ordinal numbers, Great Sov. Encycl., 2nd ed., Vol. 34, 1955, p. 227.
Statistical acceptance sampling, Great Sov. Encycl., 2nd ed., Vol. 34, 1955, pp. 498–499.
Slutskiĭ, Evgeniĭ Borisovich, Great Sov. Encycl., 2nd ed., Vol. 39, 1956, p. 378.
Smirnov, Nikolaĭ Vasil'evich, Great Sov. Encycl., 2nd ed., Vol. 39, 1956, p. 406.
Sufficient statistic, Great Sov. Encycl., 2nd ed., Vol. 51, 1958, p. 106.
Information, Great Sov. Encycl., 2nd ed., Vol. 51, 1958, pp. 129–130.
Cybernetics, Great Sov. Encycl., 2nd ed., Vol. 51, 1958, pp. 149–151.

Quantity, Great Sov. Encycl., 3rd ed., Vol. 4, 1971, pp. 456–457.
Wiener, Norbert, Great Sov. Encycl., 3rd ed., Vol. 5, 1971, p. 72.
Hilbert, David, Great Sov. Encycl., 3rd ed., Vol. 6, 1971, p. 519.
Measurement, Great Sov. Encycl., 3rd ed., Vol. 10, 1972, pp. 220–221.
Integral, Great Sov. Encycl., 3rd ed., Vol. 10, 1972, pp. 300–302.
Method of exhaustion, Great Sov. Encycl., 3rd ed., Vol. 10, 1972, p. 586.
Continuum, Great Sov. Encycl., 3rd ed., Vol. 13, 1973, p. 64.
Markov, Andreĭ Andreevich, Great Sov. Encycl., 3rd ed., Vol. 15, 1974, p. 379.
Mathematics, Great Sov. Encycl., 3rd ed., Vol. 3, 1982, pp. 560–564.
Mathematical statistics, Great Sov. Encycl., 3rd ed., Vol. 15, 1974, pp. 480–484, jointly with Yu. V. Prokhorov.
Multidimensional space, Great Sov. Encycl., 3rd ed., Vol. 16, 1974, p. 372.
Orientation, Great Sov. Encycl., 3rd ed., Vol. 18, 1974, pp. 509–510.
Statistical acceptance sampling, Great Sov. Encycl., 3rd ed., Vol. 20, 1975, pp. 572–573, jointly with Yu. K. Belyaev.
Infinity, Mat. Encycl., Vol. 1, 1977, pp. 455–458.
Quantity, Mat. Encycl., Vol. 1, 1977, pp. 651–653.
Probability, Mat. Encycl., Vol. 1, 1977, pp. 667–669.
Mathematics, Mat. Encycl., Vol. 3, 1982, pp. 560–564.
Mathematical statistics, Mat. Encycl., Vol. 3, 1982, pp. 576–581.

III. Articles by Kolmogorov in the journal "Matematika v Shkole"

1. *Properties of inequalities and the concept of approximate computations,* **1941**, no. 2, 1–12, jointly with P. S. Aleksandrov.
2. *Irrational numbers,* **1941**, no. 3, 1–15, jointly with P. S. Aleksandrov.
3. *The body of mathematical knowledge for eight-year schools* (Committee on Mathematical Education of the Mathematical Section of the USSR Academy of Sciences), **1965**, no. 2, 21–24.
4. *Geometric transformations in a school geometry course,* **1965**, no. 2, 24–29.
5. *On the content of a school mathematics course,* **1965**, no. 4, 53–62, jointly with I. M. Yaglom.
6. *Functions, graphs, and continuous functions,* **1965**, no. 6, 12–21.
7. *On textbooks for the* 1966–1967 *academic year,* **1966**, no. 3, 26–30.
8. *On a school definition of identity,* **1966**, no. 2, 33–35.
9. *Introduction to S. B. Suvorova's article "An experiment on introducing the fundamentals of differential calculus at an early stage",* **1966**, no. 4, 23.
10. *On textbooks for the* 1966–1967 *academic year,* **1966**, no. 6, 31–37.
11. *On textbooks for the* 1966–1967 *academic year,* **1967**, no. 1, 43–48.
12. *On elective studies in mathematics,* **1967**, no. 2, 2–3.
13. *New curricula and some basic questions on improving the program of mathematics in secondary schools,* **1967**, no. 2, 4–13.
14. *The content of elective studies in mathematics for the academic years of* 1967–1968 *and* 1968–1969, **1967**, no. 2, 33–38.
15. *Curricula of special courses in mathematics,* **1967**, no. 3, 73–75.
16. *Curricula of special courses in mathematics,* **1967**, no. 4, 58–59.
17. *On changes in the text of A. N. Barsukov's algebra book for grades* 6–9, **1967**, no. 6, 22–24.
18. *Curricula project for secondary schools in mathematics,* **1967**, no. 1, 4–23, jointly with A. I. Markushevich, I. M. Yaglom, et al.
19. *A generalization of the concept of a power and the exponential function,* **1968**, no. 1, 24–32.
20. *A mathematics curriculum for secondary schools,* **1968**, no. 2, 5–20.
21. *On the study of the exponential function and logarithms in eight-year schools,* **1968**, no. 2, 23–25.
22. *On new curricula in mathematics,* **1968**, no. 2, 21–22.
23. *Introduction to probability theory and combinatorial analysis,* **1968**, no. 2, 63–72.
24. *A supplement to Yu. A. Shikhanovich's review (of A. A. Stolyar's book),* **1968**, no. 3, 92.
25. *Letter to the editor (about inaccuracies in B. E. Veĭts' article),* **1969**, no. 2, 93.
26. *Scientific foundations for a school course in mathematics. First lecture: Contemporary views on the nature of mathematics,* **1969**, no. 3, 12–17.
27. *Scientific foundations for a school course in mathematics. Second lecture: Natural numbers,* **1969**, no. 5, 8–17.
28. *Scientific foundations for a school course in mathematics. Third lecture: Generalizing the notion of number. Nonnegative rational numbers,* **1970**, no. 2, 27–32.
29. *On a trial geometry book for* 6*th grade,* **1970**, no. 4, 21–34, jointly with A. F. Semenovich.
30. *Educational materials on geometry for* 5*th grade,* **1970**, no. 5, 30–45, jointly with A. F. Semenovich and R. S. Cherkasov.
31. *On a system of basic concepts and notation for a school course in mathematics,* **1971**, no. 2, 17–22.
32. *A trial geometry book for* 7*th grade,* **1971**, no. 3, 9–17, jointly with A. F. Semenovich, F. F. Nagibin, and R. S. Cherkasov.
33. *Summary of the report "Elements of logic in secondary school" (to the Committee on Mathematics of the Academic Methods Council of the USSR Ministry of Education),* **1971**, no. 3, 91–92.
34. *On a new edition of the trial geometry book for* 6*th grade,* **1971**, no. 4, 23–35, jointly with A. F. Semenovich, F. F. Nagibin, and R. S. Cherkasov.

35. *On a new edition of the trial geometry book for 6th grade,* **1971**, no. 5, 25–28, jointly with A. F. Semenovich, F. F. Nagibin, and R. S. Cherkasov.
36. *From the new geometry textbook for 6th grade (Geometric constructions),* **1971**, no. 6, 13–21, jointly with A. F. Semenovich, F. F. Nagibin, and R. S. Cherkasov.
37. *Contemporary mathematics, and mathematics in contemporary schools,* **1971**, no. 6, 2–3.
38. *From a new geometry textbook for 6th grade,* **1972**, no. 1, 22–31, jointly with A. F. Semenovich, F. F. Nagibin, and R. S. Cherkasov.
39. *Boris Vladimirovich Gnedenko,* **1972**, no. 1, 85–86, jointly with R. S. Cherkasov.
40. *On V. Ya. Vilenkin's letter* (the letter "Equality or congruence", in same issue), **1972**, no. 6, 34–35.
41. *On procedures for studying "parallel translation" in a 7th-grade geometry course,* **1973**, no. 1, 24–29, jointly with A. F. Semenovich and R. S. Cherkasov.
42. *On the structure of the new geometry book for 7th grade,* **1973**, no. 2, 17–29, jointly with A. F. Semenovich and R. S. Cherkasov.
43. *Ivan Georgievich Petrovskiĭ,* **1973**, no. 4, 81–86, jointly with P. S. Aleksandrov and O. A. Oleĭnik.
44. *Methodological comments on the 9th-grade trial textbook "Algebra and the elements of analysis",* **1973**, no. 5, 64, jointly with B. E. Veĭts and I. T. Demidov.
45. *The boarding school of the university. What is it?* (in the section "Ten years of the boarding school of Moscow State University", **1974**, no. 2, 56–60.
46. *Anna Maksimilianovna Fisher* [obituary], **1974**, no. 2, 87, jointly with A. F. Semenovich.
47. *Algebra and the elements of analysis. Mathematical induction,* **1975**, no. 1, 8–14, jointly with S. I. Shvartsburd.
48. *The elements of combinatorial analysis,* **1975**, no. 2, 16–25.
49. *Real numbers, infinite sequences, and their limits,* **1975**, no. 2, 25–35, jointly with O. S. Ivashev-Musatov.
50. *The trigonometric functions, their graphs and derivatives, in a 10th grade textbook,* **1976**, no. 1, 10–25, jointly with S. I. Shvartsburd.
51. *Thirty-eighth Moscow Mathematics Olympiad (February–March* 1975), **1976**, no. 4, 68–72, jointly with G. A. Gal′perin.
52. *The integral in a 10th-grade textbook,* **1976**, no. 6, 15–17.
53. *What is a function?,* **1978**, no. 2, 27–29 (in connection with G. V. Dorofeev's article in the same issue: "The concept of function in mathematics and in school").
54. *On cultivating a dialectic-materialistic world view in mathematics and physics lessons,* **1978**, no. 3, 6–9.
55. *Draft of a mathematics curriculum for secondary schools,* **1978**, no. 4, 7–32.
56. *Sergeĭ L′vovich Sobolev,* **1978**, no. 6, 67–73, jointly with O. A. Oleĭnik.
57. *The new curricula in French secondary schools,* **1978**, no. 6, 74–78, jointly with A. M. Abramov.
58. *On the textbook "Geometry 6–8",* **1979**, no. 3, 38–42, jointly with A. F. Semenovich and R. S. Cherkasov.
59. *Alekseĭ Ivanovich Markushevich* [obituary], **1979**, no. 5, 77–78, jointly with V. D. Belousov, V. G. Boltyanskiĭ, et al.
60. *The exponential and logarithmic functions,* **1979**, no. 6, 22–27, jointly with A. N. Abramov, O. S. Ivashev-Musatov, B. M. Ivlev, and S. I. Shvartsburd.
61. *On the textbook "Algebra and the elements of analysis 9–10",* **1980**, no. 4, 18–21, jointly with A. M. Abramov, O. S. Ivashev-Musatov, B. M. Ivlev, and S. I. Shvartsburd.
62. *On the concept of vector in the secondary school curriculum,* **1981**, no. 3, 7–8.
63. *On giving the first lessons on the topic "Vectors",* **1981**, no. 3, 8–11, jointly with A. M. Abramov.
64. *Review of L. S. Pontryagin's book "The analysis of infinitesimals",* **1981**, no. 5, 73–74.
65. *Izabella Grigor′evna Bashmakova,* **1981**, no. 1, 73–74, jointly with P. S. Aleksandrov, B. V. Gnedenko, S. S. Demidov, S. S. Petrov, K. A. Rybnikov, and A. P. Yushkevich.

66. *Boris Vladimirovich Gnedenko,* **1982**, no. 1, 72–73, jointly with R. S. Cherkasov.
67. *Leonid Vital′evich Kantorovich* (*On his seventieth birthday*), **1982**, no. 2, 77–78, jointly with V. A. Zalgaller.
68. *On the concept of limit in schools of general education,* **1982**, no. 5, 56.
69. *Newton and contemporary mathematical thought,* **1982**, no. 5, 58–64.
70. *Pavel Sergeevich Aleksandrov,* **1983**, no. 1, 47–48, jointly with B. V. Gnedenko.
71. *On the textbook "Geometry", by A. V. Pogorelov,* **1983**, no. 2, 45–46.
72. *Remarks on the concept of a set in school mathematics courses,* **1984**, no. 1, 52–53.
73. *S. L. Sobolev and contemporary mathematics,* **1984**, no. 1, 73–77, jointly with O. A. Oleĭnik.
74. *On scalar quantities,* **1986**, no. 3, 32–33.

IV. Articles by Kolmogorov in the journal "Kvant"

1. *What is a function?*, Kvant **1970**, no. 1, 27–36.
2. *A parquet of regular polygons*, Kvant **1970**, no. 3, 24–27.
3. *Problem M3*, Kvant **1970**, no. 1, 52–53.
4. *What is the graph of a function?*, Kvant **1970**, no. 2, 3–13.
5. *On the solution of Hilbert's tenth problem*, Kvant **1970**, no. 7, 39–44, jointly with F. L. Varpakhovskiĭ.
6. *Foreword to the article: The epistemology of Lenin and the concepts of mathematics, by V. G. Boltyanskiĭ and N. Kh. Rozov*, Kvant **1970**, no. 7, 2.
7. *Semilogarithmic and logarithmic coordinate scales*, Kvant **1973**, no. 3, 2–7.
8. *Mathematics as a profession*, Kvant **1973**, no. 4, 12–18.
9. *The sieve of Eratosthenes*, Kvant **1974**, no. 1, 77.
10. *Groups of transformations*, Kvant **1976**, no. 10, 2–5.
11. *The physics-mathematics school of Moscow State University*, Kvant **1977**, no. 1, 56–57, jointly with V. V. Vavilov.
12. *The physics and mathematics school of Moscow State University after* 15 *years*, Kvant **1979**, no. 1, 55–57, jointly with V. V. Vavilov and I. T. Tropin.
13. *The dialectic-materialistic world view in school mathematics and physics courses*, Kvant **1980**, no. 4, 15–18.
14. *Introduction to probability theory*, "Kvant" Library series, no. 23, "Nauka", Moscow, 1982, jointly with I. G. Zhurbenko and A. V. Prokhorov.
15. *Mathematics: science and profession*, "Kvant" Library series, no. 64, "Nauka", Moscow, 1988, jointly with G. A. Gal′perin.

V. Articles by Kolmogorov on the theory of poetry and the statistics of text

1. *The rhythmics in Mayakovskiĭ's poems*, Vopr. Yazykoznaniya **1962**, no. 3, 62–74, jointly with A. M. Kondratov. (Russian)
2. *On the study of Mayakovskiĭ's rhythmics*, Vopr. Yazykoznaniya **1963**, no. 4, 64–71. (Russian)
3. *On the dolnik in contemporary Russian poetry: general characteristics*, Vopr. Yazykoznaniya **1963**, no. 6, 84-95, jointly with A. V. Prokhorov. (Russian)
4. *Statistics and probability theory in the study of Russian versification*, Abstracts and Annotations, Sympos. on the Comprehensive Study of Artistic Creativity, "Nauka", Leningrad, 1963, p. 23, jointly with A. V. Prokhorov. (Russian)
5. *On the dolnik in contemporary Russian poetry: a statistical characterization of the dolnik of Mayakovskiĭ, Bagritskiĭ, and Akhmatova*, Vopr. Yazykoznaniya **1964**, no. 1, 75–94, jointly with A. V. Prokhorov. (Russian)
6. *On the meter in Pushkin's "Song of the Western Slavs"*, Russk. Literatura **1966**, no. 1, 98–111. (Russian)
7. *Remarks on an analysis of the rhythm in Mayakovskiĭ's "Poem about a Soviet passport"*, Vopr. Yazykoznaniya **1965**, no. 3, 70–75. (Russian)
8. *On the roots of the classical Russian metrics*, Cooperation of the Sciences and the Mysteries of Creativity, Iskusstvo, Moscow, 1968, pp. 397–432, jointly with A. V. Prokhorov. (Russian)
9. *An example of the study of meter and its metric variants*, The Theory of Poetry, "Nauka", Leningrad, 1968, pp. 145–167. (Russian)
10. *An analysis of the metric structure of A. S. Pushkin's poem "Arion"*, Problems in the Theory of Poetry, "Nauka", Leningrad, 1984, pp. 118–120. (Russian)
11. *A model of the rhythmic structure of Russian speech suitable for studying the metrics of classical Russian poetry*, Russian Versification. Traditions and the Problems of Evolution, "Nauka", Moscow, 1985, pp. 113–134. (Russian)

VI. Articles by Kolmogorov on popular science

1. *Present-day controversies on the nature of mathematics*, Nauchn. Slovo **1929**, no. 6, 41–54. (Russian)
2. *Contemporary mathematics*, Front Nauki i Tekhniki **1934**, no. 5/6, 25–28. (Russian)
3. *The Institute of Mathematics and Mechanics of Moscow State University*, Front Nauki i Tekhniki **1934**, no. 5/6, 75–78. (Russian)
4. *The Second All-Union Mathematics Congress*, Sotsialisticheskaya Rekonstruktsiya i Nauka **1934**, no. 7, 142–145. (Russian)
5. *Contemporary mathematics*, Collection of Articles on the Philosophy of Mathematics, ONTI, Moscow, 1936, pp. 7–13. (Russian)
6. *Theory and practice in mathematics*, Front Nauki i Tekhniki **1936**, no. 5, 39–42. (Russian)
7. *Probability theory and its applications*, Mathematics and the Natural Sciences in the USSR, GONTI, Moscow–Leningrad, 1938, pp. 51–61. (Russian)
8. *Nikolaĭ Ivanovich Lobachevskiĭ.* 1793–1843, Gostekhizdat, Moscow–Leningrad, 1943, jointly with P. S. Aleksandrov. (Russian)
9. *Newton and contemporary mathematical thought*, Moscow University: In Memory of Isaac Newton, 1643–1943, Izdat. Moskov. Gos. Univ., Moscow, 1946, pp. 27–43. (Russian)
10. *The role of Russian science in the development of probability theory*, The Role of Russian Science in the Development of World Science and Culture, vol. 1, book 1, Izdat. Moskov. Gos. Univ., Moscow, 1947, pp. 53–64. (Russian)
11. *Probability theory*, Thirty years of Mathematics in the USSR, 1917–1947, Gostekhizdat, Moscow–Leningrad, 1948, pp. 701–727, jointly with B. V. Gnedenko. (Russian)
12. *Ivan Georgievich Petrovskiĭ*, Uspekhi Mat. Nauk **6** (1951), no. 3, 161–164. (Russian)
13. *Introductory article and comments*, Complete Collected Works of N. I. Lobachevskiĭ, vol. 5, Gostekhizdat, Moscow–Leningrad, 1951, pp. 329–332; 342–348, jointly with A. N. Khovanskiĭ. (Russian)
14. *Mathematics as a profession: a guide to matriculants of institutions of higher learning*, Sov. Nauka, Moscow, 1952. (Russian)
15. *Probability theory*, Mathematics: its Content, Methods, and Meaning, vol. 2, Izdat. Akad. Nauk SSSR, Moscow, 1956, pp. 252–284. (Russian)
16. *Sergeĭ Mikhaĭlovich Nikol'skiĭ (on his fiftieth birthday)*, Uspekhi Mat. Nauk **11** (1956), no. 2, 239–244, jointly with S. B. Stechkin. (Russian)
17. *On a justification of the theory of real numbers*, Mat. Prosveshchenie **2** (1957), 169–173. (Russian)
18. *Mathematics as a profession*, 2nd ed., Izdat. Moskov. Gos. Univ., Moscow, 1958. (Russian)
19. *Probability theory*, Forty Years of Mathematics in the USSR, 1917–1957, vol. 1, Fizmatgiz, Moscow, 1959, pp. 781–795. (Russian)
20. *On the work of N. V. Smirnov on mathematical statistics (on his sixtieth birthday)*, Teor. Veroyatnost. i Primenen. **5** (1960), 436–440, jointly with B. V. Gnedenko, Yu. V. Prokhorov, and O. V. Sarmanov. (Russian)
21. *Mathematics as a profession*, 3rd ed., Izdat. Moskov. Gos. Univ., Moscow, 1960. (Russian)
22. *Automata and life*: notes for the report at the methodology seminar of the Mechanics and Mathematics Department of Moscow State University, 5 April 1961), Mash. Per. i Prikl. Lingvist. **1961**, no. 6, 3–8. (Russian)
23. *Automata and life*, Tekhnika—Molodezhi **1961**, no. 10, 11, 16–19, 30–33. (Russian)
24. *Properties of inequalities and the concept of approximate computations. Irrational numbers*, Questions on Teaching Mathematics in Secondary School, Uchpedgiz, Moscow, 1961, jointly with P. S. Aleksandrov. (Russian)
25. *How I became a mathematician*, Ogonek **1963**, no. 48, 12–13. (Russian)
26. *Natural numbers and positive scalar quantities*, Mathematics School. Lectures and Problems, nos. IV–V, Izdat. Moskov. Gos. Univ., Moscow, 1965, pp. 19–35. (Russian)
27. *Geometry on the sphere and geology*, Nauka i Zhizn' **1966**, no. 2, 32. (Russian)
28. *A problem in the theory of curves*, Mathematics School. Lectures and Problems, no. VIII, 1966, p. 35. (Russian)
29. *New developments in school mathematics*, Nauka i Zhizn' **1969**, no. 3, 62–66. (Russian)
30. *A letter to P. L. Kapitsa*, Voprosy filosofii **1972**, no. 9, 127–128. (Russian)
31. *The science instructor*, in: The Joy of Discovery, Detsk. Literatura, Moscow, 1972. (Russian)

32. *Coordination of the teaching of mathematics and physics*, Abstracts, All-Union Scientific and Practical Conference on the Problem of Education in Schools and Classes with More Intensive Study of Individual Subjects (mimeographed notes), Nauchn. Issled. Inst. SIMO Akad. Pedagog. Nauk, Moscow, 1972. (Russian)
33. *Scientific foundations for a school course in mathematics*, in: Curricula for Pedagogical Institutes, Prosveshchenie, Moscow, 1973. (Russian)
34. *Materials for discussion in the Committee on School Terminology and Notation* (mimeographed notes), Acad. Methods Council, Ministry of Education of the USSR, Moscow, 1973. (Russian)
35. *On ensuring worthy reserves*, Vestnik Vyssh. Shkoly **1974**, no. 6, 26–33, jointly with I. T. Tropin and K. V. Chernyshev. (Russian)
36. *New programs: specialized schools*, Mathematical Education Today, Prosveshchenie, Moscow, 1974, pp. 5–12. (Russian)
37. *Some considerations on the structure of mathematical textbooks*, School Textbook Problems, no. 3, Prosveshchenie, Moscow, 1974, pp. 14–17. (Russian)
38. *How I became a mathematician. What is mathematics? Science in your profession*, Znanie **1978**, no. 11, 5–9. (Russian)
39. *On the formation of a dialectic-materialistic ideology in schoolchildren through lessons in mathematics and physics*, The Role of Educational Literature in the Formation of an Ideology in Schoolchildren, Pedagogika, Moscow, 1978, pp. 69–74. (Russian)
40. *Introduction to Probability theory*, "Kvant" Library Series, no. 23, "Nauka", Moscow, 1982, jointly with I. G. Zhurbenko and A. V. Prokhorov. (Russian)

VII. Newspaper articles by Kolmogorov

1. *A great Russian scholar and innovator*: *the* 150th *birthday of N. I. Lobachevskiĭ*, Izvestiya, 2/Nov/43.
2. *The water and air must be clean*! (on the cleanness of the Klyaz'ma River), Lit. Gazeta, 30/Jun/53, jointly with P. S. Aleksandrov.
3. *The young forces in science* (on the scientific research of Moscow State University students), Komsomol'skaya Pravda, 15/May/55.
4. *School and the training of scientific workers* (a discussion of the propositions of the Central Committee of the Communist Party of the Soviet Union on schools), Trud, 10/Dec/58.
5. *You have chosen science*, Moskov. Univ., 1/Sep/61.
6. *Science demands enthusiasm*, Izvestiya, 21/Feb/62.
7. *Are scientific schools needed*?, Izvestiya, 27/Oct/62.
8. *Simplicity to complex*, Izvestiya, 31/Dec/62.
9. *An ordinary profession*: *mathematician*, Komsomol'skaya Pravda, 2/Mar/63.
10. *The search for talent* (on the work of young Soviet mathematicians), Izvestiya, 7/Apr/63.
11. *Off to a good journey* (on the development of mathematics and the training of personnel), Pravda, 8/Sep/63.
12. *Yes, mathematics is thriving* (the round-table of "Nedelya"), Nedelya, no. 3, 1964, a report of V. Kornilov and V. Yankulin.
13. *The physics and mathematics school*, Uchit. Gazeta, 11/Feb/64.
14. *The Moscow Mathematical Society*, Pravda (Moscow edition), 19/Oct/64, jointly with A. G. Kurosh.
15. *What is a cultured person*? (An evening at Moscow State University, conducted by P. S. Aleksandrov and A. N. Kolmogorov), written by V. Yankulin, Nedelya, no. 1, 1965.
16. *Algorithms for love of life*, Komsomol'skaya Pravda, 9/Jul/65.
17. *The waste under respectful covers* (about publications on economical-mathematical methods), Izvestiya, 24/Oct/65.
18. *The grand prospects ahead*, Moskov. Univ., 10/Dec/65.
19. *The keys to a forest lodge*, Lit. Gazeta, 16/Aug/66.
20. *A step into science*, Moskov. Univ., 10/Dec/66.
21. *Knowledge, skills, abilities, and competitive examinations* (on erudition of students in physics and mathematics schools), Lit. Gazeta, 11/Jan/67.
22. *Revision of the school mathematics program*, Uchit. Gazeta, 14/Feb/67.
23. *Experience, problems, prospects*, Moskov. Univ., 8/Dec/67.
24. *The joy of learning about the world* (to the beginning of the school year), Pravda, 1/Sep/68.
25. *Congratulations on the fifth birthday* (the Physics and Mathematics School of Moscow State University), Moskov. Univ., 4/Dec/68.
26. *A reference book for researchers* (a nomination for the Lenin Prize), Izvestiya, 21/Mar/70.
27. *School solves its main problem*, Moskov. Komsomolets, 14/Dec/71.
28. *The profession of continual youth* (to the beginning of the school year), Pravda, 1/Sep/71.
29. *How to cultivate talent* (on the work of the Physics and Mathematics School of Moscow State University), Uchit. Gazeta, 28/Oct/71.
30. *Talent requires attention* (the Physics and Mathematics School of Moscow State University), Moskov. Komsomolets, 22/Oct/71.
31. *A teacher cannot be replaced* (how to make children actively interested in knowledge), Komsomol. Pravda, 19/Jan/72.
32. *Fostering intuitive thinking* (on a new program in mathematics), Uchit. Gazeta, 22/Aug/72.
33. *The probability is not zero* (on contact with extraterrestrial civilizations), Lit. Gazeta, 20/Sep/72.
34. *From the old to the new geometry* (on a new course in geometry), Uchit. Gazeta, 21/Dec/72.
35. *The joy of a scientific quest*, Smena, no. 16, 1972, p. 14.
36. *The boarding school of the university. What is it for*?, Moskov. Univ., 30/Nov/73.
37. *A dialogue on mathematics* (a dialogue between Academician A. N. Kolmogorov and a teacher, V. A. Sadchikov, at the Moscow School No. 317), Uchit. Gazeta, 12/Jan/74, jointly with V. A. Sadchikov.
38. *Everything begins with a road* (on the interviewer, an older friend), Lit. Gazeta, 13/Oct/76.

39. *What are grades X-XI?* (a collective letter from ten academicians with a proposal for a project of school reform, Izvestiya, 26/Jan/84, jointly with B. V. Gnedenko, N. P. Dubinin, I. K. Kikoin, et al.
40. *How to cultivate enthusiasm* (I. Prelovskaya, A meeting of graduates of the Physics and Mathematics School of Moscow University who are Doctors of Sciences, with A. N. Kolmogorov), Izvestiya, 28/Jan/84.
41. *What "Kvant" gives, or could give, to a teacher of mathematics*, Uchit. Gazeta, 22/Jan/85.
42. *A life in the name of science* (response to his students, 26/Apr/86), Uchit. Gazeta, 26/Nov/87.
43. *A student on his teacher* (an interview with Kolmogorov in an issue devoted to N. N. Luzin: V. A. Uspenskiĭ was the interviewer), Put′ v Nauku (a newsletter of Kemerovo University), 7/Sep/83.
44. *A response to my students*, Uchit. Gazeta, 26/Nov/87.

VIII. Addresses by Kolmogorov at meetings of the Moscow Mathematical Society

8 October 1922. Example of a Fourier–Lebesgue trigonometric series that diverges almost everywhere.

5 April 1925. On the possibility of a general definition of derivative, integral, and sum of a series.

16 November 1926. On an everywhere divergent Fourier series.

7 December 1926. The double negation principle and the definition of a function.

6 December 1927. A generalization of Chebyshev's theorem.

20 March 1928. On a general scheme in probability theory.

18 December 1928. A new interpretation of intuitionistic logic.

11 December 1932. On geometric ideas of Plücker and Klein.

11 November 1934. On contingencies, jointly with I. Ya. Verchenko.

11 December 1934. A supplementary communication (see 11/Nov/34).

5 January 1935. Markov chains and reversibility of laws of nature.

16 February 1937. The statistical theory of crystallization of hardening metals.

22 April 1937. On differentiability properties of functions of two variables.

10 November 1937. The development in the USSR of mathematical methods for learning about nature (differential and integral equations and probability theory).

22, 28 November 1937. Meetings devoted to a discussion of a plan to put together a new elementary algebra textbook (by P. S. Aleksandrov and A. N. Kolmogorov). Two meetings.

10 or 16 March 1938 (exact date uncertain). Meeting devoted to a discussion of the draft of Kolmogorov's article "Mathematics" for the *Great Soviet Encyclopedia.*

22 December 1938. Contemporary questions in set-theoretic geometry (survey report).

16 February 1939. On principles for estimating the reliability of statistical hypotheses (survey report).

22 March 1939. On the extrapolation of stationary series in dependence on the nature of their spectra.

4 June 1939. Stationary sequences of elements of a Hilbert space.

22 November 1939. On the definition of the stationary property for an individual function.

22 March 1940. A mathematical study of deterministic and random processes.

18 March 1941. On two forms of an axiomatic method (first survey report).

1 April 1941. On two forms of an axiomatic method (second survey report).

16 April 1941. On measures invariant under a group of transformations.

7 March 1942. On a variational problem.

3 November 1943. Report at the meeting dedicated to the memory of Lobachevskiĭ.

2 February 1944. Unitary representations of infinite groups.

3 October 1944. A mathematical theory of turbulence.

31 October 1944. The contemporary state of the theory of Markov chains and unsolved problems in this area (survey report).

11 December 1944. Problems in probability theory.

4 December 1945. Computable sequences and their significance in investigations of the foundations of mathematics (survey report).

7 May 1946. The work of P. S. Aleksandrov on set theory and the theory of functions (on his fiftieth birthday).

10 June 1947. On some new work in probability theory (survey report).

17 June 1947. An address at the meeting dedicated to the memory of Alekseĭ Konstantinovich Vlasov.

9 December 1947. The structure of complete metric Boolean algebras.

17 February 1948. On a draft of curricula for secondary schools.

24 February 1948. Best approximation of complex functions.

30 March 1948. A local limit theorem for Markov chains.

28 September 1948. On Ostrogradskiĭ's critique of the work of Lobachevskiĭ.

30 November 1948. Measures and probability distributions in function spaces (survey report).

4 October 1949. A presentation of the foundations of Lebesgue measure theory.

18 October 1949. "Axiom", "Infinitesimal" (articles for a new edition of the *Great Soviet Encyclopedia*).

6 December 1949. Basic types of Markov processes.

23 December 1949. The development of mathematics during the Soviet epoch.

9 January 1951. On some mathematical problems connected with control of production (survey report).

15 May 1951. The work of V. V. Stepanov on the theory of functions.

27 November 1951. Two-valued functions of two-valued variables and their application to relay-contact networks.

20 May 1952. D. E. Men′shov's work on the theory of functions of a real variable (on his sixtieth birthday).

30 September 1952. On the spectra of dynamical systems on a torus.

25 February 1953. The research direction of the Probability Theory Branch.

17 March 1953. On the concept of algorithm.

26 May 1953. On almost periodic motions of a heavy rigid body about a fixed point.

8 December 1953. The concept of "information" in mathematical statistics and in the Shannon theory of information transmission.

27 April 1954. Estimates of the minimal number of elements in ε-nets in various function classes and their application to the question of representability of functions of several variables as superpositions of functions of a smaller number of variables.

25 May 1954. On stability of conditionally periodic motions in conservative dynamical systems.

28 September 1954. On the International Congress of Mathematicians in Amsterdam.

25 October 1955. A few words about a trip to Stockholm, jointly with P. S. Aleksandrov.

28 February 1956. On a scientific trip to France, the German Democratic Republic, and Poland.

17 April 1956. The representation of continuous functions of several variables by several continuous functions of a smaller number of variables.

29 May 1956. "Éléments de mathématique" by Nicolas Bourbaki.

5 June 1956. On some asymptotic characteristics of a totally bounded metric space.

18 December 1956. Uniform limit theorems for sums of independent terms.

25 December 1956. On D. F. Egorov's work in the theory of functions and integral equations (on the twentieth anniversary of his death).

2 April 1957. What is cybernetics? (on a draft of an article for the *Great Soviet Encyclopedia*).

24 September 1957. On the representation of a continuous function of several variables by means of continuous functions of a single variable and addition.

17 October 1957. On the approximate representation of functions of several variables by superpositions of functions of a smaller number of variables and ε-entropy of function classes.

7 October 1958. On the training of workers for the physical and mathematical sciences and the new technologies.

13 January 1959. "Small denominators" in problems of mechanics and analysis (survey report).

21 April 1959. On some features of the contemporary stage of development of mathematics.

17 May 1960. Solved and unsolved problems associated with Hilbert's thirteenth problem.

13 December 1960. A. Ya. Khinchin's work in the theory of functions.

27 December 1960. Mathematical methods in the study of Russian poetry (survey report).

4 April 1961. What is 'information'?

16 May 1961. An estimation of the difficulty of defining and computing finite sequences and discrete functions.

21 November 1961. Self-constructing apparatuses.

26 December 1961. N. K. Bari's work in the representation of superposition functions.

26 March 1963. On uniform limit theorems for sums of independent terms.

25 April 1963. Sharing work experience.

28 May 1963. An introductory address to the specialized meeting on the theory of random processes.

19 November 1963. Computable functions and the foundations of information theory and probability theory.

24 October 1964. On the influence of the ideas of information theory on the development of mathematics.

15 December 1964. The asymptotic behavior of the complexity of finite segments of an infinite sequence.

18 May 1965. Experiment and mathematical theory in the study of turbulence.

7 December 1965. The calculus of finite problems of Yu. T. Morozov.

3 March 1966. On a draft of a mathematics curriculum for secondary school.

10 May 1966. An introductory word dedicated to P. S. Aleksandrov's seventieth birthday.

14 February 1967. On elective studies in mathematics in secondary school for 1967/68.

28 March 1967. Logic, intuitionism, and the foundations of mathematics in the work of L. E. J. Brouwer.

31 October 1967. Some theorems on algorithmic entropy and the algorithmic amount of information.

24 February 1970. The statistical hydrodynamics of the ocean.

23 November 1971. The complexity of determining and the complexity of constructing mathematical objects.

14 December 1971. On P. L. Chebyshev's work in probability theory.

25 April 1972. On D. E. Men′ov's work in the theory of orthogonal series.

16 April 1974. The complexity of algorithms and an objective definition of randomness.

15 February 1977. An introductory word dedicated to the sixtieth birthday of G. E. Shilov.

18 January 1978. Remarks on statistical solutions of the Navier–Stokes equations.

IX. Contents of Selected Works of Kolmogorov, Vols. I–III

Vol. I. Mathematics and Mechanics (Kluwer, 1991)

48. Equations of turbulent motion in an incompressible fluid
49. A remark on Chebyshev polynomials deviating least from a given function
50. On the breakage of drops in a turbulent flow
51. On dynamical systems with an integral invariant on a torus
52. On the preservation of conditionally periodic motions under small variations of the Hamiltonian function
53. The general theory of dynamical systems and classical mechanics
54. Some fundamental problems in the aproximate and exact representation of functions of one or several variables
55. On the representation of continuous functions of several variables as superpositions of continuous functions of a smaller number of variables
56. On the representation of continuous functions of several variables as superpositions of continuous functions of one variable and addition
57. On the linear dimension of topological vector spaces
58. A refinement of the concept of the local structure of turbulence in an incompressible viscous fluid at large Reynolds numbers
59. P. S. Aleksandrov and the theory of δs-operations
60. A qualitative study of mathematical models of population dynamics

Commentary

On the papers on the theory of functions and set theory (*A. N. Kolmogorov*)
Trigonometric and orthogonal series (*P. L. Ul'yanov*)
Descriptive set theory (*I. I. Parovichenko*)
Measure theory and theory of the integral (*V. A. Skvortsov*)
Points of discontinuity of functions (*E. P. Dolzhenko*)
Theory of approximation (*S. A. Telyakovskiĭ and V. M. Tikhomirov*)
Inequalities for derivatives (*V. M. Tikhomirov and G. G. Magaril-Il'yaev*)
Rings of continuous functions (*E. A. Gorin*)
Curves in a Hilbert space (*Yu. A. Rozanov*)
On the papers on intuitionistic logic (*A. N. Kolmogorov*)
Intuitionistic logic (*V. A. Uspenskiĭ and V. E. Plisko*)
On the papers on homology theory (*A. N. Kolmogorov*)
Homology theory (*G. S. Chogoshvili*)
On the paper "On open mappings" (*A. N. Kolmogorov*)
Topology (*A. V. Arkhangel'skiĭ*)
Axiomatics of projective geometry (*A. V. Mikhalev*)
On the paper on the diffusion equation (*A. N. Kolmogorov*)
The diffusion equation (*G. I. Barenblatt*)
On the papers on turbulence (*A. N. Kolmogorov*)
Turbulence (*A. M. Yaglom*)
On the papers on classical mechanics (*A. N. Kolmogorov*)
Classical mechanics (*V. I. Arnol'ld*)
On the papers on superpositions (*A. N. Kolmogorov*)
Superpositions (*V. I. Arnol'ld*)

List of works by A. N. Kolmogorov

Vol. II. Probability Theory and Mathematical Statistics (Kluwer, 1992)

Series Editor's Preface
From the publishers of the Russian edition
A few words about A. N. Kolmogorov (*P. S. Aleksandrov*)

1. On convergence of series whose terms are determined by random events
2. On the law of large numbers
3. On a limit formula of A. Khinchin
4. On sums of independent random variables
5. On the law of the iterated logarithm
6. On the law of large numbers
7. General measure theory and probability calculus
8. On the strong law of large numbers

9. On analytical methods in probability theory
10. The waiting problem
11. The method of the median in the theory of errors
12. A generalization of the Laplace–Lyapunov theorem
13. On the general form of a homogeneous stochastic process
14. On computing the mean Brownian area
15. On the empirical determination of a distribution law
16. On the limit theorems of probability theory
17. On the theory of continuous random processes
18. On the problem of the suitability of forecasting formulas found by statistical methods
19. Random motions
20. Deviations from Hardy's formulas under partial isolation
21. On the theory of Markov chains
22. On the statistical theory of metal crystallization
23. Markov chains with a countable number of possible states
24. On the reversibility of the statistical laws of nature
25. Solution of a biological problem
26. On a new confirmation of Mendel's laws
27. Stationary sequences in Hilbert space
28. Interpolation and extrapolation of stationary random sequences
29. On the logarithmic normal distribution of particle sizes under grinding
30. Justification of the method of least squares
31. A formula of Gauss in the method of least squares
32. Branching random processes
33. Computation of final probabilities for branching random processes
34. Statistical theory of oscillations with continuous spectrum
35. On sums of a random number of random terms
36. A local limit theorem for classical Markov chains
37. Solution of a probabilistic problem relating to the mechanism of bed formation
38. Unbiased estimators
39. On differentiability of transition probabilities of time-homogeneous Markov processes with a countable number of states
40. A generalization of Poisson's formula for a sample from a finite set
41. Some recent work on limit theorems in probability theory
42. On A. V. Skorokhod's convergence
43. Two uniform limit theorems for sums of independent terms
44. Random functions and limit theorems
45. On the properties of P. Lévy's concentration functions
46. Transition of branching processes to diffusion processes and related genetic problems
47. On the classes $\Phi^{(n)}$ of Fortet and Blanc-Lapierre
48. On conditions of strong mixing of a Gaussian stationary process
49. Random functions of several variables almost all realizations of which are periodic
50. An estimate of the parameters of a complex stationary Gaussian Markov process
51. On the approximation of distribution of sums of independent terms by infinitely divisible distributions
52. Estimators of spectral functions of random processes
53. On the logical foundations of probability theory

Comments

On the papers on probability theory and mathematical statistics (*A. N. Kolmogorov*)
Analytical methods in probability theory (No. 9) (*A. D. Venttsel'*)
Markov processes with a countable number of states (No. 10) (*B. A. Sevast'yanov*)
Homogeneous random processes (No. 13) (*V. M. Zolotarev*)
Homogeneous Markov processes (No. 39) (*A. A. Yushkevich*)
Branching processes (Nos. 25, 32, 33, 46) (*B. A. Sevast'yanov*)
Stationary sequences (No. 27) (*Yu. A. Rozanov*)
Stationary processes (No. 48) (*V. A. Statulyavichus*)
Statistics of processes (No. 50) (*A. N. Shiryaev*)
Spectral theory of stationary processes (No. 34) (*A. M. Yaglom*)

Vol. III. Information Theory and the Theory of Algorithms (Kluwer, 1993)

X. Publications about Kolmogorov

1. A. M. Abramov, *On the pedagogical heritage of A. N. Kolmogorov*, Uspekhi Mat. Nauk **43** (1988), no. 6, 39–74; English transl. in Russian Math. Surveys **43** (1988).
2. P. S. Aleksandrov and A. Ya. Khinchin, *Andreĭ Nikolaevich Kolmogorov (On his fiftieth birthday)*, Uspekhi Mat. Nauk **8** (1953), no. 3, 178–200. (Russian)
3. P. S. Aleksandrov and B. V. Gnedenko, *Kolmogorov as pedagogue*, Uspekhi Mat. Nauk **18** (1963), no. 5, 115–120; English transl. in Russian Math. Surveys **18** (1963).
4. P. S. Aleksandrov, *Some words about A. N. Kolmogorov*, Uspekhi Mat. Nauk **38** (1983), no. 4, 7–9; English transl. in Russian Math. Surveys **38** (1983).
5. P. S. Aleksandrov, E. B. Dynkin, and I. G. Petrovskiĭ, *Laureate of the Balzan Prize (Academician A. N. Kolmogorov)*, Moskovskaya Pravda, June 9, 1963. (Russian)
6. V. I. Andrianov, *Preface*, The Flaming Bolshevist Word. Underground Literature in Yaroslavl Province in 1902–1917, Verkhne-Volzhsk. Knizhn. Izdat., 1975. (Russian)
7. D. V. Anosov, *On the works of A. N. Kolmogorov and V. I. Arnol′ld, recipients of the* 1965 *Lenin Prize*, Mat. v Shkole **1966**, no. 1, 7–12. (Russian)
8. G. Anfilov, *Scholar and teacher*, Ogonek **1963**, no. 48, 11–13. (Russian)
9. V. I. Arnol′ld, *A few words about Andreĭ Nikolaevich Kolmogorov*, Uspekhi Mat. Nauk **43** (1988), no. 6, 37; English transl. in Russian Math. Surveys **43** (1988).
10. N. N. Bogolyubov, B. V. Gnedenko, and S. L. Sobolev, *Andreĭ Nikolaevich Kolmogorov (On his eightieth birthday)*, Uspekhi Mat. Nauk **38** (1983), no. 4, 11–26; English transl. in Russian Math. Surveys **38** (1983).
11. A. A. Borovkov, S. L. Sobolev, and V. V. Yurinskiĭ, *Depth of investigations, breadth of problems*, Nauka v Sibiri, May 12, 1983. (Russian)
12. A. Butyagin, *Outstanding scholars at Moscow State University (On the awarding of the Stalin Prizes)*, Uchit. Gazeta, March 16, 1941. (Russian)
13. *The last journey*, Pravda, October 24, 1987. (Russian)
14. *The last journey*, Izvestiya, October 25, 1987. (Russian)
15. *Outstanding successes of Soviet mathematicians (On the awarding of the Stalin Prizes)*, Pravda, April 14, 1941. (Russian)
16. I. M. Gel′fand, S. M. Nikol′skiĭ, S. L. Sobolev, B. V. Gnedenko, M. A. Prokof′ev, V. G. Razumovskiĭ, N. A. Ermolaeva, R. S. Cherkasov, and A. M. Abramov, *A life in the name of science*, Uchit. Gazeta, October 26, 1987 (no. 141). (Russian)
17. B. V. Gnedenko and N. V. Smirnov, *A. N. Kolmogorov's work in probability theory (On his sixtieth birthday)*, Teor. Veroyatnost. i Primenen. **8** (1963), 167–174; English transl. in Theory Probab. Appl. **8** (1963).
18. B. V. Gnedenko, *Andreĭ Nikolaevich Kolmogorov*, Mat. v Shkole **1963**, no. 2, 67. (Russian)
19. B. V. Gnedenko, *Andreĭ Nikolaevich Kolmogorov (On his seventieth birthday)*, Uspekhi Mat. Nauk **28** (1973), no. 5, 5–15; English transl. in Russian Math. Surveys **28** (1973).
20. B. V. Gnedenko, *Scholar and pedagogue (On the seventieth birthday of Andreĭ Nikolaevich Kolmogorov)*, Mat. v Shkole **1973**, no. 2, 88–89. (Russian)
21. B. V. Gnedenko, *A mathematician (The creative path of A. N. Kolmogorov)*, Moskov. Komsomolets, May 6, 1973. (Russian)
22. B. V. Gnedenko, *Scholar, pedagogue, reformer (On the seventy-fifth birthday of A. N. Kolmogorov)*, Mat. v Shkole **1978**, no. 2, 93–94. (Russian)
23. B. V. Gnedenko, *Andreĭ Nikolaevich Kolmogorov (On his eightieth birthday)*, Mat. v Shkole **1983**, no. 2, 76–78. (Russian)
24. N. Gorbachev, *The algorithm of a remarkable person*, Uchit. Gazeta, March 6, 1977. (Russian)
25. N. Gorbachev, *I look for enthusiasts*, Uchit. Gazeta, April 25, 1978. (Russian)
26. N. Gorbachev, *Academician Kolmogorov*, Smena **1978**, no. 12; reprinted, *Soviet scholars. Sketches and remembrances*, Izdat. Akad. Pedag. Nauk, Moscow, 1982, pp. 277–293. (Russian)
27. D. G. Kendall, G. K. Batchelor, and K. Moffat, *Obituary. Dr. Andrei Kolmogorov. Giant of mathematics*, The London Times, October 26, 1987.
28. V. Lomsargis, *At nineteen I discovered mathematics (An interview with A. N. Kolmogorov)*, Komsomol′skaya Pravda (organ of the Central Committee of LKSM of Lithuania), July 23, 1977. (Russian)
29. I. Mikhaĭlova, *Words about a scientist (On the film "Stories about Kolmogorov")*, Sovetsk. Ekran **1984**, no. 22, 11. (Russian)
30. *Andreĭ Nikolaevich Kolmogorov* [obituary], Mat. v Shkole **1987**, no. 6, 3–5. (Russian)

31. *Andreĭ Nikolaevich Kolmogorov* [obituary], Pravda, October 23, 1987. (Russian)
32. *Andreĭ Nikolaevich Kolmogorov* [obituary], Izvestiya, October 24, 1987. (Russian)
33. G. Nemchuk, *Power over numbers* (*On the life and activities of A. N. Kolmogorov*), Lit. Gazeta, August 16, 1966. (Russian)
34. S. M. Nikol′skiĭ, S. L. Sobolev, et al., *A life in the name of science*, Uchit. Gazeta, November 26, 1987. (Russian)
35. S. P. Novikov, *Remembrances of A. N. Kolmogorov*, Uspekhi Mat. Nauk **43** (1988), no. 6, 35–36; English transl. in Russian Math. Surveys **43** (1988).
36. *Greetings to A. N. Kolmogorov from the Moscow Mathematical Society and the Editorial Board of "Uspekhi Matematicheskikh Nauk" on his sixtieth birthday*, Uspekhi Mat. Nauk **18** (1963), no. 5, 3; English transl. in Russian Math. Surveys **18** (1963).
37. *Greetings from the Presidium of the USSR Academy of Sciences on the seventieth birthday of A. N. Kolmogorov*, Uspekhi Mat. Nauk **28** (1973), no. 5, 3; English transl. in Russian Math. Surveys **28** (1973).
38. *Greetings to A. N. Kolmogorov on his seventy-fifth birthday from the Mathematics Section of the USSR Academy of Sciences, the Moscow Mathematical Society, and the Editorial Board of "Uspekhi Matematicheskikh Nauk"*, Uspekhi Mat. Nauk **33** (1978), no. 2, 212–213; English transl. in Russian Math. Surveys **33** (1978).
39. *Greetings to A. N. Kolmogorov from the Moscow Mathematical Society on his eightieth birthday*, Uspekhi Mat. Nauk **38** (1983), no. 4, 3–5; English transl. in Russian Math. Surveys **38** (1983).
40. *Greetings to A. N. Kolmogorov on his eightieth birthday*, Teor. Veroyatnost. i Primenen. **28** (1983), 208; English transl. in Theory Probab. Appl. **28** (1983).
41. *Soviet physics and mathematics*, Izvestiya, January 30, 1946. (Russian)
42. V. M. Tikhomirov, *The life and creative work of Andreĭ Nikolaevich Kolmogorov*, Uspekhi Mat. Nauk **43** (1988), no. 6, 3–33; English transl. in Russian Math. Surveys **43** (1988).
43. V. M. Tikhomirov, *Selfless service to science*, Moskov. Univ., May 26, 1983. (Russian)
44. A. Shiryaev, *The first Head* (*of the Probability Theory Branch*), Moskov. Univ., December 11, 1985. (Russian)
45. A. N. Shiryaev, *On the scientific heritage of A. N. Kolmogorov*, Uspekhi Mat. Nauk **43** (1988), no. 6, 209–210; English transl. in Russian Math. Surveys **43** (1988).
46. A. Yusin, *There, around the bend, in Komarovka*, Sovetsk. Sport, July 3, 1970. (Russian)
47. A. P. Yushkevich, *A. N. Kolmogorov: Historian and philosopher of mathematics. On the occasion of his 80th birthday*, Historia Math. **10** (1983), no. 4, 383–395.
48. V. L. Yanin, *Kolmogorov as historian*, Uspekhi Mat. Nauk **43** (1988), no. 6, 189–195; English transl. in Russian Math. Surveys **43** (1988).
49. V. Yankulin, *What is a cultured person?* (*Discussion between Moscow University students and Academicians P. S. Aleksandrov and A. N. Kolmogorov*), Nedelya, January 2, 1965. (Russian)
50. V. Yankulin, *On mathematics* (*Concerning A. N. Kolmogorov*), Nedelya, August 9, 1970. (Russian)
51. V. Yankulin, *The mathematician*, Ogonek **1983**, no. 20, 26. (Russian)
52. Ya. G. Sinaĭ, *About A. N. Kolmogorov's work on the entropy of dynamical systems*, Ergodic Theory Dynamical Systems **8** (1988), 501–502.
53. Ya. G. Sinaĭ, *Kolmogorov's work on ergodic theory*, Ann. Probab. **17** (1989), 833–839.
54. J. L. Doob, *Kolmogorov's early work on convergence theory and foundations*, Ann. Probab. **17** (1989), 815–821.
55. E. B. Dynkin, *Kolmogorov and the theory of Markov processes*, Ann. Probab. **17** (1989), 822–832.
56. T. M. Cover, P. Gacs, and R. M. Gray, *Kolmogorov's contributions to information theory and algorithmic complexity*, Ann. Probab. **17** (1989), 840–865.
57. A. N. Shiryaev, *Kolmogorov: Life and creative activities*, Ann. Probab. **17** (1989), 866–944.
58. W. R. van Zwet, *Andrei Nikolaevich Kolmogorov*, Annals of the Netherlands Royal Academy of Sciences **1989**, 167–171.
59. R. Hasminskii and I. Ibragimov, *On density estimation in the view of Kolmogorov's ideas*, Ann. Statist. **18** (1990), 999–1010.
60. A. L. Rukhin, *Kolmogorov's contributions to mathematical statistics*, Ann. Statist. **18** (1990), 1011–1016.

61. R. M. Dudley, S. L. Cook, J. Llopis, and N. P. Peug, *A. N. Kolmogorov and statistics: A citation bibliography*, Ann. Statist. **18** (1990), 1017–1031.
62. N. N. Chentsov, *The unfathomable influence of Kolmogorov*, Ann. Statist. **18** (1990), 987–998.
63. D. G. Kendall, *Kolmogorov: The man and his work*, Bull. London Math. Soc. **22** (1990), 31–47.
64. G. K. Batchelor, *Kolmogorov's work on turbulence*, Bull. London Math. Soc. **22** (1990), 47–51.
65. N. H. Bingham, *Kolmogorov's work on probability, particularly limit theorems*, Bull. London Math. Soc. **22** (1990), 51–61.
66. J. M. E. Hyland, *Kolmogorov's work on logic*, Bull. London Math. Soc. **22** (1990), 61–64.
67. G. G. Lorentz, *Superpositions, metric entropy, complexity of functions, widths*, Bull. London Math. Soc. **22** (1990), 64–71.
68. H. K. Moffat, *KAM-theory*, Bull. London Math. Soc. **22** (1990), 71–73.
69. W. Parry, *Entropy in ergodic theory—the initial years*, Bull. London Math. Soc. **22** (1990), 73–79.
70. A. A. Razborov, *Kolmogorov and the complexity of algorithms*, Bull. London Math. Soc. **22** (1990), 79–82.
71. C. A. Robinson, *The work of Kolmogorov on cohomology*, Bull. London Math. Soc. **22** (1990), 82–83.
72. P. Whittle, *Kolmogorov's contributions to the theory of stationary processes*, Bull. London Math. Soc. **22** (1990), 83–85.
73. D. G. Kendall, *Kolmogorov as I remember him*, Statist. Sci. **6** (1991), no. 3, 303–312.
74. A. N. Shiryaev, *Everything about Kolmogorov was unusual ...*, Statist. Sci. **6** (1991), no. 3, 313–318.
75. B. V. Gnedenko, *The Probability Theory Branch of Moscow University*, Teor. Veroyatnost. i Primenen. **34** (1989), 119–127; English transl. in Theory Probab. Appl. **34** (1989).
76. N. H. Bingham, *The work of A. N. Kolmogorov on strong limit theorems*, Teor. Veroyatnost. i Primenen. **34** (1989), 152–164; English transl. in Theory Probab. Appl. **34** (1989).
77. V. M. Zolotarev, *Limit theorems as stability theorems*, Teor. Veroyatnost. i Primenen. **34** (1989), 178–189; English transl. in Theory Probab. Appl. **34** (1989).
78. Ya. G. Sinaĭ and A. N. Shiryaev, *On the fiftieth anniversary of the creation by A. N. Kolmogorov of the Probability Theory Branch of the Mechanics and Mathematics Department of Moscow State University*, Teor. Veroyatnost. i Primenen. **34** (1989), 190–191; English transl. in Theory Probab. Appl. **34** (1989).
79. N. N. Vakhaniya, *A. N. Kolmogorov and the development of the theory of probability distributions in linear spaces*, Teor. Veroyatnost. i Primenen. **34** (1989), 197–202; English transl. in Theory Probab. Appl. **34** (1989).
80. V. S. Korolyuk, *The boundary layer in asymptotic analysis of random walks*, Teor. Veroyatnost. i Primenen. **34** (1989), 208–215; English transl. in Theory Probab. Appl. **34** (1989).
81. V. A. Uspenskiĭ, *Our great contemporary Kolmogorov*, Mathematics and Its Historical Development (by A. N. Kolmogorov), "Nauka", Moscow, 1991, pp. 11–20. (Russian)
82. A. N. Shiryaev, *Andreĭ Nikolaevich Kolmogorov (April* 25, 1903–*October* 20, 1987), Teor. Veroyatnost. i Primenen. **34** (1989), 5–118; English transl. in Theory Probab. Appl. **34** (1989).
83. D. G. Kendall, *Andrei Nikolaevich Kolmogorov*, Biographical Memoirs of Fellows of the Royal Society **37** (1991), 301–319.

XI. Publications (by various authors) cited in the biographical sketch

1. P. S. Aleksandrov and A. Ya. Khinchin, *Andreĭ Nikolaevich Kolmogorov (on his fiftieth birthday)*, Uspekhi Mat. Nauk **8** (1953), no. 3, 178–200. (Russian)
2. T. V. Arak, *On the convergence rate in Kolmogorov's uniform limit theorem,* I, II, Teor. Veroyatnost. i Primenen. **26** (1981), 225–245, 449–463; English transl., Theory Probab. Appl. **26** (1981), 219–239, 437–451.
3. T. V. Arak and A. Yu. Zaĭtsev, *Uniform limit theorems for sums of independent random variables*, Trudy Mat. Inst. Steklov. **174** (1986); English transl. in Proc. Steklov Inst. Math. **1988**, no. 1.
4. ———, *On the rate of convergence in Kolmogorov's second uniform limit theorem*, Teor. Veroyatnost. i Primenen. **28** (1983), 333–353; English transl., Theory Probab. Appl. **28** (1983), 351–374.
5. T. V. Arak, *On the approximation of n-fold convolutions of distributions having nonnegative characteristic functions with accompanying laws*, Teor. Veroyatnost. i Primenen. **25** (1980), 225–246; English transl., Theory Probab. Appl. **25** (1980), 221–243.
6. K. B. Athreya and P. E. Ney, *Branching processes*, Springer-Verlag, New York–Berlin, 1972.
7. S. N. Bernstein, *An experiment in giving an axiomatic justification of probability theory,* Soobshch. Kharkovsk. Mat. Obshch. **15** (1917), 209–274. (Russian)
8. ———, *Principes de la théorie des équations différentielles stochastiques*, Trudy Mat. Inst. Akad. Nauk SSSR **5** (1934), 95–114. (Russian)
9. ———, *Probability theory*, Gostekhizdat, Moscow, 1946. (Russian)
10. A. Blanc-Lapierre and R. Fortet, *Théorie des fonctions aléatoires*, Masson, Paris, 1953.
11. G. Bohlmann, *Die Grundbegriffe der Wahrscheinlichkeitsrechnung in ihrer Anwendung auf die Lebensversicherung*, Atti del IV Congresso Internazionale dei Matematici (Rome, 6–11 April 1908), Vol. 3, 1909.
12. A. A. Borovkov, *Boundary-value problems, the invariance principle and large deviations*, Uspekhi Mat. Nauk **38** (1983), no. 4, 227–254; English transl., Russian Math. Surveys **38** (1983), no. 4, 259–290.
13. D. R. Brillinger, *Time series: Data analysis and theory*, Rinehart, Holt and Winston, New York, 1975.
14. N. Wiener, *Extrapolation, interpolation, and smoothing of stationary time series*, MIT Press, Cambridge, 1949.
15. V. A. Volkonskiĭ and Yu. A. Rozanov, *Some limit theorems for random functions,* I, Teor. Veroyatnost. i Primenen. **4** (1959), 186–207; English transl., Theory Probab. Appl. **4** (1959), 178–197.
16. ———, *Some limit theorems for random functions,* II, Teor. Veroyatnost. i Primenen. **6** (1961), 202–215; English transl., Theory Probab. Appl. **6** (1961), 186–198.
17. I. I. Gikhman, *On some differential equations with random functions*, Ukrain. Mat. Zh. **2** (1950), 45–69. (Russian)
18. ———, *On the theory of differential equations of stochastic processes*, Ukrain. Mat. Zh. **2** (1950), 37–63; **3** (1951), 317–339. (Russian)
19. ———, *On a theorem of A. N. Kolmogorov*, Nauchn. Zap. Kievsk. Univ. Mat. Sb. **7** (1953), 76–94. (Russian)
20. I. I. Gikhman and A. V. Skorokhod, *An introduction to the theory of stochastic processes*, 2nd rev. ed., "Nauka", Moscow, 1977; English transl., Saunders, Philadelphia, PA, 1969; reprint, Dover, 1996.
21. ———, *Stochastic differential equations and their applications*, "Naukova Dumka", Kiev, 1982. (Russian)
22. V. Glivenko, *Sulla determinazione empirica delle leggi di probabilita*, Giron. Inst. Ital. Attuari **4** (1933), 92–94.
23. A. Dvoretsky, J. Kiefer, and J. Wolfowitz, *Asymptotic minimax character of the sample distribution function and of the classical multinomial estimator*, Ann. Math. Statist. **27** (1956), 642–669.
24. M. Donsker, *An invariance principle for certain probability limit theorems*, Mem. Amer. Math. Soc. **6** (1951).
25. ———, *Justification and extension of Doob's heuristic approach to the Kolmogorov–Smirnov theorems*, Ann. Math. Statist. **23** (1952), 23, 227–281.

26. J. L. Doob, *Heuristic approach to the Kolmogorov–Smirnov theorems*, Ann. Math. Statist. **20** (1949), 393–403.
27. ———, *Stochastic processes*, Wiley, New York, 1953.
28. J. Jacod and A. N. Shiryaev, *Limit theorems for stochastic processes*, Springer-Verlag, Berlin–New York–Heidelberg, 1987.
29. I. G. Zhurbenko, *The spectral analysis of time series*, Moskov. Gos. Univ., Moscow, 1982; English transl., North Holland, Amsterdam, 1986.
30. A. Yu. Zaĭtsev, *On the accuracy of approximation of distributions of sums of independent random variables which are nonzero with small probability by means of accompanying laws*, Teor. Veroyatnost. i Primenen. **28** (1983), 625–636; English transl., Theory Probab. Appl. **28** (1983), 657–669.
31. K. Itô, *Differential equations determining a Markov process*, J. Pan-Japan Math. Coll. 1077 (1942).
32. ———, *Stochastic integral*, Proc. Imper. Acad. **20** (1944), 519–524.
33. ———, *On a stochastic integral equation*, Proc. Japan Acad. **22** (1946), 32–35.
34. ———, *On a formula concerning stochastic differentials*, Nagoya Math. J. **3** (1951), 55–65.
35. ———, *On stochastic differential equations*, Mem. Amer. Math. Soc. **4** (1951), 1–51.
36. ———, *Selected papers*, Springer-Verlag, Berlin, 1986.
37. F. M. Kagan, *On a limit theorem of Yu. V. Prokhorov*, Limit theorems of probability theory, "Fan", Tashkent, 1963, pp. 38–42. (Russian)
38. F. P. Cantelli, *Sulla determinazione empirica delle leggi di probabilita*, Giorn. Inst. Ital. Attuari **4** (1933), 421–424.
39. T. M. Cover and R. C. Ring, *A convergent gambling estimate of the entropy of English*, IEEE Trans. Inform. Theory **IT-24** (1978), 413–421.
40. I. P. Kornfel′d, Ya. G. Sinaĭ, and S. V. Fomin, *Ergodic theory*, "Nauka", Moscow, 1980; English transl., Springer-Verlag, New York, 1982.
41. H. Cramér and M. R. Leadbetter, *Stationary and related stochastic processes*, Wiley, New York, 1967.
42. H. Cramér, *On the composition of elementary errors, Statistical applications* (2 papers), Scand. Aktuarietidskr. **1** (1928), 141–180.
43. ———, *On harmonic analysis in certain function spaces*, Ark. Mat. Astr. Fys. **28B** (1942), 1–7.
44. ———, *Mathematical methods of statistics*, Princeton Univ. Press, Princeton, NJ, 1974.
45. N. Krylov and N. O. Bogolyubov, *La théorie générale de la mesure dans son application à l'étude des systèmes dynamiques de la mécanique non linéaire*, Ann. Math. **38** (1937), 65–113.
46. P. Lévy, *Sur les intégrales dont les éléments sont des variables aléatoires indépendantes*, Ann. Scuola Norm. Sup. Pisa Ser. 2 **3** (1934), 337–366.
47. V. P. Leonov and A. N. Shiryaev, *On a method of calculation of semi-invariants*, Teor. Veroyatnost. i Primenen. **4** (1959), 342–355; English transl., Theory Probab. Appl. **4** (1959), 319–329.
48. V. P. Leonov, *Some applications of higher cumulants to the theory of stationary stochastic processes*, "Nauka", Moscow, 1964. (Russian)
49. M. A. Leontovich, *Zur Statistik der kontinuierlichen Systeme und des zeitlichen Verlaufes der physikalische Vorgange*, Physik Z. Sowjetunion **3** (1933), 35–63.
50. ———, *Fundamental equations of kinetic gas theory from the standpoint of the theory of stochastic processes*, Zh. Èksper. Teoret. Fiz. **5** (1935), 211–231. (Russian)
51. L. Li and P. M. B. Vitanyi, *Two decades of applied Kolmogorov complexity*, Proc. IEEE 3rd Structure in Complexity Theory Conference (Washington, D.C., June 1988), pp. 1–23; Preprint TR-02-08, Harvard Univ., Center for Research in Computing Technology.
52. R. Sh. Liptser and A. N. Shiryaev, *Statistics of random processes*, "Nauka", Moscow, 1974; English transl., Springer-Verlag, New York, 1977.
53. ———, *The theory of martingales*, "Nauka", Moscow, 1986. (Russian)
54. D. Loveland, *The Kleene hierarchy classification of recursively random sequences*, Trans. Amer. Math. Soc. **125** (1966), 497–510.
55. ———, *A new interprctation of the von Mises concept of random sequences*, Z. Math. Logik Grundlag. **12** (1966), 279–294.
56. A. Łomnicki, *Nouveaux fondements du calcul des probabilités*, Fund. Math. **4** (1923), 34–71.

57. M. Loève, *Fonctions aléatoires du second ordre*, Appendix to: P. Lévy, *Processes stochastiques et mouvement brownien*, Gauthier–Villard, Paris, 1948.
58. P. Martin-Löf, *On the concept of a random sequence*, Teor. Veroyatnost. i Primenen. **11** (1966), 198–200; English transl., Theory Probab. Appl. **11** (1966), 177–179.
59. ——, *The definition of random sequence*, Inform. and Control **9** (1966), 602–619.
60. G. Maruyama, *The harmonic analysis of stationary stochastic processes*, Mem. Fac. Sci. Kyushu Univ. Ser. A **4** (1949), 45–106.
61. ——, *Continuous Markov processes and stochastic equations*, Rend. Circ. Mat. Palermo **4** (1955), 1–43.
62. L. D. Meshalkin, *A lower bound for the rate of convergence of distributions of sums to a set of infinitely divisible laws*, Dokl. Akad. Nauk SSSR **132** (1960), 766–768; English transl., Soviet Math. Dokl. **1** (1960), 648–654.
63. ——, *On approximations of distribution functions of sums by infinitely divisible laws*, Teor. Veroyatnost. i Primenen. **6** (1961), 257–275; English transl., Theory Probab. Appl. **6** (1961), 233–252.
64. R. von Mises, *Grundlagen der Wahrscheinlichkeitsrechnung*, Math. Z. **5** (1919), 52–99.
65. ——, *Wahrscheinlichkeit, Statistik und Wahrheit*, Springer-Verlag, Vienna, 1928.
66. ——, *Probability and statistics*, Russian transl. of German, Gosizdat, Moscow–Leningrad, 1930.
67. ——, *Mathematical theory of probability*, Academic Press, New York, 1964.
68. A. L. Miroshnikov and B. A. Rogozin, *Inequalities for the concentration function*, Teor. Veroyatnost. i Primenen. **25** (1980), 178–183; English transl., Theory Probab. Appl. **25** (1980), 176–180.
69. ——, *Remarks on an inequality for the concentration function of sums of independent random variables*, Teor. Veroyatnost. i Primenen. **27** (1982), 787–789; English transl., Theory Probab. Appl. **27** (1982), 848–850.
70. A. S. Monin and A. M. Yaglom, *Statistical fluid mechanics*, Parts I, II, "Nauka", Moscow, 1965, 1967; English transl., MIT Press, Cambridge, MA, 1971, 1975.
71. S. V. Nagaev, *On the rate of convergence in a boundary problem*, I, II, Teor. Veroyatnost. i Primenen. **15** (1970), 179–199, 419–441; English transl., Theory Probab. Appl. **15** (1970), 163–186, 403–429.
72. J. von Neumann, *Zur Operatorenmethode in der klassischen Mechanik*, Ann. Math. **33** (1932), 587–642.
73. ——, *Mathematische Grundlagen der Quantenmechanik*, Springer, Berlin, 1932.
74. O. Nikodým, *Sur une généralisation des intégrales de M. J. Radon*, Fund. Math. **15** (1930), 168.
75. S. M. Nikol′skiĭ, *Aleksandrov and Kolmogorov in Dnepropetrovsk*, Uspekhi Mat. Nauk **38** (1983), no. 4, 37–49; English transl., Russian Math. Surveys **38** (1983), no. 4, 41–55.
76. A. M. Obukhov, *Some specific features of atmospheric turbulence*, J. Fluid Mech. **13** (1962), 77–81.
77. ——, *Kolmogorov flow and its laboratory simulation*, Uspekhi Mat. Nauk **38** (1983), no. 4, 101–111; English transl., Russian Math. Surveys **38** (1983), no. 4, 113–126.
78. D. S. Ornstein, *Ergodic theory, randomness, and dynamical systems*, Yale Univ. Press, New Haven, CT, 1974.
79. I. G. Petrovskiĭ, *Selected works. Differential equations. Probability theory*, "Nauka", Moscow, 1987. (Russian)
80. R. G. Piotrovskiĭ, *Measurement of the information in language*, "Nauka", Leningrad, 1968. (Russian)
81. L. S. Pontryagin, A. A. Andronov, and A. A. Vitt, *On the statistical analysis of dynamical systems*, Zh. Èksper. Teoret. Fiz. **3** (1933), 165–180. (Russian)
82. P. S. Aleksandrov et al. (eds.), *Hilbert's problems*, "Nauka", Moscow, 1969. (Russian)
83. A. V. Prokhorov, *Teoria prawdopodobeństwa w badaniach rytmu wiersza*, Pamiętnik Literacki **61** (1970), 113–127. (Polish)
84. ——, *On random versification.* (*The question of theoretical and speech models of poetry*), Problems in the theory of verse, "Nauka", Leningrad, 1984, pp. 89–98. (Russian)
85. Yu. V. Prokhorov, *Probability distributions in function spaces*, Uspekhi Mat. Nauk **8** (1953), no. 5, 165–167. (Russian)

86. ———, *On sums of identically distributed variables*, Dokl. Akad. Nauk SSSR **105** (1955), 645–647. (Russian)
87. ———, *Convergence of random processes and limit theorems of probability theory*, Teor. Veroyatnost. i Primenen. **1** (1956), 177–238; English transl., Theory Probab. Appl. **1** (1956), 157–214.
88. *On a uniform limit theorem of A. N. Kolmogorov*, Teor. Veroyatnost. i Primenen. **5** (1960), 103–113; English transl., Theory Probab. Appl. **5** (1960), 98–106.
89. B. A. Rogozin, *An estimate for concentration functions*, Teor. Veroyatnost. i Primenen. **6** (1961), 103–105; English transl., Theory Probab. Appl. **6** (1961), 94–97.
90. Yu. A. Rozanov, *Stationary random processes*, Fizmatgiz, Moscow, 1963; English transl., Holden-Day, San Francisco, 1967.
91. ———, *The theory of innovation processes*, "Nauka", Moscow, 1974; English transl., Halsted Press, New York, 1977.
92. N. G. Rychkova, *Linguistics and mathematics*, Nauka i Zhizn' **9** (1961), 76–77. (Russian)
93. A. P. Savchuk, *On estimates for the entropy of a language according to Shannon*, Teor. Veroyatnost. i Primenen. **9** (1964), 154–157; English transl., Theory Probab. Appl. **9** (1964), 138–141.
94. A. I. Sakhanenko, *On the rate of convergence in a boundary problem*, Teor. Veroyatnost. i Primenen. **19** (1974), 416–420; English transl., Theory Probab. Appl. **19** (1974), 399–403.
95. B. A. Sevast′yanov, *Branching processes*, "Nauka", Moscow, 1971; German transl., Akademie Verlag, Berlin, 1974.
96. Ya. G. Sinaĭ, *On higher-order spectral measures of ergodic stationary processes*, Teor. Veroyatnost. i Primenen. **8** (1963), 463–470; English transl., Theory Probab. Appl. **8** (1963), 429–436.
97. ———, *The theory of phase transitions*, "Nauka", Moscow, 1980; English transl., Pergamon Press, New York, 1982.
98. A. V. Skorokhod, *Studies in the theory of random processes*, Izdat. Kiev. Univ., Kiev, 1961; English transl., Addison-Wesley, Reading, MA, 1965.
99. N. V. Smirnov, *Approximation of distribution laws of random variables from empirical data*, Uspekhi Mat. Nauk **10** (1944), 179–206. (Russian)
100. J. F. Steffenson, *Deux problèmes du calcul des probabilités*, Ann. Inst. H. Poincaré **3** (1933), 331–334.
101. D. W. Stroock and S. R. S. Varadhan, *Diffusion processes with continuous coefficients,* I, II, Comm. Pure Appl. Math. **22** (1969), 345–400, 479–530.
102. B. de Finetti, *Le funzione caratteristiche de legge instantanea*, Atti Accad. Naz. Lincei Rend. **12** (1930), 278–282.
103. R. A. Fisher, *The genetical theory of natural selection*, Oxford Univ. Press, Oxford, 1930.
104. R. Fortet and E. Mourier, *Convergence de la répartition empirique vers la théorique*, Ann. École Norm. Supér. **70** (1953), 267–285.
105. P. Halmos and J. von Neumann, *Operator methods in classical mechanics,* II, Ann. Math. **43** (1942), 332–338.
106. T. E. Harris, *Theory of branching processes*, Prentice-Hall, Englewood Cliffs, NJ, 1963.
107. A. Ya. Khinchin, *Asymptotic laws of probability theory*, ONTI, Moscow–Leningrad, 1936; Russian transl. of German version, *Asymptotische Gesetze der Wahrscheinlichkeitsrechnung*, Springer-Verlag, Berlin, 1933.
108. ———, *A new derivation of P. Lévy's formula*, Byull. Moskovsk. Univ. Ser. Mat. **1** (1937). (Russian)
109. ———, *The concept of entropy in probability theory*, Uspekhi Mat. Nauk **8** (1953), no. 3, 3–20. (Russian)
110. ———, *On the fundamental theorems of information theory*, Uspekhi Mat. Nauk **11** (1956), no. 1, 17–75; English transl., Morris D. Friedman (MR 18, 630h).
111. S. Chapman, *On the Brownian displacements and thermal diffusion of grain suspended in a non-uniform fluid*, Proc. Roy. Soc. London Ser. A **119** (1928), 35–54.
112. A. Church, *On the concept of a random sequence*, Bull. Amer. Math. Soc. **46** (1940), 130–135.
113. C. E. Shannon, *Prediction and entropy of printed English*, Bell System Tech. J. **30** (1951), 50–64.

114. A. N. Shiryaev, *Some problems in the spectral theory of higher-order moments,* I, Teor. Veroyatnost. i Primenen. **5** (1960), 293-313; English transl., Theory Probab. Appl. **5** (1960), 265–284.
115. ———, *On conditions for ergodicity of stationary processes in terms of higher-order moments*, Teor. Veroyatnost. i Primenen. **8** (1963), 470–473; English transl., Theory Probab. Appl. **8** (1963), 436–439.
116. ———, *Statistical sequential analysis. Optimal stopping rules*, 2nd ed., "Nauka", Moscow, 1976; English transl. of 1st ed., Amer. Math. Soc., Providence, RI, 1972.
117. E. Schrödinger, *Über die Umkehrung der Naturgesetze*, Sitzungsber. Preuss. Akad. Wiss. Phys.-Math., 12 Marz, 1931, 144–153.
118. A. Einstein and M. Smoluchowski, *Brownian motion*, ONTI, Moscow–Leningrad, 1936 (Russian translation of a collection of papers).
119. P. Erdős and M. Kac, *On the number of positive sums of independent random variables*, Bull. Amer. Math. Soc. **53** (1947), 1011–1021.
120. A. M. Yaglom and I. M. Yaglom, *Probability and information*, "Nauka", Moscow, 1973; English transl., Reidel, Dordrecht, 1983.
121. A. M. Yaglom, *Correlation theory of stationary stochastic processes*, Gidrometeoizdat, Leningrad, 1981; English transl., Springer-Verlag, New York–Berlin–Heidelberg, 1987.